The On-line Electric Vehicle

Nam P. Suh · Dong Ho Cho
Editors

The On-line Electric Vehicle

Wireless Electric Ground Transportation Systems

Springer

Editors
Nam P. Suh
Department of Mechanical Engineering
Massachusetts Institute of Technology
Cambridge, MA
USA

Dong Ho Cho
School of Electrical Engineering
Korea Advanced Institute of Science and Technology
Daejeon
Korea (Republic of)

ISBN 978-3-319-51182-5 ISBN 978-3-319-51183-2 (eBook)
DOI 10.1007/978-3-319-51183-2

Library of Congress Control Number: 2016963660

Printed on acid-free paper

This Springer imprint is published by Springer Nature
The registered company is Springer International Publishing AG
The registered company address is: Gewerbestrasse 11, 6330 Cham, Switzerland

To
Moon Soul Chung
A. Neil Pappalardo
B.J. Park
Hock Tan

For their generous support of higher education and technology innovation

Acknowledgements

> OLEV and SMFIR received international recognitions from the World Economic Forum (top 10 Emerging Technologies of 2013), The TIME (50 Best Inventions of 2010), and the International Union of Railways (Innovation Award of 2014 given to KRRI). Reuters ranked KAIST as the 6th most Innovative University in the world.

The ultimate goal of EGTS is to electrify all ground transportation systems using OLEV, SMFIR, and other related technologies. When EGTS is adopted worldwide, it should significantly reduce CO_2 emission worldwide. The technologies described in this book are a result of dedicated work by many at KAIST, and the financial support and goodwill of many people, worldwide.

The OLEV project was the largest R&D project undertaken by KAIST in its 35-year history. From 2009 to 2011, the Korean government provided approximately US$50 million to KAIST for this project. We are most grateful to President M.B. Lee and his senior staff and cabinet members Dr. J.W. Park, J.S. Yoon, and J.K. Choi for their support.

The idea for OLEV was germinated at KAIST because we had undertaken the EEWS (energy, environment, water, and sustainability) project in 2007. Prime Ministers D.S. Han and S.S. Han provided the initial funding for the EEWS project.

The OLEV system was first installed and commercially used in the Seoul Grand Park in 2010 to transport people around the 14-mile loop around the park. OLEV buses have been in operation on the KAIST campus since 2011. At the 2012 World's Expo in Yeosu in Korea, OLEV was used to transport visitors, thanks to the support of Minister D.S. Kang and Vice Minister H.K. Yeo of the Ministry of Land, Infrastructure, and Transport. Also Director General J.S. Kim of Frequency Policy in the Korea Communications Commission helped in expediting the commercial use of OLEV. In 2013, Gumi, an industrial city in Korea, adopted the first commercial use of regular OLEV buses in the city. We are grateful to Mayor Y.J. Nam of Gumi City and Director General K.C. Kim of Seoul City government.

The work presented in this book was possible because of the outstanding professors, staff, and students at KAIST, who participated in the OLEV project under the able leadership of Prof. D.H. Cho, vice president in charge of the ICC Campus. Professor Joseph D.J. Joo, vice president for External Affairs, made invaluable contributions to the success of the project by securing the support of various government agencies and the national assembly. In addition to the authors of various chapters of this book, many other faculty and research staff members contributed to the project: Profs. S.J. Jun, J.H. Kim, K.S. Kim, H.K. Lee, B.K. Park, G.W. Moon, and C.T. Rim. Professors N. Kim of Choong-Buk University and Y.M. Kim, of Dan Guk University, also participated in the OLEV project.

Many outside organizations worked with KAIST on this project: Hyundai Heavy Industries, Korea Railroad Research Institute, Korea Electrical Safety Corporation, Korea Rail Network Authority, and Korea Automobile Testing & Research Institute. We are particularly grateful to Dr. S.M. Hong, president of KRRI, for his exceptional leadership in implementing SMFIR technology to trams and high-speed trains. We are also indebted to K.J. Yong, president of Korea Automobile Testing & Research Institute; S.K. Lee, president of KESCO and Dr. W.K. Han of KESCO; S.K. Kim, vice chairman of the Korea Rail Network Authority; Chairman Y.J. Cho of Hankuk Fiber Group and Chairman S.H. Baek of Daewoo Bus Company; Dr. G.S. Min, chairman of Hyundai Heavy Industries; and Dr. S.B. Yoo of Korea Automobile Testing & Research Institute.

To educate and conduct research on ground, air, and ocean green transportation systems of the twenty-first century, KAIST established the CCS Graduate School for Green Transportation with the generous financial support of chairman and Mrs. C.S. Cho. The first dean of the CCS Graduate School was Prof. D.H. Cho.

KAIST granted licenses to Dong Won Group for the Korean and Asian markets, and the OLEV Technologies, Inc., Boston, Massachusetts, for the American markets. Both of these companies have licenses for the European market. The support of Chairman J.C. Kim of Dong Won Group is deeply appreciated. We would also like to acknowledge the support of I.K Park, vice chairman of Dong Won Group. Mr. Y.W. Jung, president, and Mr. T.S. Chung, the former president of Dong Won OLEV Co., led the OLEV effort in Korea.

The OLEV Technologies, Inc., based in Boston, Massachusetts, USA., was established in 2012, because two major benefactors of KAIST and MIT, Dr. A. Neil Pappalardo and Dr. B.J. Park, wanted to see the OLEV technology used in the USA. Dr. and Mrs. Park also donated the B.J. and Chung Park KI Building, the largest building on the KAIST campus. Dr. and Mrs. Pappalardo established the Pappalardo Medical Center at KAIST, one of the best university medical facilities. The Pappalardos and the Parks also gave major gifts to the MIT Mechanical Engineering Department. Bryan Wilson, Roger Burns, and Dr. Hikyu Lee of OLEV Technologies, Inc. have played a pioneering role to commercialize OLEV in the USA.

In preparing this book, it has been our pleasure to work with Oliver Jackson, the engineering editor of Springer. His guidance and advice in preparing this book are gratefully acknowledged. Dr. Dan Penrice edited parts of the manuscript. We are indebted to Ms. Jin Hui Lee for thoroughly reviewing the final manuscript to make them ready for publication.

Finally, we are indebted to Prof. Gang Chen, Head of the MIT Department of Mechanical Engineering, for supporting this book-writing effort, for providing funds needed to complete this book, and for inviting Prof. D.H. Cho as a visiting professor of MIT during the final stages of preparing the manuscript of this book.

September 2016

Nam P. Suh
Dong Ho Cho

Personal Note of Appreciation by Nam P. Suh

In 2006, Dr. W.S. Kim, deputy prime minister of Korea, approached me about joining KAIST as its president. I accepted the kind offer hoping to help KAIST to become one of the leading universities in the world.

There were many challenges in transforming KAIST into a globally competitive top leading university. Changes are difficult to embrace, especially in a strong institution with proud tradition and well-established culture. However, changes were necessary, since the world around us changed. With inputs from the faculty and staff, we developed an ambitious strategic plan for KAIST. A few years later, thanks to the effort of many, Reuter ranked KAIST as one of the top 10 most innovative universities of the world.

A university president cannot function without the support and leadership of the board of trustees. KAIST had an exceptional leader in Dr. Moon Soul Chung, chairman of the Board of Trustees of KAIST. His leadership and support were indispensible to the success of KAIST's transformation. There were other exceptional board members: Dr. Young Gil Kim, president of HanDong University; Mr. Dong-Soo Huh of chairman of GS Caltex Corporation; Ms. Y.S. Chang, chairwoman of Aekyung Group; Mr. Soon Tak Cho of vice chairman of SK Group; and Mr. Ju-Myung Hwang of founding partner of HMP, a leading law firm. Their support was essential at KAIST. I am also indebted to Dr. Donald C.H. Kim for his advice and counsel.

I had the pleasure of working with exceptional colleagues at KAIST. During the first term of my presidency, the team consisted of Profs. Soon Heung Chang, provost; Ji-Won Yang, VP for external affairs; Sang Soo Kim, VP for research; Kwang Hyung Lee, dean for Academic Affaires; Kyung Wook Paik, dean for Student Affairs; Yong Hoon Lee, dean of Engineering; Young Kyu Do, dean of Science; Do-Kyung Kim, dean of Admissions; and Sang Yup Lee, dean of Biological Science. Their intimate knowledge of KAIST and Korea guided me through complicated pathways.

During the second term of my presidency, I had the pleasure of working with truly outstanding colleagues. I cannot name them all, but the dream team was led by Provost Yong Hoon Lee; Vice Presidents Joseph D.J. Joo, Kyung Wook Paik, and

Dong Ho Cho; Deputy Vice President Chang Dong Yoo; and Deans Yong Man Han, Seung-Bin Park, Jin Koo Kang, Dong Soo Kim, Hee Kyang Park, Ook Jun Yoo, and Tae Eok Lee. We also had strong intellectual leaders as department heads and directors of the KI Research Institutes. Professor Sun Chang Kim made his KI Center on Biological Systems a model for an outstanding interdisciplinary research. Finally, I would like to thank Jurist Sung Hee Lee, an outstanding lawyer, who provided much needed counsel and assistance.

The period 2006–2013 was a unique period of transition at KAIST. The faculty size increased from 400 to 625 with the addition of 350 outstanding young professors. The tenure process was strengthened. Department heads became the major decision makers. The seniority system was replaced by a merit-based system. The budget and the research volume were increased by a factor of approximately 2.7. KAIST accepted students from ordinary high schools, while maintaining its special relationship with science high schools as conduits to KAIST for exceptionally bright students. We increased the number of women and international students. We built 14 new buildings and changed the language of instruction from Korean to English. We initiated major large-scale research projects. We also initiated new ways of educating students by eliminating formal lectures—*Education 3.0*—under the leadership of Prof. D.H. Cho and Prof. T.O. Lee. The international ranking of KAIST literally shot up the fastest in the world, according to ranking agencies. In 2015, Reuter ranked KAIST as the only non-US university in the top 10 of the most innovative universities of the world.

Professor Dong-Ho Cho, one of the most respected professors at KAIST, directed the OLEV project. He assembled an interdisciplinary team of professors, researchers, staff, and graduate students. He held staff meetings at 10 pm every day to review the progress made and identify issues to be solved. The leaders of the entire OLEV team met with me once a week for one or two hours to discuss problems that we need to solve. It was a true collaboration of the highest order.

In addition to many KAIST professors and staff, there were other contributors to OLEV project: Dr. Soon Man Hong, president of the Korea Railroad Research Institute, where he developed trams and high-speed trains using the SMFIR technology in collaboration with KAIST. Dr. Byung Kyun Park contributed to the commercialization of OLEV.

In addition to OLEV, we concurrently initiated the Mobile Harbor (MH) project, an equally ambitious large project to deal with the lack of large harbors in the world that can accommodate large containerships. The central idea of MH is to have harbors go out to the ship moored in deepwaters instead of having these ships come into crowded harbors. The use of these MHs will be much cheaper than enlarging existing harbors or building new harbors, which will harm the environment and damage shore lands. I would like to express my special thanks to Profs. B.M. Kwak, Kyung-Soo Kim, and Soo-Hyun Kim, and many other professors and research staff for their contributions. They demonstrated the viability of Mobile Harbor with a 1/3 scale MH (note: the height of the model was about 20 m) by unloading containers in open sea. In executing the OLEV project as well as another major project at KAIST, the Mobile Harbor (described in Chap. 3), we had two outstanding system architects, Profs. T.S. Lee and G.J. Park, who monitored the design of projects. They constructed the design matrix by getting the design information from engineers and designers in order to spot possible coupled designs. When they identified coupling, they made suggestions for appropriate changes.

A major development at KAIST was the establishment of a joint research center with Saudi Aramco for conversion of CO_2 into some other useful materials. I am personally grateful to Dr. Khalid A. Al-Falih, chairman of Saudi Aramco and minister of Energy, Industry, and Minerals Resources of the Kingdom of Saudi Arabia. His leadership and vision in establishing this center should have a long-term impact on KAIST and hopefully also on Saudi Aramco and the world.

OLEV project received strong support of the KAIST President Advisory Council (PAC). PAC members were leaders of industry, academia, and government from many nations: Dr. A. Neil Pappalardo, CEO of MEDITECH, and a member of the MIT Corporation; Dr. B.J. Park, founder and former CEO of MTL, Inc.; Dr. Ki Joon Lee, former president of Seoul National University (SNU); Dr. Lars Pallisen, rector of Denmark Technical University; H.K. Lee, LG president; H. Yoshikawa, former president of the University of Tokyo; Dr. Arden Bement, former director of the US National Science Foundation; H. Tan, CEO of Avago Technologies, Inc.; Dr. W.S. Kim, former deputy prime minister for Science and Technology and former president of Yonsei University; Dr. Adnan Akay, provost, Bilkent University, Turkey; Prof. Gunnar Sohlenius, vice president, Royal Institute of Technology, Sweden; and Prof. Jorg Steinbach, rector of Technical University of Berlin.

I am personally indebted to many benefactors of KAIST. Dr. and Mrs. B.J. Park donated the Park KI Building. Dr. and Mrs. A. Neil Pappalardo gave the Pappalardo Medical Center. These two major gifts inspired Koreans to give, changing the giving culture of Korea. Dr. and Mrs. B.H. Kim gave a major gift for the Electrical Engineering Building. Mrs. Ewon Oh established the endowment for Ewon professorships. Chairman C.S. Cho endowed the Graduate School of Green Transportation. Dr. Donald Kim of Hawaii gave support for the Science Building. Chairwoman Y.S. Chang gave the Chang Student Activities Center at KAIST. CEO Hock E. Tan, in addition to his leadership in PAC, established the Nam Pyo Suh Professorship of Mechanical Engineering at MIT.

KAIST administration had exceptional staff members, who made outstanding contributions to KAIST. I had the pleasure of working closely with directors Dong-Hyuck Won, Jae Suk Jang, Ki Hyun Kim, and Won Soo Doo. I am also indebted to Boram Cho, Yong Sub Kang, and Yoonju Hong for their dedication and personal support. They are exceptional people and deserve special recognitions for their important contributions to KAIST.

Finally, I would like to thank Young Ja, my dear wife, who shared her ideas and ideals to make KAIST an outstanding institution. She has worked extremely hard throughout her life, especially during the seven years we spent at KAIST, to make the world of ours a better place for all. She is an extraordinarily gifted and dedicated person. Finally, during our stay at KAIST, Jung-Woo Surh, a distinguished jurist in Korea, provided valuable pro bono service for KAIST and me.

Sudbury, MA, USA
August 2016

Nam P. Suh

Contents

Contributors

Seungyoung Ahn The Cho Chun Shik Graduate School for Green Transportation KAIST, Korea Advanced Institute of Science and Technology (KAIST), Yuseong-Gu Daejeon, South Korea

Dong Ho Cho School of Electrical Engineering KAIST, Korea Advanced Institute of Science and Technology (KAIST), Yuseong-gu, Daejeon, South Korea

Amro M. Farid Thayer School of Engineering, Dartmouth College, Hanover, NH, USA; MIT Mechanical Engineering Department, Cambridge, MA, USA

Soon Man Hong Korea Railroad Corporation, Daejeon, South Korea

Young Jae Jang Industrial and Systems Engineering, Korea Advanced Institute of Science and Technology (KAIST), Daejeon, South Korea

Yong Hoon Jeong Nuclear and Quantum Engineering, Korea Advanced Institute of Science and Technology (KAIST), Daejeon, South Korea

Gu Ho Jung Wireless Power Transfer Research Center KAIST, Korea Advanced Institute of Science and Technology (KAIST), Yuseong-Gu Daejeon, South Korea

Byung Song Lee Korea Railroad Research Institute, Uiwang-si, Gyeonggi-do, Korea

Kon Fah Loh Independent Environmental Evangelist, Ipoh, Perak, Malaysia

Jong Han Park Dongwonolev, Seoul, South Korea

In-Soo Suh Korea Advanced Institute of Science and Technology (KAIST), Troy, MI, USA; General Motors, Warren, MI, USA

Nam P. Suh Department of Mechanical Engineering, Massachusetts Institute of Technology (MIT), Cambridge, MA, USA

Uooyeol Yoon Wireless Power Transfer Research Center KAIST, Korea Advanced Institute of Science and Technology (KAIST), Yuseong-Gu Daejeon, South Korea

Synopsis of Book

This book is a culmination of our effort at KAIST to solve the long-term global problems humanity needs to address. Initially, we limited our attention to the issues related to EEWS (energy, environment, water, and sustainability). We began this effort in late 2006. For energy and environment, we decided to replace internal combustion (IC) engines in ground transportation systems with electric drives so as to reduce CO_2 emission. In executing this project, there were two major goals: overcoming the concerns of non-believers in global warming and showing how to design and implement innovative large system in an academic environment. We received a great deal of support from all corners of Korean society, those in government, industry, universities, and most of all, ordinary citizens who believed in us. We attribute what is in this book to these supporters.

This book is concerned with presenting the technology of On-line Electric Vehicle (OLEV), the details of its various applications, and analyses of the technology in terms of performance cost, and safety, for the benefit of researchers and technologists working in the fields of electric vehicles and, in particular, on the use of wireless dynamic power transfer for EVs and systems for electrification of ground transportation systems (EGTS). From a technical standpoint, OLEV represents, in our view, the best technology yet invented for EGTS, with all that this implies for efforts to develop emissions-free vehicles so as to dramatically reduce the amount of CO_2 and other GHGs currently being poured into the atmosphere. Furthermore, the use of a minimum number of batteries, unlike all battery-powered electric vehicles (EVs), will eliminate immense problems associated with manufacturing and disposing batteries such as lithium-ion batteries. Secondly, the design and successful implementation of OLEV in the space of a mere two years (2009–2011) illustrate the power of Axiomatic Design, a method for the design of large engineered systems that offers a much more efficient and cost-effective process for the design of such systems than the methods still typically taught in many engineering schools and implemented in large, enormously expensive, publicly and privately funded projects. Thirdly, the development of OLEV at KAIST also contains valuable lessons about how institutions involved in teaching and research in engineering and design, designing and building large systems, and funding these activities need to be organized and managed in order to successfully carry out projects aimed at solving the many serious and urgent problems now confronting humankind.

The book consists of four parts. Part I has three chapters. In Chap. 1, we present the background information on the environmental issue OLEV is designed to address, i.e., CO_2 emission by internal combustion engines used in transportation systems, by explaining the history and shortcomings of internal combustion engines. Then, the rationale for replacing them with electric drives was presented, following short historical development of early technologies that led to current electric vehicles. Then, as a means of reducing the need for a large bank of batteries currently used in EVs, the invention and introduction of OLEV were presented. Chapter 2 provides a basic description of OLEV and its enabling technology of SMFIR. It then briefly compares OLEV/SMFIR in terms of environmental impact, performance, and cost with both vehicles using IC engines and other electric vehicles (a subject that is further explored in Part IV of the book); explains the potential benefits of electrifying ground transportation systems with a technology like OLEV and connecting these systems to smart electric grids; and describes efforts to date to commercialize OLEV and lessons from these efforts. Chapter 3 explains how the design and implementation of OLEV were accomplished so expeditiously by introducing readers to Axiomatic Design and, specifically, how it can be used to remove "complexity" from the design of even the kinds of very large, complicated engineered systems that are so often subject to massive delays and cost overruns.

Parts II–IV of the book provide detailed technical explanations and analyses of, respectively, the design and technology of OLEV, SMFIR, and their components; applications for OLEV and SMFIR other than those for which they were originally designed; and performance, cost, and safety factors with respect to these technologies. This section is intended for specialists in electric vehicles, wireless heavy electric power transfer (SMFIR), control of EMF and EMI, and regulatory issues. Most of the knowledge needed to implement OLEV is given here in order to help those who are interested in solving the environmental problems caused by vehicles that use internal combustion engines and petroleum.

The OLEV project was a large project involving an interdisciplinary team of professors, researchers, staff, and students. They designed the system and collaborated with industrial firms that provided vehicles, which were converted to OLEV. KAIST research staff and faculty members installed the first OLEV system in the Seoul Grand Park during the bitter cold winter of 2010. Some of the leading faculty and research staff members of the OLEV team are the authors of Chaps. 5–23. Professor Dong-Ho Cho, who was vice president of the KAIST ITC campus, led the OLEV project. He was also a visiting professor at MIT in 2015 to collaborate in writing and editing this book. Professor Nam Pyo Suh, who was the president of KAIST from 2006 to 2013, conceived the idea for OLEV and launched the project after securing the initial funding from the Korean government through a special supplementary budgetary process in 2009. In May 2016, he and Prof. Cho, together with other visitors, had the pleasure of riding the latest commercial OLEV bus installed and operating in Gumi, Korea. Fare-paying passengers seemed to enjoy the ride on the quiet OLEV bus that serves one of the commercial routes of the city. Our hope is that this book will provide sufficient fundamental knowledge for implementation of OLEV in all countries interested in improving their quality of life and the global environment for all humankind.

Part I
Synergy of Diverse Ideas Behind OLEV

Chapter 1
Making the Move: From Internal Combustion Engines to Wireless Electric Vehicles

Nam P. Suh and Dong Ho Cho

Abstract This first chapter presents the background information on the environmental issues the On-Line Electric Vehicle (OLEV) system—the electric vehicle that receives electric power wirelessly from the underground electric power supply—is designed to address. It then reviews CO_2 emission by internal combustion (IC) engines after providing a brief history of IC engines. Following a review of the history of electric technologies that led to current electric vehicles, the rationale for electrifying ground transportation systems (EGTSs) is presented. Many of the current electric vehicles (EVs) use a large bank of lithium batteries, which has shortcomings associated with its cost, weight, bulkiness, and fire hazard. To overcome these shortcomings, OLEV was created.

1.1 Introduction

Global warming is perhaps the most urgent issue that humanity must deal with in the twenty-first century. As is well known, one of the biggest causes of global warning is our use of automobiles, which, while driving economic growth in twentieth century, has also created catastrophic levels of CO_2 in the atmosphere, threatening the planet and our way of life. To deal with these concomitant environmental problems, we must look to other innovative solutions to transportation. One such solution is to electrify our ground transportation systems (EGTSs), which is the subject of this book. In chap. 1, a historical perspective is given on IC engines and electric cars after reviewing the relevant information on climate change.

N.P. Suh (✉)
Department of Mechanical Engineering, Massachusetts Institute of Technology (MIT), 77 Massachusetts Avenue, Cambridge, MA 02139, USA
e-mail: npsuh@mit.edu

D.H. Cho
School of Electrical Engineering, Korea Advanced Institute of Science and Technology (KAIST), 291 Daehak-ro, Yuseong-gu, Daejeon 34141, South Korea
e-mail: dhcho@kaista.ac.kr

N.P. Suh and D.H. Cho (eds.), *The On-line Electric Vehicle*,
DOI 10.1007/978-3-319-51183-2_1

1.2 CO_2 Emissions and Climate Change

On December 12, 2015, 195 nations at the United Nations Conference on Climate Change in Paris (Conference of Parties COP21) agreed to reduce their CO_2 emissions. The 29 articles in the agreement adopted by COP21 require each nation to choose its own specific means of reducing its greenhouse gas emissions, but the ultimate goal is to limit the rise in the earth's temperature to well below 2 °C above pre-industrial levels, in line with the limits advocated by the Intergovernmental Panel on Climate Change (IPCC). IPCC has presented overwhelming evidence that the buildup since the Industrial Revolution of the greenhouse gases CO_2, methane, water vapor, nitrous oxide, and ozone, which absorb and emit radiation in thermal infrared range, will, if left unchecked, raise the temperature of the earth's atmosphere from its current range to one that will change weather patterns, sea levels, forestation, fisheries, and many other phenomena central to maintaining life on earth, and to the life of our species in particular. CO_2 is the second most prevalent greenhouse gas (contributing 9–26% of the total greenhouse effects[1]) after water vapor and has increased the most due to human activities since the Industrial Revolution.[2] The IPCC's 2014 report for policymakers states:

> Emissions scenarios leading to CO_2-equivalent concentrations in 2100 of about 450 ppm or lower are *likely* to maintain warming below 2 °C over the twenty-first century relative to pre-industrial levels. These scenarios are characterized by 40–70% global anthropogenic GHG [greenhouse gas] emissions reduction by 2050 compared to 2010, and emissions levels near zero or below in 2100.[3]

Without dramatic reductions in CO_2 emissions in short order we will face dire consequences, many of which have already become manifest: desertification of arable land; more frequent flooding, tornadoes, wildfires, and typhoons; a rise in sea level of 2.5–6.5 ft (0.8–2 m) by 2100 caused by the melting of polar ice; complete disappearance of certain islands in the Pacific Ocean due to rising sea level; drought in California and elsewhere due to melting of mountain snow; erosion of coastal

---

[1]Kiehl et al. [1].

[2]According to the U.S. Environment Protection Agency (EPA), CO_2 accounted for 82% of greenhouse gas emission in the USA in 2013. http://www3.epa.gov/climatechange/ghgemissions/gases/co2.html.

[3]IPCC, Climate Change 2012 Synthesis Report Summary for Policymakers (https://www.ipcc.ch/pdf/assessment-report/ar5/syr/AR5_SYR_FINAL_SPM.pdf), p. 20. Two recent studies suggest that the IPCC report was conservative in its estimates. A joint paper by NASA and the University of California at Irvine, based on forty years of data collected via satellites and aircraft, predicts that the continued melting of a major section of the ice sheet in west Antarctica will raise the sea level by four feet (1.2 m). A similar conclusion was reached in a study from the University of Washington, which predicted that if the current trend continues, the sea level may rise as much as 12 feet (3.7 m). See Eric Rignot, NASA–University California at Irvine research report, May 13, 2014. http://www.nytimes.com/2014/05/13/science/earth/collapse-of-parts-of-west-antarctica-ice-sheet-has-begun-scientists-say.html?_r=0; Joughin et al. [2] The State of Florida may become a much smaller state, since its mean elevation is only about one hundred feet.

areas; and acidification of the oceans, negatively affecting marine life. It is particularly unfair that, although people living in developed nations have produced most of the CO_2, all the inhabitants of the world will share the major adverse consequences equally. Predictions of the consequences of increases in the level of CO_2 in the atmosphere come from both theoretical modeling and measurements[4] [5]. The IPCC prediction is deemed credible because many outstanding scientists have collaborated to establish the science of global warming.[5]

There are three major sources of anthropogenic CO_2 emissions: electricity generation, transportation, and industry. The two largest of these are electric power plants that burn fossil fuels (coal, oil, or natural gas, although coal-fired plants produce significantly more emissions than those burning oil or natural gas) and ground, air, and ocean transportation vehicles that use petroleum. These two sources account for roughly two-thirds of all anthropogenic CO_2.[6] Power plants and automobiles are also key contributors to two other major environmental problems: pollution of the atmosphere and the rapid depletion of certain raw materials such as rare earth elements. The first of these two problems is location specific as well as global: because automobiles and other gasoline-powered vehicles account for 31% of US CO_2 emissions as well as for emissions of particulates, air quality is worse in big cities than in other locations because of the large number of automobiles choking their highways and roads. It is also well known that large harbors (e.g., Long Beach, California) are highly polluted because of the use of trucks and automatically guided vehicles to transport containers unloaded from container ships with capacities as large as 13,000 TEUs.

Thus, from the standpoint of both global warming and the other environmental hazards posed by current provision for energy needs, electricity generation and transportation are the two major areas on which society must focus its attention.

---

[4]Scientific knowledge about historical levels of CO_2 in the atmosphere is deduced from measurement of the air trapped in ice [3]. Such measurements indicate that cyclic variations of incident solar energy began at least 800,000 years ago, with each cycle lasting roughly 100,000 years. Until the beginning of the Industrial Revolution (around 1750), concentrations of CO_2 in the atmosphere had not deviated from this cyclic variation. However, since the Industrial Revolution, with the invention of the steam engine by James Watt and the subsequent development of other energy-intensive technology, the CO_2 concentration in the atmosphere has continuously increased from about 280 ppm in 1750 to about 392 ppm in 2012 and is heading higher every year [3, 4].

[5]A summary of the IPCC findings is given in "Climate Change 2014, Synthesis Report Summary for Policy Makers"; https://www.ipcc.ch/pdf/assessment-report/ar5/syr/AR5_SYR_FINAL_SPM.pd.

[6]The source of CO_2 generation is, of course, country specific. In the USA, according to the U.S. Environmental Protection Agency (EPA), the combustion of fossil fuels to generate electricity is the largest source of CO_2 emissions, accounting for about 37% of the total CO_2 emission and 31% of total greenhouse gas emitted in 2013 in the USA. The type of fuel used to generate electricity emits different amounts of CO_2, coal producing more CO_2 than natural gas or oil. Transportation is the second largest source of CO_2 in the USA. The combustion of fossil fuels in transportation—primarily gasoline and diesel by automobiles, kerosene by airplanes, and diesel by ships, and trains—generates about 31% of the total US CO_2 emission and 26% of total US greenhouse gas emissions in 2013.

To avoid global calamity, advanced economies need, most especially, to take prudent steps *now* to reduce greenhouse gas emissions and CO_2 emissions in particular, produced by electricity generation and fossil-fueled transportation. To deal with the first source of emissions, they must replace coal-burning electric power plants with ones using nuclear power and/or renewable energy sources such as solar, wind, and hydropower.

One of the most promising potential means for arresting global warming is the elimination of internal combustion engines from automobiles.

1.3 The Problem of the Internal Combustion Engine

In the mid-to-late nineteenth century, many different versions of the internal combustion (IC) engine were developed. Jean Joseph Lenoir invented the IC engine in 1858, but could get only 4% efficiency from it. Nikolaus August Otto tried to improve the Lenoir engine, and by 1876 he and Eugen Langen created the first IC engine that compressed the fuel mixture prior to combustion, yielding far higher efficiency than any other engine to that point. This Otto engine was a large, stationary one. The credit for the small Otto engine goes to Gottlieb Daimler, Otto's employee and manager, who disagreed with Otto's view that there was no need for small engines for transportation and left Otto's company to form his own in 1883. Daimler was able to get around Otto's patents by discovering the work of Alphonse Beau de Rochas, who had invented the four-stroke engine in 1861.[7]

The explosive growth of the automobile market, in turn, created large industrial complexes in many parts of the world, including not just automobile companies and their suppliers but firms in whole new industrial fields such as mass production and automation technologies, materials, and, of course, petroleum. The fields of mechanical engineering and chemical engineering were created to deal with issues arising primarily from automobiles, power generation, and petroleum. Together with its associated industries, the automotive industry has created more jobs than any except construction and raised the standard of living of people in many nations; by the mid-twentieth century, it had made USA the world's leading economic power. Moreover, the emergence of gasoline-powered automotive technology has affected not only economic and technological development but also culture, demographics, and politics in the USA and throughout the world. The automobile industry has not only created vast wealth but also spurred the creation of networks of superhighways and speeded the migration of people from farms to cities and from cities to suburbs. Today it is estimated that about one to two billion cars are on the road and the demand for automobiles continues to increase as the world

[7]For a more complete listing of inventions related to IC engines, see https://en.wikipedia.org/wiki/History_of_the_internal_combustion_engine#Further_reading.

population continues to grow and more people become able to afford them; China, in particular, was one of the most rapidly growing markets for automobiles in 2015.[8]

When oil was first discovered in Titusville, Pennsylvania, in the USA in 1859, no one could have predicted that petroleum would become the principal energy source and one of the biggest economic engines in the world. In 2015, the world consumed approximately 96 million barrels of oil and other liquid fuel per day.[9] In 2014, the USA consumed an average of about 19.11 million barrels per day (including about 0.93 million barrels of biofuels[10]), about 71% of which (13.6 million barrels) was used for transportation.[11] This (the amount used for transportation) is roughly equivalent to 0.5 gallons of oil consumed per person per day. According to the US Energy Information Administration, about 63% of the projected growth in petroleum use between 2010 and 2040 will be for transportation.[12] It is no wonder, then, that in the twentieth century, oil became one of the most strategic of resources, influencing geopolitics and the use of military power as well as economics and transportation. Or that reliance on this fuel has made oil consumption and CO_2 emissions two of the biggest geopolitical issues of the twenty-first century.

The invention of IC engines created a market for the petroleum following the invention of the automobile by Karl Benz of Germany in 1886. Nearly all ground transportation vehicles today use the IC engine, with either reciprocating pistons or gas turbines. Over the past 120 years, IC engines with pistons have been refined and improved, and overall they perform well and reliably. However, the 17–20% efficiency of an IC engine with pistons is low. It is unfortunate that a technology as widespread as the reciprocating piston-type IC engine has such an inherently low level of energy conversion efficiency—and results in such serious emission problems. About 68–72% of the chemical energy in an IC engine is lost in the engine itself.[13] Thermodynamic efficiency increases with the temperature of the combusted gas, but the engine block must be cooled with water to enable lubrication of sliding interfaces, and the cold wall of the engine cylinder delays the evaporation of gasoline droplets that hit the wall. Additionally, the gap between the piston and

---

[8]London-based IHS Automotive puts light vehicle (passenger car and light-duty truck) production in 2013 at 84.7 million, up from 81.5 million in 2012. (Organisation Internationale des Constructeurs d'Automobiles), Paris in 1919.

[9]Data from the International Energy Agency. http://www.iea.org/aboutus/faqs/oil/.

[10]Data from the U.S. Energy Information Administration. https://www.eia.gov/tools/faqs/faq.cfm?id=33&t=6.

[11]As of 2013, the USA was the largest oil-consuming nation in the world, using about 19% of the world's supply. Korea was tenth, at 2.4% [Data from U.S. Energy Information Administration (EIA)].

[12]Twenty-eighty percent of the energy the US consumes, and 71% of the oil, is used for transportation: 48% for automobiles, 16% for diesel vehicles, and 7% for airplanes (National Academies 2008).

[13]In terms of well-to-wheel efficiency, only about 17% of the energy in crude oil goes to the wheel, since 15% of the energy is consumed in the refining process (Wang 2006).

cylinder can retain some of the fuel, leading to emission of partially oxidized hydrocarbons [6]. Energy is also lost in the many mechanical components between the engine and the wheels, including the transmission, tires, and braking system. In all cases, about 25% of the energy is exhausted in high-temperature exhaust gas [7].

As a result of these deficiencies, IC engines use an enormous amount of petroleum and are major polluters of our environment[14] and emitters of CO_2—about 23% of the CO_2 emitted in the world comes from combustion of the petroleum to propel vehicles with IC engines.[15] (Only slightly more CO_2 is generated from electric power plants that burn coal and natural gas.) IC engines with gas turbines are more efficient at high speeds[16] than reciprocating piston-type IC engines, but automobile companies have so far not been able to adopt them for automobiles because of their high rotational speeds. To significantly reduce the emissions of CO_2 and other environmentally damaging consequences of automobile use today, we must develop new technologies that can overcome their major shortcomings.

1.4 The Promise of Electric Vehicles

Although most vehicles used for ground transportation today have IC engines running on either gasoline or diesel fuel (with natural gas now also commonly used in commercial vehicles such as buses and taxis), subways, trains, and streetcars[17]

---

[14]The health consequences of air pollution from automobiles with IC engines fueled by petroleum are an important factor to consider. People who are exposed to ultrafine particulates (UFP), black carbon particles (BC), nitrogen oxides (NO_x), and carbon monoxide (CO) emitted by automobiles with internal combustion engines have higher incidences of asthma and cancer, and the death rate due to the heart and lung disease of people living near roads with heavy traffic is almost twice the norm and 1.4 times higher than overall death rate. For every increase in particulate exposure of 10 mcg/m^3, risk increased by 4% for mortality, 6% in cardiopulmonary mortality, and 8% in lung cancer mortality [8–12]. Modern automobile engines are tightly controlled for complete combustion and outfitted with catalytic converters that further reduce partially combusted hydrocarbon, and today's fuels are much cleaner than in the past in terms of sulfur content and other harmful chemicals. Yet 50% of the nitrogen oxides in air have been attributed to traffic. The World Health Organization (WHO) states the following: "… in parts of Europe where stricter standards and regulations for vehicles have been enforced, ambient air pollution levels are stable or continue to rise [sic, due to more vehicles]. Per kilometer of travel, diesel vehicles also typically emit more particulate emissions than gasoline, gas-powered, or electric vehicles of comparable size and age – so that increased reliance on diesel vehicles in the vehicle fleet may be a contributing factor to health-harmful air pollution in many cities. Diesel emissions have also been defined by WHO's International Agency for Research on Cancer as a carcinogen." http://www.who.int/phe/health_topics/outdoorair/databases/faqs_air_pollution.pdf.

[15]IPCC, Climate Change 2014: Mitigation of Climate Change, Chap. 8, p 603.

[16]Gas turbine efficiency ranges from 40% (single cycle) to 60% (combined cycle). http://www.cospp.com/articles/print/volume-11/issue-3/features/gas-turbines-breaking.html.

[17]The electric streetcar is not a recent invention, the first one in the USA having been installed in 1885 in South Bend, Indiana.

are powered by electricity. Meanwhile, electric cars have captured the imagination of the public as companies from Tesla, General Motors, and Ford, to technology giants such as Google and Apple, to obscure startups promising "connected" vehicles pursue the promise of all electric vehicles for the consumer market.

Even though most people today think that electric automobiles are a recent development, battery-powered electric cars have actually been around since the mid-nineteenth century. In fact, modern electric cars owe their invention, in part, to a series of important scientific advances made in the nineteenth century in Europe and the USA.[18]

One of the major difficulties in developing electric cars in the past was the lack of rechargeable batteries satisfying the requirements for automotive applications. Today the most commonly used battery in electric automobiles is the lithium-ion battery. The fact that lithium is the lightest metallic element and has the greatest electrochemical potential and the largest energy density per unit of weight led to the development of lithium batteries.[19] However, lithium batteries could not be made rechargeable because of the possibility of explosion. Sony replaced lithium batteries with lithium-ion batteries in 1991. Lithium-ion batteries have slightly lower energy density than lithium batteries but twice that of nickel-cadmium batteries, which are also rechargeable. They have a cell voltage of 3.6 v, three times that of nickel-based cells. Lithium batteries are also lighter, hold higher charges, and have a much greater rate of charge and discharge than the lead-cell batteries most commonly used in automobiles. They are 80–90% efficient.[20] Lithium-ion batteries are low maintenance and low self-discharge. They can provide very high current and are easy to dispose of when they have to be replaced.

[18]In 1828, Ányos Jedlik of Hungary invented an electric motor and created a small model car powered by it. In 1834, Moritz Jacobi of Germany created a rotating electric motor, which was followed by the work of Sibrandus Stratingh of the Netherlands and his assistant Christopher Becker, who built an electric motor that powered a small model car in 1835. These engineering breakthroughs followed upon prior contributions by many others: the invention of the electric battery by Alessandro Volta, the Italian physicist and chemist, in 1800; the scientific work of Hans Christian Øersted, a Danish physicist, who in 1820 demonstrated the generation of a magnetic field around electric wire when current flows in it; the creation of the electromagnet by the English physicist William Sturgeon in 1825; and the theoretical contributions to the understanding of electromagnetism by Heinrich Friedrich Emil Lenz, a Russian physicist, in 1833. In the USA, Thomas Davenport received the first patent for an electric motor in 1837. In 1859, the French physicist Gaston Planté invented the lead–acid battery, which is still used in automobiles to operate auxiliary equipment although it cannot be used to propel the vehicle because it lacks sufficient energy to provide reasonable driving range and because its discharge rates (i.e., current density) are too slow to accelerate heavy vehicles quickly.

[19]The first commercial lithium batteries were sold in 1970, although the experiments with lithium batteries had begun in 1912.

[20]Lead–acid (Pb–acid) batteries are 50–92% efficient, and nickel–metal hydride (NIMH) batteries are 66% efficient. Given the cumulative efficiencies of generating electricity, grid loss, charge and discharge losses, the well–wheel efficiency of the lithium-ion battery is 45%, while Pb–acid battery is 0.26–0.48%, and NiMH battery is 0.34%. See Sun [13].

Tesla Motors is so far the most visible and successful of the companies producing electric cars with lithium-ion batteries. The battery cost for a 60 kWh unit is about $12,000 and for an 80 kWh unit $17,000. The driving range is about 200–260 miles, and charging time is in the range of 75 min. While these performance factors are subject to improvement, there are inherent shortcomings to lithium-ion batteries that may limit the prospects for these electric vehicles (EVs), notwithstanding the appeal of expensive high-end electric cars such as Tesla today. Lithium-ion batteries are fragile, require a protection circuit, and need to be monitored for temperature within a range of 1–2 °C. One of their major disadvantages is their relatively short life of three to five years (depending on the frequency of charging and discharging and the level of discharge), although some claim a much longer battery life of up to twelve years. One solution to the long charging times is quick swapping of batteries. As for their use in battery-powered electric cars, one of the fundamental shortcomings of lithium-ion batteries is the very necessity of charging them, especially since provision for charging infrastructure is still in a relatively early state; many consumers also name "range anxiety" (the fear that an electric car will run out of charge before a charging station can be reached) as a major misgiving about driving an all electric vehicle. Yet another problem is cost: the cost of lithium batteries in the large bank of batteries in an electric car can amount to half the cost of the entire vehicle, depending on the price of the car, although prices for lithium batteries may come down significantly in the near term.

Lithium-ion batteries have other significant limitations. A Tesla Model S weighs 4647 lb,[21] 1700 lb of which is the weight of the 80 kWh battery. The weight of 16 gallons of gasoline is about 100 lb. Weight matters in automobiles [14]. The volume occupied by the batteries is also large, and the battery pack in an electric car using a lithium-ion battery can account for a third of the weight of the vehicle. Since the car uses an electric motor, its power and torque depend on the electric current delivered by its battery pack. In the case of a bus, the weight of the total battery package is as much as that of the bus itself; this weight increases energy consumption, especially in cities where stop-and-go driving is typical. There is also the issue of safety with lithium-ion batteries: all lithium batteries have both the oxidizer and the fuel in the same package, as is the case with explosives. These batteries are quite safe under normal operating conditions, but lithium is a highly reactive material that may ignite in the presence of water, and explode if the lithium battery is punctured in an accident.

Other kinds of rechargeable batteries besides lithium-ion ones are being researched or are under development: for example, zinc–air, lithium–sulfur, and lithium–air batteries. Of these technologies, lithium–air rechargeable batteries are currently believed to be the best potential alternative. This metal–air battery uses the oxidation of lithium at the anode and reduction in oxygen at the cathode to generate electric current. It is a solid-state battery that does not use any electrolyte, getting its oxygen from air rather than storing the oxidizer in an electrolyte. The lithium–air

[21]https://www.teslamotors.com/support/model-s-specifications, accessed January 20, 2016.

battery has a high specific energy density per unit of weight, approaching that of the energy density of gasoline (13 kWh/kg). Lithium–air batteries may have five to fifteen times the specific energy of lithium-ion batteries. However, there are still many technical challenges to realizing the potential of the lithium–air battery, an achievement that may be decades away at this point. In any event, we believe that regardless of the state of battery technology, it is best to use a minimum number of batteries to reduce the vehicle weight, the cost of the vehicle, and the use of lithium.

Another model for powering electric cars—hydrogen fuel cells, which convert petroleum into hydrogen so that the engine can convert the chemical energy in hydrogen into mechanical energy—is attractive because hydrogen is the lightest element and its combustion product is water. Yet this technology has several major disadvantages that have hampered its penetration of the automobile market. Like cars with IC engines using petroleum or electric cars with batteries that need to be recharged periodically, cars running on hydrogen fuel cells need to be refueled. Hydrogen is a highly reactive substance, and the safety and reliability of fuel cells in a moving vehicle are a major issue. Most notably, from an environmental standpoint, hydrogen fuel cells use either petroleum or natural gas and generate CO_2. Although many automotive companies, including Ford, invested heavily on fuel cells in the past, and Toyota and other auto manufacturers are investing in this technology now, these problems remain to be addressed before hydrogen fuel cells become practical.

All told, electric motors represent an extremely promising alternative to the IC engine—even given that we must then account not only for the additional demand for electricity but also for the environmental implications of the way in which that electricity is generated. Focusing, for the moment, only on efficiency, the energy conversion efficiency of electric power generation plants is around 40% for those using coal, and it can be as high as 60% when a combined cycle is used. The efficiency of electric motors is also high—more than 90%[22]—and electric motors do not themselves produce any CO_2. Rao estimated the well-to-wheel efficiency of electric drives as 60% higher than that of IC engines[23] [15]. The higher efficiency of electric drives could reduce ground-transport energy consumption by 50% or more,[24] with a corresponding reduction in CO_2 emissions. Further reduction in CO_2 is possible, since it is easier to capture CO_2 at a central electric power plant than from individual IC engines. (With electric vehicles, we can eliminate the direct emission of greenhouse gases from automobile tailpipes, but we will still need to reduce or eliminate CO_2 emissions from electric power plants.) This being the case, the electrification of ground transportation systems should be one of the options on

[22]"Electric Motor Efficiency and Power factor," University of Alabama, Note from ME 416/516.

[23]It is also conceivable that a new kind of automobile engine could be built using gas turbines. Gas turbine-based IC engines that have a much higher efficiency than reciprocating piston-type IC engines can generate electric power, which is then transmitted wirelessly to an electric motor or motors. The motor(s) can be in wheels that receive electric power wirelessly.

[24]Rough estimate, assuming 40% electric power plant energy conversion efficiency, 20% electric transmission loss, 90% electric motor efficiency, 10% loss in inverters, etc.

the table for dramatically reducing emissions of CO_2 from automobiles and other surface transportation vehicles. Moreover, improvements in battery technology of the kind being attempted today are not essential for this objective, as there is another kind of technology altogether available for it.

1.5 The Online Electric Vehicle (OLEV) and the Electrification of Ground Transportation Systems

In 2006, the Korea Advanced Institute of Science and Technology (KAIST) developed a strategic plan to become a world's major university in science and technology.[25] One of the thrust areas we created for research and education was in energy, environment, water, and sustainability (EEWS). With special support of the Korean government for the EEWS program, we undertook many projects in energy and environment. In the field of environment, we identified the reduction in CO_2 emission from automobiles through electrification of ground transportation systems (EGTS) as an urgent task. In 2009, KAIST received a major funding for its OLEV project to create electric vehicles that receive electric power wirelessly while in motion. We installed the first OLEV system[26] in Seoul Grand Park in 2010, a rather short time for such complicated and potentially complex project. We attributed this achievement to two factors: outstanding researchers and staff, and the use of axiomatic design, [16, 17] which is discussed in greater detail in chap. 3.

In this book, we present a new solution to the CO_2 problem generated by ground transportation vehicles: the electric vehicle or OLEV, which our team at the Korea Advanced Institute of Science and Technology (KAIST) has developed as a bus with an electric motor powered by an underground power supply system, embedded in the road that generates a magnetic field above the ground for pickup by a unit mounted on the bottom of the vehicle [18–21] (Note that although the name OLEV indicates a vehicle unit alone, it actually refers to a system comprising a vehicle or fleet of vehicles combined with a charging infrastructure that takes the form of a set of power supply systems buried in the road. In this book we will sometimes use the term to refer to the technology and sometimes to vehicles using the technology.).

[25]In 2016, Thomson Reuters ranked KAIST as one of the ten highest-ranked most innovative universities in the world together with MIT, Stanford, and Harvard, the only non-US-based university in the top 10.

[26]The OLEV technology has received major awards from prominent international organizations. The World Economic Forum selected OLEV as one of the top ten emerging technologies in the world in 2013. *Time* magazine named it one of the 50 best inventions of 2010. The Korea Railroad Research Institute, together with technologists at KAIST, has applied SMFIR technology to trams and high speed trains, winning the 2014 Innovation Award from the International Union of Railways. These are rare distinctions to be conferred by non-academic organizations on a technology developed at a university.

This power supply is augmented by a small battery on board the vehicle (although a capacitor could be used for the same purpose) that allows it to travel on non-powered roadways. When the bus returns to the powered section of the roadway, the power transmitted from that source both propels the vehicle and recharges the battery. Without the large battery bank, the vehicle is lighter and smaller, increasing its energy efficiency. To protect pedestrians from the magnetic field, the power supply in the road is segmented, and sensors turn on the field in each segment only when an OLEV is directly above it. An OLEV bus does not emit any harmful gases and avoids the need to carry a large bank of batteries for long-distance travel, which makes it lighter than vehicles that do. It is also quite reliable and safer than any existing surface vehicles.

The core technology of OLEV is shaped magnetic field in resonance, or SMFIR, a technology for the *wireless* transfer of heavy electric power over a large distance. The electric power is transmitted wirelessly from an underground cable to the moving or stationary vehicle using an alternating magnetic field. Ideally, electric power to the underground cable will in the future be supplied by smart electric grid systems. OLEV is a first step toward what we envision as the ultimate goal of the electrification of ground transportation systems (EGTS), to provide an integrated systems solution for all ground transportation including cars, buses, trains, streetcars, and even ships that sail in shallow waters, receiving their power from smart electric grids.

OLEV is designed to achieve three important goals: reducing the emission of CO_2 into the earth's atmosphere by ground transportation systems; improving the energy efficiency of ground transportation vehicles; and ultimately contributing to reducing global warming. Through EGTS, OLEV overcomes the shortcomings of traditional ground transportation systems and of all battery-powered electric vehicles (BPEVs). OLEV systems and conventional vehicles with IC engines have many differences. The first difference is in the energy consumption and efficiency: OLEV, having greater efficiency than vehicles with IC engines, uses much less energy. The second difference is in impact on the environment: OLEV does not emit any greenhouses gases. Third, the overall cost of installation and operation of OLEV over a period of ten years is much less than for vehicles that use fossil fuels. Fourth, OLEV will be part of EGTS, which allows ready integration into the overall electric grid system; currently, gas stations requiring their own infrastructure for oil delivery and storage for transportation systems must be maintained for vehicles with engines, while charging stations for use by many customers or individual charging pads are required for BPEVs. Fifth, because of its underground power system and its capability of being connected to the grid, OLEV lends itself to many applications including automated, driverless vehicles, systems for charging individual customers for power and/or road usage, and—perhaps most significantly, given that the electricity for powering OLEVs must be generated just like the electricity for any other application, including other EVs, thus increasing the overall load on the electric grid—use on electric grids drawing significant amounts of power from renewables.

OLEV is a natural step in the history of technology development. Human beings once had literally to carry energy, mass, and information with them in order to act upon the world in certain ways. Through technological innovation, however, we have successfully reduced the need to carry these things around physically. For example, we replaced human messengers with the telegraph, the telephone, and the Internet. We learned to use animals and machines to move heavy objects. Instead of physically carrying energy sources like wood, we send electrical energy over wires and natural gas through pipelines. OLEV is designed to reduce the transport of both mass and energy by some of our machines—i.e., ground transportation vehicles—in order to make them more safe and efficient.

1.6 Concluding Remarks

In this chapter, we presented the environmental issue OLEV is designed to address, i.e., CO_2 emission by internal combustion engines used in transportation systems. It gives a historical perspective of the evolution of automobile technologies as well as the rationale for replacing them with electric drives. The currently used EVs require a large number of batteries on board, which may cause a huge future problem with their disposal as well as affecting the sustainability of natural resources. OLEV overcomes the shortcomings of current commercial electric vehicles.

References

1. Kiehl JT, Kevin E Trenberth (1997) Earth's annual global mean energy budget (PDF). Bull Am Meteorol Soc **78** (2): 197–208. Bibcode:1997BAMS.0.78.197K. doi:10.1175/1520-0477(1997)078<0197:EAGMEB>2.0.CO;2
2. Joughin I, Smith BE, Medley B (2014) Marine ice sheet collapse potentially under way for the Thwaites Glacier Basin, West Antarctica. Science 344(6185):735–738
3. Dieter L, Martine LF, Bernhard B, Thomas B, Barnola JM, Siegenthale U, Raynaud D, Jouzel J, Fischer H, Kawamura K and Stocker TF (2008) High-resolution carbon dioxide concentration record 650,000–800,000 years before present. Nature **453**, 379–382. doi:10.1038/nature06949
4. Sachs JD (2015) The age of sustainable development. Columbia University Press, ISBN-10: 0231173156
5. Cox PM, Betts RA, Jones CD, Spall SA, Totterdell IJ (2000) Acceleration of global warming due to carbon-cycle feedbacks in a coupled climate model. Nature 408:184–187
6. Heywood J (1988) Internal combustion engine fundamentals. McGraw-Hill
7. Arcoumanis C (ed) (1988) Internal combustion engines. Academic Press, London
8. Hoek G, Brunekreef B, Goldbohm S, Fischer P, van den Brandt PA (2002) Association between mortality and indicators of traffic-related air pollution in the Netherlands: a cohort study. Lancet 360(9341):1203–1209
9. Pope CA III, Burnett RT, Thun MJ, Calle EE, Krewaski D, Ito K, Thjurston GD (2002) Lung cancer, cardiopulmonary mortality, and long-term exposure to fine particulate air pollution. JAMA 287:1132–1141

10. Brugge D, Durant JI, Rioux, C (2007) Near-highway pollutants in motor vehicle exhaust: a review of epidemiologic evidence of cardiac and pulmonary health risks. Environ Health 6:23
11. Ballantine C (2010) Air pollution and health. Lecture at the Center for the Environment, Catawba College. http://www.centerfortheenvironment.org/speaker-video-clay-ballantine.html
12. Gan WQ, Tamburic L, Davies HW, Demers PA, Koehoorn M, Brauer M (2010) Changes in residential proximity to road traffic and the risk of death from coronary hear disease. Epidemiology 21(5):642–649
13. Sun J (2010) Car battery efficiencies. Stanford University Course Work, Physics 240, Fall 2010
14. Ricardo Inc. (2008) Impact of vehicle weight reduction on fuel economy for various vehicle architectures. Research Report of Ricardo for Aluminum Association
15. Rao V (2013) Shale gas: the promise and the peril. RTI Publication, p 100
16. Suh NP (1990) Principles of design. Oxford University Press, New York
17. Suh NP (2001) Axiomatic design: advances and applications. Oxford University Press, New York
18. Suh NP, Cho DH, and Rim CT (2010) Design of On-Line Electric Vehicle (OLEV). In: CIRP design conference, Nantes, France
19. Suh NP, Chang SH, Cho DH, Rim CT, Jung GC, Kim KS, Jeong YH, Yoon YS, and Jeong GH (2012) Power supply device, power acquisition device, and safety system for electromagnetic induction-powered electric vehicle. US Patent Application Publication No. US. 2012/0186927 A1
20. Ahn SY, Suh NP, Cho DH (2013) The all-electric car you never plug. In: IEEE spectrum
21. Suh NP, Cho DH, Rim CT, Jeon J, Kim JH, and Ahn SY (2013) Method and device for designing a current supply and collection device for a transportation system using an electric vehicle. US Patent Application Publication No. US 2013/0304443 A1

Chapter 2
Wireless Power Transfer for Electric Vehicles

Nam P. Suh and Dong Ho Cho

Abstract This chapter provides a basic description of OLEV and its enabling technology of SMFIR (Shaped Magnetic Field In Resonance). It then briefly compares OLEV/SMFIR with other vehicles that use IC engines and other electric vehicles in terms of environmental impact, performance, and cost, which are explored further in Part IV of the book. It also explains the potential benefits of electrifying ground transportation systems (EGTS) with a technology like OLEV and connecting these systems to smart electric grids. Finally, it describes efforts made to commercialize OLEV and lessons from these efforts.

2.1 Introduction to OLEV Technology

The key features of the On-Line Electric Vehicle (OLEV) are shown in Fig. 2.1. The OLEV wireless charging system consists of the following components:

1 A road-side *Power Inverter* to bring electricity from the electric grid system to the road embedded power tracks
2 A *Roadway Infrastructure* consisting of *Road-embedded Power Tracks* installed at selected locations of the bus route
3 A *Pick-Up and Regulator* kit for the wireless power transfer installed in or under the electric vehicle.

The OLEV road-embedded power tracks can be deployed in variable lengths on the route to meet the vehicle operating needs for recharging the battery to maintain

N.P. Suh (✉)
The Department of Mechanical Engineering, Massachusetts Insitute of Technology, 77 Masachusetts Avenue, Cambridge, MA 02139, USA
e-mail: npsuh@mit.edu

D.H. Cho
School of Electrical Engineering KAIST, Korea Advanced Institute of Science and Technology (KAIST), 291 Daehak-ro, Yuseong-gu Daejeon 34141, South Korea
e-mail: dhcho@kaista.ac.kr

N.P. Suh and D.H. Cho (eds.), *The On-line Electric Vehicle*,
DOI 10.1007/978-3-319-51183-2_2

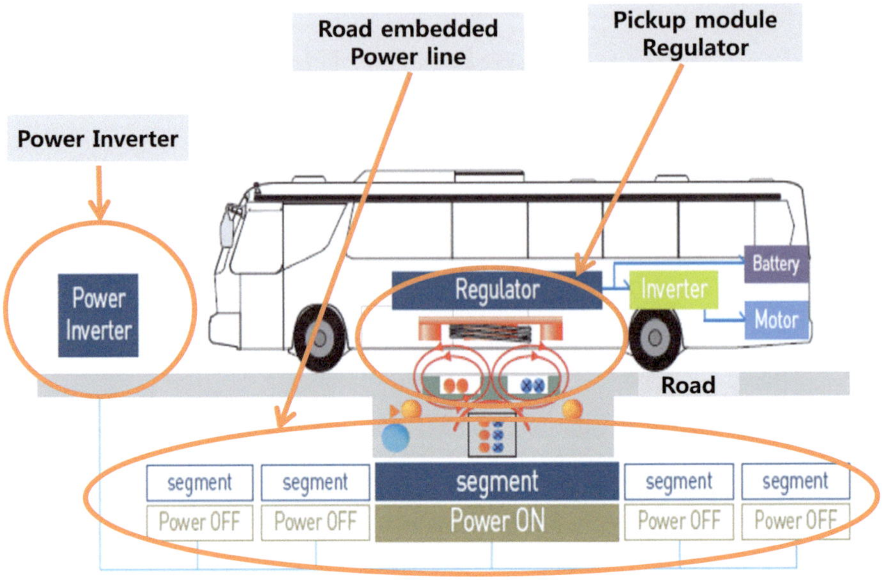

Fig. 2.1 Schematic diagram of an OLEV system

the energy balance level or providing sufficient power for its operation. The OLEV road-embedded power tracks are deployed in multiple segments so that only the segment over which the vehicle is moving above is turned on. The underground power supply system generates a two-dimensional magnetic field (perpendicular to the vehicle moving direction), which is picked up by the moving vehicle. It satisfies all international regulations, including safety.

The On-Line Electric Vehicle (OLEV) avoids the problems of conventional, battery-powered EVs—not just the weight but also the volatility of lithium-ion batteries, and the need to recharge the battery—by receiving its electric power dynamically (i.e., while in motion) and wirelessly via roads equipped with an underground power supply system. It is virtually, though not entirely, battery-less, for to provide autonomous mobility on roads not equipped with the underground power supply, OLEV carries a small battery on board. The externally supplied electric power both propels the OLEV and recharges the battery. The OLEV bus system is designed so that the electric charge in the battery returns to its original charge state when the vehicle makes a complete round trip. This charging occurs automatically when the vehicle is moving on roads with the underground power supply, without any input from the vehicle's operator. The battery size for an OLEV bus is 5–20% the size needed for an all-battery-powered EV bus. (In the case of private vehicles that use plug-in charging, the battery must be large enough to supply electric energy until the vehicle can be plugged-in at home or at a charging station. OLEV batteries are typically charged to about 50% of capacity, which allows for the storage of excess energy and helps preserve battery life.) The length of the underground power supply system, meanwhile, is optimized to minimize the overall cost of the OLEV system—which includes the cost of installing the underground power supply, the cost of the battery, and the cost of the vehicles themselves—while maximizing the speed of the vehicle and the range of autonomous mobility.

The new wireless electric power transfer technology is named as the shaped magnetic field in resonance (SMFIR).[1] SMFIR transfers heavy electric power—

---

[1]Attempts to use wireless technology to charge electric vehicles have a long history. In 1894, Nikola Tesla, who spent many years experimenting with the wireless transmission of electricity, received a U.S. patent for his invention of an electric railway system using inductive coupling, although his idea was never put to use. In the 1990 s, General Motors and Toyota experimented with inductive charging in their first electric vehicles, and in 2002 a system of inductive charging (manufactured by the German firm Conductix-Wampfler) was implemented on city buses in Turin and Genoa, Italy. The Italian buses, however, use stationary charging, meaning that the vehicles can be charged only when stopped over induction coils installed in the road. See Markkus Rovito, "OLEV Technologies' dynamic wireless inductive system charges vehicles while in motion," *Charged*, 5/1/14, https://chargedevs.com/features/olev-technologies-dynamic-wireless-inductive-system-charges-vehicles-while-in-motion/ (originally in *Charged* Issue 12–FEB 2014).

There are many ways of sending electric power over a large distance. One of oldest technologies for this is the electromagnetic radio signal transmitted by TV or radio antennas. The signal is propagated over a long distance, but because it is weak it must be picked up and amplified by supplying external electric power to the TV or radio. The technologies for wireless power transfer to EVs that have been tried may be classified into the following three types:

1. Magnetic induction

AC current flowing through a circular ring will generate three-dimensional magnetic waves. This field is picked up by a pickup unit mounted on a vehicle. This inductive technology has a basic limitation for EVs: the three-dimensional shaped magnetic field it generates cannot be picked up by another coil (on top of it) unless the two coils are in close proximity. Therefore, in some wireless power transfer technologies that use this technology, the top pickup unit is lowered to bring it into the requisite proximity to the bottom coil align precisely to obtain acceptable efficiency. This technology cannot be applied to moving vehicles.

2. Magnetic resonance

WiTricity Corp. has developed a technology based on an invention by Professor Marin Soljačić of MIT. In this technology, the coils generate electromagnetic waves that are picked up by a resonator at a large distance. This technology cannot be used to transmit the high power needed for transportation, however.

3. Shaped Magnetic Field in Resonance (SMFIR)

This technology, pioneered by OLEV, is different from the above technologies in several important ways:

(a) It uses ferrite cores to shape the two-dimensional magnetic field in order to create a "magnetic field path" from the bottom ferrite core to the core attached to a moving vehicle. The high-intensity field is confined in a relatively well-defined space between the ground and the vehicle. This is equivalent to creating a loop from the poles of the underground ferrite core (think of the top two ends of the letter U as the two poles) through the poles of the top ferrite core (an inverse-shaped U) of the pickup unit attached to the vehicle. As the magnetic field oscillates through these ferrite "loops," we pick up the energy associated with the magnetic more than 100 kW—and overcomes the limitations of conventional EVs that have to recharge its batteries. (EVs need about 500–1250 kWh of electric energy, or 100–250 kW of electric power for about five hours of operation). The gap between the ground surface and the pickup unit attached to an OLEV bus is typically about 20–25 cm. The maximum efficiency of power transfer achieved to date is 85%, which depends on the size of the gap and the vehicle's alignment with the magnetic field. (See Fig. 2.2) As the gap decreases—e.g., when vehicles are moving on rails—the efficiency increases significantly. In current installations, the underground power supply system is connected to a conventional electric grid, which will eventually be replaced by smart electric grids to better optimize the power distribution, generation, and storage. The resonance and the high quality factor (i.e., Q factor) are the key requirements of SMFIR.

The input power to the underground system is 200 A at 440 V and 20 kHz. The magnetic field above the ground is shaped to reach the vehicle. To maximize the power pickup, the pickup unit mounted on the vehicle is tuned to the frequency of the magnetic field. The power picked up is supplied at 60 Hz in AC to the electric motor that drives the wheels of the OLEV, and to the battery in DC to recharge it. Two kinds of electromagnetic field (EMF) shielding are deployed to protect people from EMF radiation emitted by the SMFIR system: one embedded underground, and a passive cancellation system mounted on the vehicle. (Sometimes an active shielding system is mounted on the vehicle.) The EMF radiation from OLEV is well below the internationally specified level of 62.5 mG at 20 kHz. The power supply is segmented so that only the segment directly below the vehicle is activated.

(Footnote 1 continued)

field using the resonance effect. In order to pick up the magnetic field, the top pickup unit must be in resonance with the field frequency of the lower unit imbedded in the ground, which creates a "continuous loop" of magnetic field. This is why we call our technology "Shaped Magnetic Field in Resonance" (SMFIR), which is a patented technology.

(b) We control the height of the heavy magnetic power transfer by changing the width of the two ends of U-shaped ferrite cores; the farther apart they are, the greater is the height the field can reach.

(c) Unlike magnetic induction, which generates a three-dimensional magnetic field, OLEV generates a two-dimensional magnetic field along the direction of the vehicle motion by having a series of U-shaped ferrite structure of the lower unit imbedded in the ground, which creates a "continuous loop" of magnetic field. This is why we call our technology "Shaped Magnetic Field in Resonance" (SMFIR), which is a patented technology.

Additional technical detail about SMFIR is provided in later chapters of this book. The technology for shaped magnetic field in resonance (SMFIR) system, a critical part of the development of OLEV, was designed and implemented by researchers at KAIST under the leadership of Professor D.H. Cho, using the theoretical design framework of axiomatic design. There are a large number of patents covering these technologies. For further discussion of wireless technology and its various uses, and of future applications for SMFIR, see Chap. 19.

		1G(Bus)	2G(Bus)	3G(Bus)	4G(Bus)
Date		July 14. 2009	Jan. 31, 2010.	Oct. 25, 2012.	Dec. 25, 2013.
Vehicle					
System Spec.		air-gap = 17cm Efficiency = 72%	air-gap = 23cm Efficiency = 81%	air-gap = 23cm Efficiency = 85%	air-gap = 23cm Efficiency = 85%
EMF		Less than 51mG	Less than 50mG	Less than 62.5mG	Less than 62.5mG
Power Rail (width)		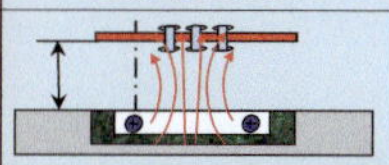	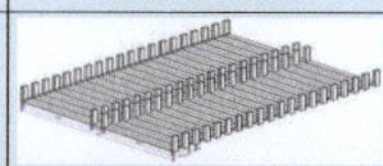	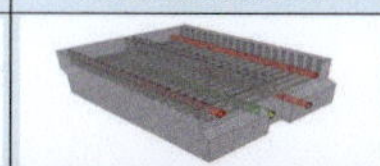	
			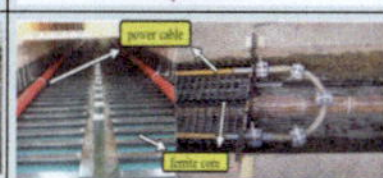	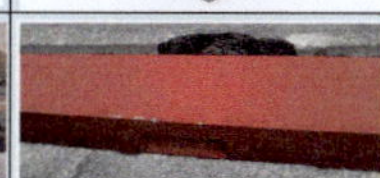	
		1400mm	800mm	800mm	800mm
Pick-up	Picture	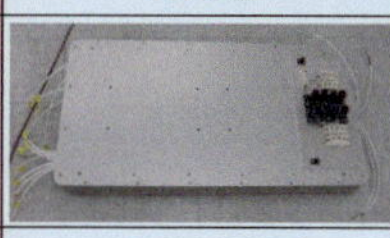			
	Power	12kW(pickup unit) 60kW(pickup total)	15kW(pickup unit) 75kW(pickup total)	33.33kW(pickup unit) 100kW(pickup total)	20kW(pickup unit) 100kW(pickup total)
	Weight	160kg(pickup unit) 800kg(pickup total)	110kg(pickup unit) 550kg(pickup total)	150kg(pickup unit) 450kg(pickup total)	90kg(pickup unit) 450kg(pickup total)
	Size (unit)	1600(W)×600(L)x110(H) mm^3	1250(W)×740(L)x177(H) mm^3	1250(W)×710(L)x165(H) mm^3	1250(W)×740(L)x165(H) mm^3

Fig. 2.2 Four generations of OLEV

OLEV's first test of its SMFIR system in 2009 was the first successful demonstration of a wireless power system for electric vehicles in motion. The OLEV system has so far been deployed at four sites in Korea. The OLEV tram system originally installed in Seoul's Grand Park as a pilot project in late 2009–early 2010 has been running as a commercial service since 2011, replacing a noisy and smelly diesel system. It travels on a 2.2 km circular path around the park, powered by an underground power supply system totaling 372.5 m in four segments (Fig. 2.3). In 2012, OLEV buses serviced the international exhibition Expo 2012 in Yeosu. Since 2012, two OLEV buses have been transporting students, staff, and faculty around the KAIST campus in Daejeon. In August 2013, OLEV was also installed in Gumi, an important industrial city in Korea with a population of 350,000. The thirty-five kilometer route is serviced by six buses and powered by six charging pads under the road (Fig. 2.4). Gumi added two more OLEV buses to its system in May 2016. The latest to install an OLEV system is Sejong, a newly developed city that houses many central government offices.

Fig. 2.3 OLEV tram installed in Seoul's Grand Park

Fig. 2.4 OLEV bus in Gumi. The *blue lines* indicate the location of the underground power supply cables

One exciting prospect arising from this technology is high-speed trains equipped with SMFIR. Europe has had high-speed electric trains for three decades, Japan for only somewhat less. Korea also has a high-speed electric train that is the primary mode of transport between major cities; the electricity cost of transporting a passenger on these high-speed trains from Seoul to Busan (about 200 miles) is extremely low. China has been building high-speed train-systems throughout the country, which may be a good investment in the long term. (Although the United States, for political reasons, has not yet been able to build high-speed railways, sooner or later it will end up building them—albeit at greater cost than necessary owing to the delay.) In Korea, KAIST and the Korea Railroad Research Institute

(KRRI) developed the proof of concept system for high-speed trains that can travel at 450–600 km/h using SMFIR transmitting one megawatt of electric power. These trains operate without the cumbersome overhead infrastructure (including sliding mechanical contacts for electric power transfer) that limits the speed of current high-speed train systems,[2] and thus require smaller (by 30%) tunnels than do current systems. Subways could also be built more cheaply with SMFIR since the overhead superstructure would not be needed. In existing subway tunnels, double-decker trains could be used at minimal additional cost. The KAIST team has also applied the SMFIR technology to a tram that runs on rails in cooperation with KRRI (Fig. 2.5).[3] Recently, in cooperation with many industries, KRRI and KAIST are developing the light rail transit with wireless charging, which will be commercialized in 2019.

In Korea, the one country in the world in which OLEV has so far been successfully commercialized, two studies have demonstrated the benefits, in terms of cost (including, in one case, the "CO_2 cost"), of OLEV technology versus more conventional technologies for city buses. It is estimated that, over a ten-year period, the cost of implementing and operating a new OLEV six-bus system (like the one currently operating in Gumi City) in another location in Korea would be about half of the cost of implementing and operating a similar route with CNG (compressed natural gas) buses [2].[4] In another study, the total cost of four different bus systems, i.e., diesel, CNG, OLEV, and plugin electric vehicles (PEVs), are compared as shown in Fig. 2.6 [2]. The total cost of the OLEV bus system is the lowest, despite a greater initial bus cost for the OLEV bus than for the diesel or CNG buses.

---

[2]Typical high-speed trains in Korea run at about 300 km/h. What limits their speed is the mechanical sliding contact between catenary electric wires and the pantograph on top of the train for transmission of electricity. Since the OLEV train runs on rails, its efficiency is expected to be up to 90% higher than that of the OLEV bus, because the pickup unit can be closer to the emitting unit. For additional discussion of wireless power transfer technology for trains, see Chap. 17.

[3]One interesting application area is the use of SMFIR to transmit electric power wirelessly to ships. Since the magnetic field is not affected by water, the SMFIR system can be deployed in water. A ship can charge its battery while in dock or while in motion in shallow waters.

[4]Implementing such a system, this study found, would cost approximately $900,000 per bus for purchase of the vehicles plus installation of the OLEV infrastructure. This compares with a cost of approximately $600,000 per bus for the purchase of six CNG buses (factoring in the government subsidy of $300,000 for each OLEV bus and $100,000 for each CNG bus; typical heavy-duty diesel buses cost from $200,000 to $600,000 [3]). However, the initial cost for the OLEV system would be offset by fuel savings over a ten-year period (during which the fuel costs for CNG buses would significantly surpass the fuel costs for OLEV buses). To make one round trip, a CNG bus uses $20.58 worth of fuel, but the OLEV bus uses just $3.92 worth of electricity. Over ten years, this adds up to $4,500,000 of fuel for the six CNG buses, and just $860,000 of electricity for the six OLEV buses. Adding in the carbon tax of $401,000 for the use of natural gas, the total operating and capital cost of the CNG buses is about $5,510,000 over ten years, while the OLEV buses cost $2,659,000—about 50% less than the CNG bus system. Even without the government subsidy for the buses, the OLEV bus system still costs about $1.5 million less. This study also estimates that over a ten-year period, the overall cost of an OLEV system is 40–60% that of an equivalent diesel bus system, because of fuel savings.

Fig. 2.5 SMFIR-based tram tested in Korea (90 kHz power supply)

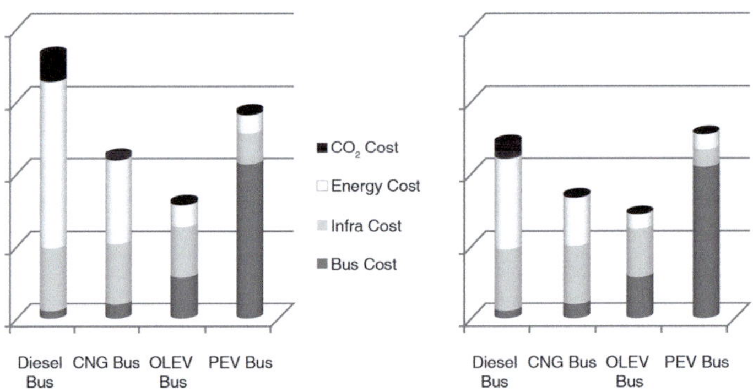

Fig. 2.6 Cost comparison of diesel, CNG, OLEV and PEV bus systems. The figure on the *left* is for one dispatch every four minutes, and the figure on the *right* is for every ten minutes. The vertical axis is millions of US dollars [2]

The PEV bus is the most expensive of the four systems because of the higher initial cost for a bus. The operating costs of diesel and CNG are higher because of the fuel cost.[5]

[5]This analysis was based on a proposal submitted to the government by a city in Korea for a bus rapid transit (BRT) system. It assumed a 20 km bus line traveled by fifteen buses running at four-minute or ten-minute intervals (except for the PEV system, which would require an additional seven buses due to the thirty-minute battery charging time). The analysis also assumed an amortization period of nine years. The total operating cost was calculated based on the costs of the buses, the infrastructure, the fuel, and the CO_2. The CO_2 cost was calculated under the assumption that 50% of the electricity was generated by nuclear power plants, via a formula advanced by a national environmental research institute in Korea.

How much would it cost to transition from a ground transportation system built mainly around vehicles with IC engines to EGTS using the SMFIR technology? To answer this question, we have to assess two major costs in addition to those for owning and operating OLEVs: the cost of generating more electric power for electrified ground transportation and the cost of laying the underground power supply system on roadways. To electrify ground transportation, additional electric power plants would obviously need to be built, although utilities and financial markets would support this because the costs would be recovered through increased electricity sales. (Although reductions of CO_2 and other GHGs achieved through EGTS would have to be balanced against increases in emissions from electricity generated from fossil fuels, capturing and sequestering emissions at power plants is more efficient than capturing it at the tailpipes of individual IC vehicles. Moreover, as we shall see, a system of wirelessly powered electric vehicles connected to the grid would facilitate the use of clean, renewable energy sources for electricity generation.)

The cost of electrifying roadways depends, of course, on what fraction of the roadway must have the underground power supply system installed in it. The three OLEV bus systems currently operating in Korea run with between 2 and 15% of their roadways electrified. Continuous, nonstop operations require longer roadway installations than those in which a bus waits at a station before the next run (since stations are equipped with infrastructure that charges the batteries automatically when the buses are stopped there). In major cities like Seoul, a maximum of only about 30% of streets might need to have the underground power supply system if all cars, buses, and trucks were OLEVs equipped with a small battery that could propel them for about ten miles between charges. Highways, where vehicles move at higher speeds, might need the underground power supply system installed in more than 50% of the roadway.[6] At any rate, as the number of OLEVs on the road increased, the cost of the infrastructure would be spread over this increased number and the cost per vehicle would decline, as the same infrastructure can be used by any number of OLEVs. This is in contrast to BEVs using stationary charging, for which additional infrastructure (charging stations or pads) must be installed as the number of vehicles on the road increases.

In Korea, which imports all of its energy, as well as in other energy-dependent countries, a commercial bus system like the one now operating in Gumi City could be cost-effectively implemented throughout the country. Korea consumes about 2.5 million barrels of oil per day, and in 2013 about 68%[7] of that oil was used for ground transportation. With the price of oil at $100 per barrel,[8] Korea's annual expenditure for fuel used in ground transportation is roughly $62 billion. If

[6]Regardless of the speed of the OLEV vehicle, the time required to recharge the battery for use on roads without the underground power supply system remains the same. Therefore, as the average speed of OLEV increases, the length of the underground power supply system must be longer to maintain the same charging time.

[7]Energy Data Handbook 2013.

[8]The cost of petroleum fluctuates. In 2015, it came down to about $50 a barrel.

adoption of EGTS reduced oil imports for ground transportation by 30%,[9] Korea could save about $18 billion annually on fuel, which it could use to install the underground power supply system on about 9000 miles of roadways per year, assuming a cost of $2 million per mile.[10] Electrification of all of Korea's 56,000 miles of paved roadways would take about six years, if all of the savings on oil imports were used for electrification (These are estimates that obviously depend on many factors including the price of oil, cost reductions via economies of scale, and the profit margins of vendors.).

All this being said, a more realistic scenario for the introduction of EGTS in most countries in the world would be not an immediate transition from IC vehicles to EGTS but, rather, a slower shift during which IC vehicles are gradually replaced (not without enormous resistance from those heavily invested in IC technology, primarily the oil industry and the old-line automobile manufacturers) with electric vehicles employing a variety of power and charging technologies that would then compete among themselves for dominance. The question we then need to answer is: Why OLEV rather than primarily battery-powered alternatives for EVs?

2.2 OLEV Versus Other EV Technologies

As noted in Chap. 1, there are several problems with using batteries as the primary source of power for EVs. The need to charge the battery periodically and the attendant limitations on vehicle range (particularly at the present stage of battery technology) create one major drawback for BEVs. The danger of electric shock posed by plug-in battery charging is another, while a lithium-ion battery—the type currently favored in BEVs—carries a flammable electrolyte that makes it essentially an explosive, as evidenced by the major fire and explosion of a Tesla car in 2013 when the lithium battery was punctured after the battery pack was hit by a "large metallic object."[11] Finally, the disposal of used lithium-based batteries is going to be a major environmental issue when more than a billion vehicles use batteries, which is already

[9]This is lower than Rao's estimate of 60%.

[10]The system installed in Gumi City cost $1.4 million for 35 km (~22 miles), or $64,000 per mile. It incorporated six charging stations, each costing about $230,000. A new installation in the same city is budgeted to cost $150,000 per charging section. The number of charging stations required increases with average vehicle speed, so highways require more of them than city streets. Two million dollars per mile is a generous allowance to cover costs such as installing electric power supply lines in Korea. In America the cost of the power supply infrastructure will be higher than $2 million per mile, and in China it will be lower, due to differences in the cost of power components and equipment in the two countries.

[11]Per Liljas, "America's Safest Car Ablaze After Fire Starts in Battery Pack," *Time*, October 3, 2013 (http://business.time.com/2013/10/03/americas-safest-car-ablaze-after-fire-starts-in-battery-pack/, accessed December 5, 2015.) In January 2016, a 2014 T Model S was destroyed after bursting into flames while charging at one of Tesla's "Supercharger" stations in Norway; see Clifford Atiyeh, "Tesla Model S Catches Fire at Supercharger Station in Norway," *Car and Driver*, January 4, 2016

the most recycled consumer product [1]. Although we do not have experience with lithium-based batteries, the recent articles on lead batteries highlight future potential problems associated with using a massive number of lithium-based batteries.[12]

The weight of electrochemical batteries also decreases a vehicle's efficiency. In urban driving, with its frequent stops and starts, pure battery-powered EVs require more energy than an OLEV due to the weight of batteries. Since OLEVs are expected to be lighter than all-battery-powered EVs by 25% or more, the energy consumption of OLEV per unit of distance driven is expected to be less than those of BPEVs by roughly 10–20%. Moreover, if the same size engine is used in an EV as in an OLEV, the OLEV is going to have much higher acceleration, since acceleration is inversely proportional to the mass of the vehicle.

Finally, electrochemical batteries store electric energy (when charged) as chemical energy, then convert chemical energy to electric energy again, with a loss of energy and efficiency with every phase change. Simply put, carrying a vehicle's energy supply on the vehicle itself is an inherently inefficient system—particularly in contrast to wireless transmission for EVs, which is weightless. This is a compelling reason for hastening the transition to wirelessly powered EVs rather than continuing to rely on battery power.[13]

While few studies are available at this point for comparing OLEVs and more conventional EVs in terms of cost, effects on emissions, and the merits and demerits of carrying a large number of Li-ion batteries on board, this book presents research on the energy efficiency of OLEV buses versus plug-in hybrid and pure battery-powered ones (further details in Chap. 21), and on the systemic impacts of both OLEV and more conventional EV concepts on the transportation-electricity nexus in relation to which the performance of any EV must ultimately be evaluated (further discussed Chap. 20). Such research should help quantify the effect on GHG emissions of implementing BEV versus wireless dynamic technology. Future research will need to examine the long-term implications for safety, cost, sustainability, and geopolitics of using an astronomical number of lithium-ion batteries with increases in the number of EVs on the road. For example, just as the need for oil for automobiles has altered and even distorted geopolitics, the need to access lithium supply could have a similar effect, especially in the regions around the lithium-rich countries.

This is all separate, of course, from the issue of the potential impact of the two technologies on how electricity is generated and electricity usage managed as EVs

(Footnote 11 continued)

(http://blog.caranddriver.com/tesla-model-s-catches-fire-at-supercharger-station-in-norway/, accessed January 5, 2016).

[12]Joshua Parlow and Joby Warrick, "A Dangerous Export—America's car-battery wastge is making Mexican communities sick", Washington Post, February 26, 2016 (http://www.washingtonpost.com/sf/national/2016/02/26/a-dangerous-export-americas-car-battery-waste-is-making-mexican-communities-sick/).

[13]As battery technology continues to evolve in the direction of greater efficiency and safety, OLEV will benefit, as it uses small batteries for free autonomous mobility on roads without the underground power supply system.

began to supplant vehicles powered by IC engines. In 2014, the United States, the world's second-largest consumer of electric power (after China), generated roughly 37% of its CO_2 emissions from electric power plants fueled by coal and natural gas,[14] while 28% of its CO_2 emissions came from vehicles burning fossil fuels.[15] Thus the transition from internal combustion vehicles to EVs, which will increase demand for electric power, will only heighten the urgency of replacing fossil fuels with nuclear power and renewable energy sources to meet future electricity needs. One advantage of using OLEV technology for EGTS, compared with relying on battery power and stationary charging, is the way in which OLEV and EGTS could be used both to manage electricity demand and to facilitate the transition to renewable sources of energy for electricity generation.

2.3 OLEVs, Smart Grids, and Renewable Energy Sources

One of the well-known issues with renewables such as solar is that electricity generation from these intermittent sources fluctuates in ways that cannot be controlled, greatly adding to the inherent complexity of managing supply and demand on an electric grid. Although the simplest system is one in which total power generation and total power consumption match exactly at all times, this rarely occurs. Some systems have time-varying power demand, which must be matched by the power input. Conversely, the power input may vary as a function of time, such as when power is generated by windmills or solar cells.

When the input to the system cannot be controlled, the output (i.e., demand) side must be modulated to consume all the power coming into the system from the input ends. This modulation is performed by a "buffer" in the system. When the input power to the system is greater than demand, the excess energy is stored in the buffer (unless the demand for power can be increased artificially). When the demand for power is greater than the supply, the grid can use energy stored in the buffer (or increase the power input at the supply side by using standby electric power generators—e.g., gas turbines). In theory, with a large number of "smart" OLEVs connected to the grid and equipped with the requisite sensing and communications devices operating in real time, their batteries could act as buffers for the grid. Their limitation for this purpose would be the smallness of the battery or battery pack on an OLEV, although precise information on the time-varying fluctuation of demand, or of the power input, should allow a "smart" grid to function with a smaller buffer than otherwise.[16]

---

[14]U.S. Energy Information Administration. http://www.eia.gov/tools/faqs/faq.cfm?id=77&t=11.

[15]U.S. Energy Information Administration. http://www.eia.gov/tools/faqs/faq.cfm?id=307&t=10.

[16]To manage the dynamic nature of an electric grid with many diverse energy sources, sinks, and storage sites (and where some individual nodes can act as either a source, sink, or storage unit at various times), information must be measured and transmitted among all the elements on the grid including central power generation plants, distributed generation and storage systems, transmission

Meanwhile, the fact that OLEVs draw power from the grid and charge their batteries in real time (when most appropriate from the standpoint of system optimization), while battery-powered EVs have to be charged at fixed intervals, points to a way in which OLEVs could lend themselves much better than conventional EVs to deployment with renewable sources of electricity. The complexities of managing supply and demand on a typical electric grid are exacerbated by the fact that the value of electricity depends on the time of the day at which it is consumed: at night, when demand is lower, electricity is more plentiful and thus cheaper than during the day. For owners of battery-powered EVs, this is currently creating an incentive to charge vehicles overnight after they have discharged their batteries during the day. According to a recent report, however, from the U.S. National Renewable Energy Laboratory that examines new technologies and systems with the potential for achieving "transformative reduction" of GHG emissions from transportation, "analysis of the electrical energy demand of electrified roadways highlights that the demand time period may coincide with renewable generation and thus lead to expanded capability to accommodate renewables in the electrical grid and reduction of the GHG impacts of transportation." In other words, a system like OLEV that draws its power in real time from electrified roadways could actually help advance the integration of renewables into electrical power supply systems—thereby increasing the impact it would already have on reducing GHG emissions because of its greater efficiency compared with vehicles that carry their power source on board.[17]

Finally, smart OLEVs connected in real time to the grid in an electrified ground transportation system could also provide accurate and dynamic data on vehicles in use for: (1) communication among vehicles for purposes including minimizing traffic congestion, collision avoidance, driverless steering, and even remote control of EV operating speeds; and (2) systems for charging drivers for electricity use, which could facilitate uses including congestion pricing and vehicle sharing.

(Footnote 16 continued)

nodes and plants, and consumers (stationary or mobile). This is the reason for making the electricity grid "smart.".

[17]As this report notes in another place, "In high renewable penetration scenarios, curtailment of renewable generation occurs seasonally when load is low and renewable generation is high. Electrified roadway scenarios move some of the load that would have occurred in the evening to the daytime periods and thus offers new load that can utilize the excess renewable generation." Laura Vimmerstedt et al., *Transformative Reduction of Transportation Greenhouse Gas Emissions: Opportunities for Change in Technologies and Systems* (Golden, CO: National Renewable Energy Laboratory, April 2015), pp. 45, 47. The author is grateful to one of the report's co-authors, Tony Markel of NREL, for answering questions about its findings.

2.4 Paving the Way for OLEV

OLEV offers considerable advantages as a technology for the ground transportation of the future, and in many ways seems far and away the most logical one so far invented and actually demonstrated. Yet its introduction and adoption on a large scale are, of course, by no means inevitable and will not happen without difficulty. Some of the major challenges are suggested by the attempts to date to commercialize the technology outside of its home country of Korea.

The commercial implementation of OLEV in Korea to date has been carried out by a company, Dongwon OLEV, that was founded in 2011 and granted an exclusive license to commercialize the technology in Asia. The Dongwon Group is a Korean conglomerate involved in marine products and logistics, food processing and distribution, industrial materials, construction, information and communication technology, IT services, and—with the establishment of Dongwon OLEV—green transportation. Dongwon OLEV installed the OLEV systems now operating in Gumi City, KAIST campus and Seoul Grand Park and will do the installations in Sejong. The central government and regional government in Korea jointly provided funding for these projects. These OLEVs operate at 20 kHz and are designed to receive 100 kW of power. The high-speed train developed by KAIST and KRRI operates at 60 kHz and transmits 1000 kW wirelessly. Meanwhile, a company called OLEV Technologies, headquartered in Boston, was also founded in 2011 with an exclusive license to commercialize OLEV in the Americas and co-exclusive rights (with Dongwon OLEV) for Europe and Africa. Two old friends in the United States who were on the President's Advisory Council at KAIST and who wanted to help KAIST realize its goal of commercializing OLEV[18] put personal funds into a venture company in the United States to launch the company.

The primary challenge OLEV Technologies has faced to date is that EVs of the kind for which OLEV, in its current state of development, is well suited have not yet penetrated the market in sufficient numbers.[19] There are currently less than 100

[18]These two angel investors were Dr. A. Neil Pappalardo, the founder and CEO of MEDITECH, Inc., and Dr. "BJ" Park, the founder and former CEO of MTL, Inc.

[19]The first president of OLEV Technologies, Dr. Hikyu Lee, attempted to sell the patented and proprietary OLEV system for use with buses in the United States. As each OLEV 20 kW power unit weighs 400 lbs., the technology is currently too heavy for smaller vehicles but still lighter than the battery packs carried on battery-powered electric buses. A regional port authority showed interest in purchasing the system for airport buses but balked at the idea of OLEV Technologies, a for-profit company, making a profit from the sale.

In 2013, with OLEV Technologies having gone through its first round of funding, the board wanted new leadership to pursue opportunities for sales to the private sector. We brought in Bryan S. Wilson, who had spent the previous twelve years developing infrastructure for the wireless communications industry. The company's two angel investors put in a second round of funding and Wilson began to looking to sell the OLEV technology to concerns including shuttle operators, ports (where electric vehicles are an attractive option for transferring cargo from ships to trucks and trains and vice versa), and mining companies. A major multinational corporation with mining interests showed interest in OLEV for electric mining vehicles that operate underground (making battery electric buses in operation in the United States. These vehicles are either battery-powered and being recharged with plug-in systems overnight, or are powered by overhead conductive systems. Both options are much less than optimal, but this is the way the vehicles are currently being built by the OEMs. For OLEV Technologies to have an adequate customer base, there will have to be a sufficient number of EVs in the marketplace, or being readied for the marketplace, so that OLEV systems could either replace existing systems or become a component of new vehicles. (The latter would require partnering with OEMs who are currently not selling enough of their own vehicles using plug-in technology or battery-swapping to be seeing any demand for online vehicles, at the cost of an OLEV system, at present.) Even at its current cost, OLEV (as one of the studies cited above makes clear) makes good economic sense in terms of total cost of ownership for bus operators using diesel buses, although this has not yet created enough demand for such a system. It also seems worth noting that two major companies selling inductive power transfer systems for electric vehicles (IPT Technology in Germany and the Canadian multinational Bombardier[20]) have installed numerous systems for electric buses in Europe, but all of these systems currently use wireless power only for stationary charging. If a deep-pocketed company like Bombardier—which has its own technology for wireless dynamic power transfer—saw a market, at present, for a wirelessly powered vehicle, it would certainly have the means to commercialize it, but so far it has not.

Given the reality, however, that the IC engine needs to be replaced by electric ones in order to reduce GHG emissions from vehicles to the necessary levels, wireless dynamic power transfer for vehicle operation (not just for charging a battery-powered vehicle) is the best technology available over the long run because of the inherent inefficiency of a vehicle carrying its own energy supply on board. In some countries, government will step in, embrace the idea of EGTS as part of its long-term energy policy, and give the technology a boost by

(Footnote 19 continued)

charging and swapping challenging), but wanted to have a product to try out; not being in the business of producing vehicles, or having relationships with OEMs that could produce vehicles equipped with OLEV pickup units, OLEV Technologies was unable to supply a product. A company that makes vehicles for airport baggage transfer expressed interest in having its vehicles equipped with OLEV technology, but the technology costs more, at present, than the vehicles it would have been installed for. A heavy equipment manufacturer making electric vehicles for which OLEV would be a good fit appeared to be another prospect, but was still trying to develop its own customer base for its products. As these sales efforts came up short, OLEV Technologies began seeking grant money from government agencies interested in trying out its system, only to have these initiatives founder on the complexities of doing business with government.

[20]For a description of Bombardier's PRIMOVE wireless power transfer system for trams, see Chap. 17.

underwriting road electrification just as it constructed today's highway systems.[21] The cost-benefit analysis for installing the infrastructure for a system like OLEV will, of course, vary from place to place and is subject to many variables including future improvements in battery technology for BEVs, in capacitors (which could replace the batteries in OLEVs if there were enough underground charging infrastructure installed), and in OLEV technology itself (e.g., reduction of the weight and size of the pickup unit).[22] China and India may prove to be the real proving grounds for wirelessly powered vehicles. China will eventually have the largest market for automobiles in the world, followed by India. These two

[21]Sweden and the U.K. are two countries where the government has been funding, or has pledged to fund, research on the feasibility of electrified roads for vehicles using wireless dynamic power transfer. In Sweden, a project launched in 2010, managed by Volvo Group Trucks Technology, and partly financed by the Swedish Energy Agency studied and proposed means of transmitting electrical energy from highways to vehicles via both conductive and inductive (i.e., wireless) technology. Scania AB, the Swedish commercial vehicle manufacturer, and Bombardier developed the project's "inductive energy transfer solution" based on Bombardier's PRIMOVE. In 2014, two organizations involved in this study, the Swedish Transport Administration and the research institute Viktoria Swedish ICT, partnered with the independent Swedish National Road and Transport Research Institute to test electrified roads employing both conductive and inductive technology in a "driving simulator." In 2015, the U.K.'s Transport Minister, Andrew Jones MP, announced that the government was allocating £500 million ($781 million) over five years to finance off-road trials for road electrification using dynamic wireless power transfer technology. Both countries have also begun implementing wirelessly charged city buses. See Viktoria Swedish ICT on behalf of Volvo GTT and Scania CV, "Slide-in Electric Road System, *Inductive project report*" (Scania CV, October 18, 2013), pdf available at https://www.viktoria.se/sites/default/files/pub/www.viktoria.se/upload/publications/slide-in_inductive_project_report_draft_phase_1_2013-10-18.pdf (accessed January 6, 2016); VTI (Swedish National Road and Transport Research Institute), "Electric roads: a solution for the future," press release January 16, 2015 (http://www.vti.se/en/news/electric-roads-a-solution-for-the-future/, accessed January 6, 2016); Highways England and Andrew Jones MP, "Off road trials for 'electric highways technology," press release August 11, 2015 (https://www.gov.uk/government/news/off-road-trials-for-electric-highways-technology, accessed January 6, 2016); Federico Guerrini, "The UK Will Be Trialling Roads That Wirelessly Recharge Your Electric Vehicle While You Drive," *Forbes*, August 13, 2015 (http://www.forbes.com/sites/federicoguerrini/2015/08/13/the-uk-will-be-trialling-roads-that-wirelessly-recharge-your-electric-car-while-you-drive/, accessed January 6, 2016); Scania Group, "Scania to test wirelessly charged city bus for the first time in Sweden," press release December 17, 2014 (http://www.scania.com/media/pressreleases/N14050EN.aspx. accessed January 6, 2016); and Neil Bowdler, "Wirelessly charged electric buses set for Milton Keynes," BBC News, January 9, 2014 (http://www.bbc.com/news/technology-25621426, accessed January 6, 2016).

[22]The size and weight of the pickup unit mounted on an OLEV vehicle should be able to be reduced so that the weight drops from 400 lb to about 100—perhaps by laminating thin layers of ferrite to get rid of eddy current fields and skin effects. (This is pure speculation at this point, not based on firm physics.) Meanwhile, we can do a rough lower-bound estimate of the cost of EVs and EGTS. We will assume that there are six billion people on earth and one billion cars on the road. If we assume that cars are used for ten years, we will have to replace 100 million cars a year. If we further assume that the cost of the battery in each EV is $1000, we will be spending $100 billion a year on batteries. If we assume that the infrastructure for EGTS lasts six years, each year we could invest $600 billion for infrastructure rather than in batteries. The need for additional cost-benefit analysis for OLEV is discussed in Chap. 16.

nations must expand their transportation systems since both are still in the process of catching up in terms of automobiles per capita. What they ultimately do for the electrification of ground transport systems will have a big impact on OLEV and other EV technologies, and the fact that their systems are in nascent state could enable them to leapfrog at least some road construction for IC vehicles and BEVs and go straight to electrified roads. If they embrace EGTS, they will become a formidable force pushing the global automotive market in this direction, stimulating innovation and demand for OLEV-type systems and helping to drive down costs for both vehicles and infrastructure including inverters, ferrite and other magnetic materials, special electric cables, capacitors, and special construction equipment.[23]

In the United States, which lags other advanced countries in public spending for transportation infrastructure, the adoption of OLEV or another OLEV-like technology is most likely to be accomplished through a gradual commercialization process that will resemble the spread of cellular phone technology, expanding outward from a small core of users as the technology is proven and gains in popularity while network effects increase its economic value. Such a process may begin with buses traveling on fixed routes in urban settings—school buses and transit buses, for example—with public-sector vehicle operators paying for cost of the infrastructure. As inverters and underground coils are installed, systems could then be extended for public uses like garbage pickup, or private ones like corporate shuttles, and expand outward from there. This process will begin in urban cores, where services such Uber and ZipCar will also benefit from it, and move outward from there. Inter-city and long-distance road electrification even in such a large country as the United States might also not prove as daunting as one might think. The National Renewable Energy Laboratory, for example, has already estimated that "[e]lectrifying 1% of U.S. Interstate Highways would cover 17% of traveling road vehicles … [e]lectrifying 5% of U.S. Interstate Highways would cover 40% of traveling road vehicles … [and] [e]lectrifying 25% of U.S. Interstate Highways would cover 80% of traveling road vehicles."[24]

---

[23]The production cost of airplanes decreases by 10% each time the production volume doubles. Assuming a similar pattern in bus production, if Korea were to produce 1000 OLEV buses a year, the cost of the buses would be reduced by 65%. The U.S. market for buses in 2012 was around 74,000 [3]. Using the same mass production algorithm, we would expect a similar reduction in manufacturing cost. Mass production of key components of the ground transportation system, including buses, inverters, and ferrite cores, could reduce the cost per component by two-thirds (Willcox 2004). Overall, I believe that the cost of the components that go into OLEV and EGTS will decrease by 50% or more if demand increases sufficiently. The price for the ferrite core, for example, should come down through redesign and more R&D when demand for the product increases.

[24]Vimmerstedt et al. "Transformative Reduction of Transportation Greenhouse Gas Emissions: Opportunities for Change in Technologies and Systems," National Renewable Energy Laboratory/U.S. Department of Transportation (NREL/TP-5400-62943 April 2015), p. 72. For an analysis of how EV penetration will be affected by choices made about plug-in versus online infrastructure, see Chap. 22.

References

1. Gaines L (2014) The future of automotive lithium-ion battery recycling: charting a sustainable course. Sustain Mater Technol 1–2:2–7
2. Park JH (2014) Cost analysis, personal communication
3. Lowe M, Aytakin B, Gereffi G (2009) Public transit buses: a green choice gets greener chapter 12 in manufacturing climate solutions. Center on globalization, governance & competitiveness. Duke University

Chapter 3
Design of Large Engineered Systems

Nam P. Suh

Abstract OLEV is a large system that was designed and implemented in two years. This chapter explains how the design and implementation of OLEV were accomplished so expeditiously by introducing readers to Axiomatic Design and complexity theory. It discusses how these theories provide the basis for designing large systems and how to remove "complexity" from the design of these systems that are so often subject to massive delays and cost overruns.

3.1 Large Systems Versus Complex Systems

We implemented the first OLEV system commercially only two years after undertaking the project. This was possible because we used the theoretical foundation that enables the development of large complicated systems rationally, quickly, and at a minimal cost by making certain that correct design decisions are made and implemented properly during the course of project execution. This chapter reviews the theoretical foundation that enabled us to achieve such a feat: Axiomatic Design (AD) and Complexity theory. These theories have been applied to a wide range of different design problems and systems in technical and non-technical areas, including hardware and software [1–4].

Because of the two design axioms and their corollaries and theorems, AD may be unique in several ways: It enables the designer to make the right and the best design decisions as the design proceeds; it prevents the possibility of creating coupling of functional requirements (FRs), which is the root cause of complexity and design failures of large as well as small systems; it enables many designers to work together regardless of the size of the system through step-by-step decomposition of the design tasks, and it allows the selection of the best design from among those proposed. When the design is completed following AD, the design is done

N.P. Suh (✉)
The Department of Mechanical Engineering, Massachusetts Insitute of Technology,
77 Masachusetts Avenue, Cambridge, MA 02139, USA
e-mail: npsuh@mit.edu

N.P. Suh and D.H. Cho (eds.), *The On-line Electric Vehicle*,
DOI 10.1007/978-3-319-51183-2_3

with no need for iteration, etc. AD differs from those system theories[1] that attempt to model and optimize a designed system. Modeling of a poorly designed system cannot overcome the wrong design decisions made during the design process. It also differs from the approach of hiring smart designers who intuitively designs based on their past experience.[2] Even smart people make wrong design decisions when they depend on their intuition alone.

A Simple Example: System Design Based on AD and Other Methodologies

To illustrate the AD concept in system design, consider the design of water faucet that must satisfy two functional requirements (FRs). The two FRs are: FR1 = control the temperature of water within 1 °C of the desired temperature and FR2 = control the flow rate of water to within 1 cm^3/min of the set flow rate. Suppose we came up with a design consisting of a water faucet with two valves, one for hot water and the other for cold water. According to the Independence Axiom of AD, we would discard this design right away and search for another, because it violates the Independence Axiom, i.e., opening the hot water valve more to increase the temperature would also changes the flow rate. If we turn the valve to change the flow rate, the temperature will change as well. Thus, this valve cannot maintain the independence of FRs, i.e., violates the Independence Axiom. Then, according to AD theory, we must look for another design that can independently satisfy the flow rate and temperature without affecting the other FR. There are many designs that can change temperature and the flow rate independently.[3] Other system theories (such as a model-based system design) may not come to the conclusion that this two-valve faucet design is not an acceptable design. Instead the designer accept the two valve design and may try to satisfy the two FRs by introducing a feedback control mechanism that links the two valves by creating a model for the relationship between the flow rate and temperature change. They may ultimately come up with a coupled system that has a set of unique solutions with sensors and feedback control, etc., which is complicated and costly. Such a system will take a long time to develop and ultimately cost a great deal, and yet may be prone to frequent failures. These coupled systems fail frequently, because they have a extremely small "sweet spots".

When a large system is developed purely based on modeling rather than starting out with right design decisions based on design axioms, coupled designs may be introduced, sometimes causing endless reiteration of design/build/test cycles. In realizing the concept of OLEV to a working real system, we had system architects whose job was to identify coupled designs and suggest a new design. Many other projects were done this way, resulting in short time for conversion of an idea to a real system in two to three years. It should be noted that an uncoupled system is easy to model mathematically because the

[1]For review of other systems theories, see Farid [4], An Engineering Systems Introduction to Axiomatic Design, (Chapter 1, Axiomatic Design in Large Systems, Farid and Suh, Eds.), Springer, ISBN 978-3-319-32387-9, DOI 10.1007/978-3-319-32388-6, Library of Congress Control Number: 2016936680, Springer International Publishing Switzerland 2016.

[2]Many industrial firms proceed with their system development by designing their system based on their past experience. Such an approach may work well if they are developing a product that is similar to existing products. Even then, they may end up going through repetitious cycles of design/build/test/ redesign/build/test. This experienced-based development process is unpredictable and costly. See Suh [1–3] and Farid and Suh [4] for additional examples.

[3]Suh [2], pp. 119–124.

multi-FR and multi-DP problem can be converted into a set of one-input/one-output problems, which makes it easy to implement and has a large "sweet spots". Perhaps some of the controversial projects such as the F-35 plane have inadvertently introduced coupled designs that have prolonged the project development that has escalated the cost of development as well as make it costly to insure reliability. Many other examples are given in Appendix 1. For more extensive discussions, see [1–4].

The On-Line Electric Vehicle (OLEV) system is a large system that involves a large number of electrical components, mechanical parts, software routines, structural elements, and people with diverse disciplinary backgrounds. It is a *complicated* system. However, we designed it to be not a *complex* system by satisfying the Independence Axiom. We satisfied the functional requirements (FRs) at all levels of decomposition as per the Independence Axiom, which required minimal trial-and-error processes and uncertainty. The important point is that a complicated system and a large system do not necessarily have to be complex if they satisfy the Independence Axiom, where *complexity is defined as a measure of uncertainty in satisfying the FRs.* This chapter presents the essence of the ideas involved in designing and implementing large complicated systems based on Axiomatic Design (AD) and the complexity theory based on AD.

For centuries, people have developed imaginative and innovative solutions to satisfy human and societal needs in fields such as energy, electric power generation, food, transportation, healthcare, education, information technology, banking, defense, environment, communications, and materials. Many of these innovations were in the form of *systems* designed to satisfy a specific set of functional requirements and constraints. These man-made systems have varied in physical size, information content, materials used, etc. Some have been large systems with sub-systems. Some have been time invariant, satisfying the same set of functional requirements (FRs) all the time (e.g., internal combustion engines, most buildings, transfer lines that make engine blocks, highways, nuclear power plants), whereas some have satisfied different sets of functional requirements as a function of time (e.g., flexible manufacturing systems, governments, defense systems). Today we face several serious problems—many of them stemming from climate change and its potential effects on both the natural and the human environment—calling for large engineered systems for their solution.

Many of the engineered systems we are familiar with are large and complicated. Well-known examples are F-35 fighter airplanes, Boston's Central Artery/Tunnel Project (the "Big Dig"), Japan's Fukushima nuclear power plants, and the software system for "Obamacare." There is a general perception that such large systems are expensive and difficult to design, develop, and implement, a perception that has been created by the long delays and cost overruns on some infamous projects. For example, Boston's Big Dig (or Central Artery/Tunnel Project, as it was formally known)—the largest and the most expensive highway project in the history of the USA—has transformed the city but took 30 years to plan and twelve years to construct, cost $18 billion instead of the $2 billion initially projected, and was saved from becoming a complete fiasco only by the deep pockets of the Federal government and the Commonwealth of Massachusetts. Another well-known large system that has greatly exceeded the original cost and development schedule is the F-35 Joint Strike Fighter airplane developed by Lockheed Martin for the US Department of Defense.[4] Meanwhile, the "futuristic" new airport currently being built for Berlin, originally scheduled to open in 2012, has had its opening date delayed to 2017 while its cost has ballooned to $6 billion—more than ten times the original budget. The delay and cost overrun in this case have been attributed to the poor design of the project's fire alarm and communications systems.[5]

However, not every large system suffers from such problems in the course of design and development. We developed many large and complicated systems such as Mixalloy and Mobile Harbor in relatively short times (see Appendix 1). The KAIST OLEV team designed the vehicle (initially a diesel bus converted to an OLEV) and the underground power supply system, tested the designs, simulated performance through modeling and analysis, had the parts manufactured by collaborating industrial firms, and assembled them to create the OLEV system. They also designed and constructed the EMF shield for the vehicle and the underground power supply system and obtained the necessary licenses and approvals from government agencies and laboratories for commercial operation of this new technology. In some cases, new standards had to be created for SMFIR, OLEV, and EGTS, because nothing like them had ever existed before. From the view point of design, OLEV could have encountered the same set of problems like the F-35, Fukushima Nuclear Power Plants, and other failed systems if we did not rigorously implement AD in designing and implementing OLEV.

When the KAIST team tackled the OLEV project, we had a narrow window and a small margin for error. Had our initial system resulted in failure, we would not have gotten a second chance because of the number of people, especially some engineering professors in leading Korean universities and a politician with a science background, who opposed our project from the beginning and would have

---

[4]The development of the Joint Strike Fighter, a fifth-generation stealth jet, illustrates how complexity due to poor design causes spiraling costs and schedule delays. The program's price tag is nearly $400 billion for 2457 planes—almost twice the initial estimate. U.S. Senator John McCain slammed the F-35 Joint Strike Fighter's troubled history on April 26, 2016, saying it "has been both a scandal and a tragedy with respect to cost, schedule and performance." He noted that the F-35 program had originally promised 1013 fighters by fiscal year 2016 but had only delivered 179. Similarly, David Francis noted that "The F-35 Joint Strike Fighter is the most expensive, and possibly the most error ridden, project in the history of the United States military. But DOD has sunk so much money into the F-35—which is expected to cost $1.5 trillion over the 55-year life of the program—that the Pentagon deemed it 'too big to fail' in 2010 … American taxpayers will pay between $148 million and $337 million per jet plane in 2015, depending on the model." (*The Fiscal Times*, July 31, 2014) http://www.thefiscaltimes.com/Articles/2014/07/31/How-DOD-s-15-Trillion-F-35-Broke-Air-Force.

[5]Joshua Hammer, "How Berlin's futuristic airport became a $6 billion embarrassment," Bloomberg Business, July 23, 2015, http://www.bloomberg.com/news/features/2015-07-23/how-berlin-s-futuristic-airport-became-a-6-billion-embarrassment (accessed 10/29/15).

condemned it at the first indication of failure.[6] Therefore, the goal was not just to create a successful design. It was to create the final, successful product in the shortest possible time, at the lowest possible cost. We performed our tasks systematically without letting the external factors drive our schedule or plan. We held extensive and intensive weekly meetings of the key members of the OLEV team, including the President of KAIST, to be sure that we do not introduce coupled designs. Our goal was to design a system that would operate reliably for many years without failure. We succeeded in achieving these goals. It took a mere two years from the decision by the Korean government to fund the OLEV project to design and implement the basic core technologies, whereas projects of similar size and scope often take four or more years. OLEV was also designed and implemented for the first time within the original budget and has been running reliably ever since.

How was OLEV designed and implemented so quickly and efficiently? While not as large and complicated as, say, Boston's Big Dig, the OLEV system is indeed large and complicated and is based on an entirely new technology. It includes many components and sub-systems, both hardware and software. It has infrastructure for the segmented underground power supply system made up of ferrite, Litz electric wires, etc.; electric power inverters; electric vehicles capable of receiving magnetic power transmitted wirelessly; EMF and EMI shields; communication systems; operators of the vehicles; and other essential parts. Yet unlike many large, complicated systems, OLEV is not *complex* in the sense in which that word is used in Axiomatic Design, the method that was used to create it. By contrast, many large systems that pose difficulties undergo delays and come in dramatically over budget do so because they *are* complex due to coupling of FRs (i.e., due to wrong design decisions), and it is their complexity that makes them difficult to design in a rational and efficient manner.

3.2 Complexity and Coupled Designs

Complexity is defined in many different ways, depending on the field; Professor Seth Lloyd[7] once stated that there are as many definitions of complexity as the number of people working in the field of complexity. What is meant by complexity

---

[6]When the idea for OLEV was first proposed for funding, a few prominent professors in Korea opposed the project, stating that heavy electric power could not be transferred over a distance of more than a couple of centimeters, even to a stationary vehicle. Not knowing the basis for their claim, we assumed that they were thinking of induction between two circular coils, similar to wireless power transmission between electric toothbrush and its charger, rather than field effects. Such a system would require that the vehicle be stationary and the coils on the vehicle be precisely on top of the power supply coil in close proximity. However, OLEV transmits heavy electric power over a large distance to a moving or stationary vehicle.

[7]Seth Lloyd is the Nam Pyo Suh Professor of Mechanical Engineering at M.I.T., Cambridge, MA, U.S.A.

in the context of Axiomatic Design? In complexity theory based on Axiomatic Design, complexity is *defined as a measure of uncertainty in satisfying the functional requirements (FRs) of the system.* According to this definition, a system that always satisfies its FRs is not complex, irrespective of its size, the number of parts or lines of software code, etc., although it may be more *complicated* to design and assemble as the number of parts increases. Conversely, according to this in some ways counterintuitive definition of complexity, a system that cannot easily satisfy its FRs is complex, regardless of its size. Even a small system can be complex when poorly designed—for example, the ignition switch in GM automobiles that led to 124 deaths and 275 injuries. The major cause of complexity in engineered systems, in turn, is the *coupling of its functional requirements* (FRs), where FRs define the specific design objectives.[8] Axiomatic design theory enables designers to avoid coupling of functional requirements (FRs) and eliminate the possibility of complexity by making proper design decisions. Just as other theories have done, once we develop a theory that explains what appear to be rather complicated incomprehensible phenomena, the solution is obvious.

The engineers and managers who worked on the Berlin airport project, the F-35 Joint Strike Force airplanes, and the Big Dig must have been intelligent and capable people, since such advanced technology projects attract some of the brightest and most experienced engineers and scientists. Such important projects also receive generous financial support. It is most likely that those involved had every intention of creating great systems on budget, on schedule, and even exceeding the original expectations. What then is the reason for the cost overruns and missed schedules that are so common in such large projects? Even those who lead such projects or participate in making important decisions related to them may not know the fundamental cause of the problems that lead to cost overruns and delays. They may know the *symptoms* of failure but not *the causes*—i.e., the basic design decisions that inevitably lead to failure.

At present, the only way to delineate the cause of system failures like the ones discussed above is to compare these projects with ones that were finished in time and within the estimated cost. The limited case studies I have been able to make[9]

---

[8]It is important to note that it is coupling of functions, not physical coupling or integration of components, that we are talking about here.

[9]The relationship between coupled designs and cost overruns and delays can also be stated as a theorem and proof as follows:

Theorem on Cost Overruns and Project Delays

The coupling of the functional requirements (FRs) of the system under development is the root cause of cost overruns and project delays.

Proof of the Theorem

Consider that two designs—Design A and Design B—with an equal number of FRs and DPs. Then, they are both ideal designs per Theorem 4 [2]. Assume that Design A is a coupled design and Design B is an uncoupled design. Then, the information content of Design A, I_A, is greater than the information content of Design B, I_B. That is, $I_A > I_B$. Therefore, the design parameters of suggest that the successful projects are those that adhered to the principles of what is called, in Axiomatic Design, the Independence Axiom—which states that FRs in a system must be satisfied independently of one another—while the projects that have encountered major problems are those that violated the Independence Axiom by means of coupled designs. Intentionally or inadvertently, the designers of these problematic systems introduced coupled designs at the system level as well as at lower levels of the design hierarchy.[10] These couplings then led to cost overruns and time delays, or worse. The Independence Axiom is a powerful theory in developing complicated systems.

One of the most catastrophic recent failures of a large technological system was the destruction of Japan's Fukushima nuclear power plants in the tsunami that hit the coastal area of Sendai, Japan, in November 2011. The human and ecological toll from this disaster can hardly be overstated. According to one research team, it was the coupling of functions of the plants' electrical systems with those of their mechanical systems that led to the failure of the entire system [5, 6]. The GM ignition key system, which was physically small with relatively few components, also had a coupled design, which resulted in its failure, making this small system a complex one that led to fatal accidents. This kind of disaster was fortunately avoided when, in 2013, the US Federal Aviation Administration grounded all Boeing 787 s following a fire initiated on a 787-8 by a lithium-ion battery in the plane's auxiliary power unit that disabled other batteries and other functions of the

(Footnote 9 continued)

Deign A must have tighter tolerances than Design B. Furthermore, Design A works only when there is a unique solution (i.e., all DPs associated with coupled FRs must have the exact values requiring tight tolerances for the system to satisfy its FRs). In such a design, all FRs are functions of other FRs. In Design B, each DP satisfies one specific FR, and thus, the tolerances can be much larger. That is, Design B can be treated as a one-output one-input system with larger tolerances for DPs. Therefore, the cost of Design B, $\$_B$, is less than the cost of Design A, $\$_A$, i.e., $\$_B < \$_A$. Similarly, the time taken to complete the project A, T_A, is longer than the time taken to complete Design B, since each task does not have to be repeated. Thus, $T_A > T_B$.

[10]The internal combustion engine, interestingly, is an example of a coupled design. When the IC engine with reciprocating pistons was first invented and then deployed in automobiles, oil was plentiful and cheap and the environmental consequences of burning petroleum were not yet understood, so that neither of these factors was treated as a constraint. Had these two constraints been factored into the design of the IC engine and the FRs had been kept independent of one another, its current version would not have been developed. In this design, a coupling of functional requirements occurs because the engine block must be cooled in order to lubricate the interface between the piston ring and the cylinder wall, although the thermodynamic efficiency of these heat engines increases with the temperature of the combustion product. The FRs of mixing fuel and air, the exact timing of ignition, and maintaining the stoichiometric ratio of fuel and air require precise tuning of all functions. Even with all the advances made over a century of continuous effort, the IC engine is not an efficient energy conversion device.

airplane.[11] Even short of such disasters or near-disasters, however, when FRs are coupled to each other, it is difficult to unravel a problem after the product or system is made, since everything may have to be changed, requiring new tooling and new components.

Lockheed Martin became aware early on of the need to adopt a more systematic approach to developing large systems in order to avoid cost overruns and project delays on its F-35 project, which began in 2000. When the company got the contract from NASA to develop the Orbital Space Plane (OSP), a program that began in 2001, its program manager, Robert Ford, had decided to use a system design methodology in developing the OSP.[12] In 2002–2003, he invited the author to teach the lead engineers and designers in the OSP program about system design based on Axiomatic Design. About 250 engineers were taught over a period of about six months in Denver, Colorado. Young engineers grasped AD relatively quickly, but the experienced engineers, who had spent years doing design and development through a repetitive cycle of design/build/test, had a difficult time learning a new way of thinking about system design; they were used to starting out with a physical embodiment of what the system should look like rather than a set of functional requirements to be satisfied. Yet if design of a large system is done based purely on experience, the probability of creating coupled FRs is likely to increase as the number of FRs increases.

Many industrial firms and organizations design their systems based primarily on the past experience of their engineers and managers, who often use ad hoc, trial-and-error processes along with intuition and their past experience. Modeling and simulation, two other traditional tools, are necessary in system development, but they are not the best methods for identifying the causes of cost overruns and missed schedules. In developing a new product, many companies repeat a cycle of design/build/test cycle until they arrive at a solution that performs to their satisfaction, a process that is time-consuming, expensive, and may lead to coupled designs. Coupled designs, in turn, are difficult to make function properly and will likely fail when one of the design parameters unexpectedly changes for a variety of

[11]The FAA ordered the grounding of the fleet when, only five days after the fire, another Boeing 787 made an emergency landing after the pilots received a warning of a battery malfunction. See Umair Irfan, in ClimateWire, December 18, 2014 "How Lithium Ion Batteries Grounded the Dreamliner," reprinted by *Scientific American*, http://www.scientificamerican.com/article/how-lithium-ion-batteries-grounded-the-dreamliner/ (accessed 11/27/15). According to the NTSB (U.S. National Transportation Safety Board), the fire was the result of a short circuit in one battery that led to "thermal runaway" in other cells when the batteries became overheated to the point where a chemical reaction was triggered that released even more heat, causing a cascading effect that caused the batteries to explode or catch fire.

[12]Mr. Robert Ford, an outstanding expert in rocket systems, realized that if the OSP project used the traditional methods of product development the mistakes of the past would simply be repeated. Ref: Robert Ford, *Orbital Space Plane (OSP) Program at Lockheed Martin*, American Institute of Aeronautics and Astronautics, September 2003, Long Beach, California, U.S.A., http://ntrs.nasa.gov/archive/nasa/casi.ntrs.nasa.gov/20030106071.pdf (accessed 12/18/15).

reasons.[13] Furthermore, designers using ad hoc methods may not be able find a combination of FRs that can be satisfied at all times. As the number of FRs increases, there are too many combinations to test them all. In a coupled design system, any time any one of the FRs is changed, all other FRs must be changed as well. These coupled systems are not robust against unexpected variations of DPs due to wear, temperature variation, etc., because the system fails if one of the FRs changes unexpectedly due to changes in DPs and constraints. In a coupled design, when one of the DPs changes all the FRs are affected and all DPs must be changed to satisfy FRs—if there is a solution at all. What we want to achieve in designing an engineered system is to enable its many FRs to be satisfied as specified at all times. An uncoupled design makes a multi-input/multi-output system operate as a set of a single-input/single output system that can be controlled or varied easily and that is more tolerant to unexpected variations of DPs.[14] The basic design of OLEV and SMIFR is an *uncoupled or decoupled system*, which is what enabled us to go from design to implementation in just two years.

Because systems are designed to produce certain outputs using a set of input parameters within a set of constraints, we should be able to design any system using the same methodology rather than a variety of ad hoc approaches, although the specific nature of the problem, physical principles, data, and acceptable variations are field-specific. And we should be able to do so without creating coupled designs. Axiomatic design is a methodology that makes these things possible.[15]

3.3 An Introduction to Axiomatic Design[16]

There are basically two different approaches to design: algorithmic or axiomatic. Most of the software used in computers is an algorithm. An algorithm is a step-by-step, "do this and then do that" approach that leads to a solution to a specific problem. Many fields have standard codes for design that are basically algorithms, such as ASME's pressure vessel codes that tell designers how to design a boiler following a set procedure described in the code. This code was created

[13]To make a coupled system function properly without failure, design parameters in the system must be exact and precise to make the system with multiple-input/multiple-output yield the desired results.

[14]If we satisfy the Independence Axiom, even a large design project with many FRs can be treated as a one-input/one-output (OI/OO) project, a real advantage in system design. Modeling, simulation, and optimization become much simpler to implement and actual physical implementation can avoid the trial-and-error process.

[15]Recently, it was revealed that jumbo jet Boeing 747 was designed and completed in 29 months under the direction of Mr. Joe Sutter. His success might have been because of his intuitive understanding of how to satisfy FRs independently based on his years of experience at Boeing (The New York Times, September 1, 2016).

[16]For more details, see [1–3].

through much empirical and analytical work to prevent boiler accidents that had previously wrecked ships and injured people when steam engines were first introduced. Apprenticeship programs for training electricians and other such technicians are often based on algorithmic models. If you are designing something that has never existed before, such as a solution to the CO_2 problem, there is no algorithmic approach established for it. In such cases, many people have used trial-and-error processes or attempted experientially based design. We chose instead to use an axiomatic design approach.

Axioms have advanced science and engineering in countless ways. Any design for a heat engine should be immediately discarded if it does not conform to the first and second laws of thermodynamics, which are axioms. Euclidian geometry is built on axioms such as "the shortest distance between two points is a straight line," and Newton's three laws are axioms that Newton created in relatively short time in order to derive Kepler's three laws of planetary motion. We cannot derive these axioms, but we believe they are correct because we cannot find exceptions or counter examples. Often they are valid within certain bounds; for example, Newton's laws are valid when an object is moving much slower than the speed of light. Axioms are just as useful in design as they are in science. Axioms have provided basic foundations for scientific and technological thought processes.

The AD design theory depends on the idea that the design world is made up of four *design domains* as shown in Fig. 3.1. The first domain is the customer domain. The second is the functional domain, where customer needs are translated into one or more FRs. The third domain is the physical domain, where design parameters (DPs) are chosen to satisfy the FRs of the functional domain. The fourth domain is the process domain, where the process variables (PVs) are chosen to satisfy the DPs.

In the customer domain of our design, we identified the need to improve the environment. We decided that electric vehicles (EVs) that receive their electric power wirelessly from an external power source while in motion should replace automobiles with IC engines.

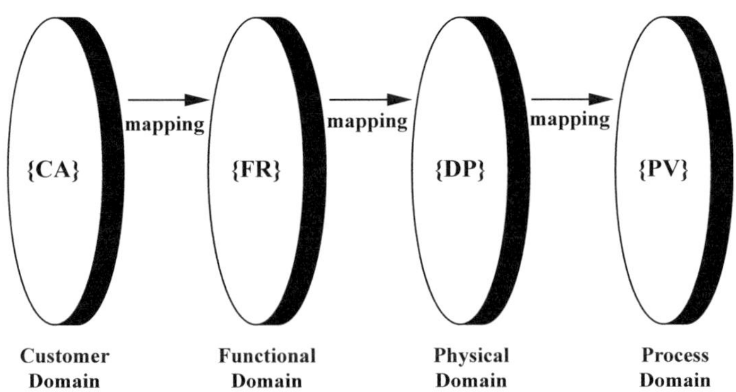

Fig. 3.1 Four domains of axiomatic design

Mathematically, FRs and DPs are vectors with many components. The relationship between these two vectors can be represented by the following *design equation*:

$$\{FR\} = [D]\{DP\} \tag{3.1}$$

$\{FR\}$ is a vector consisting of components FR1, FR2, etc.; $\{DP\}$ is a vector with components DP1, DP2, etc.; and [D] is the *design matrix*, i.e., a second-order tensor, that defines the relationship between $\{FR\}$ and $\{DP\}$.

AD has two *design axioms*, from the features always present in good designs. The first is the Independence Axiom, which states that *a system must maintain the independence of FRs*. In other words, when we choose a DP to satisfy a specific FR, we have to make sure that other FRs are not affected. These axioms were confirmed by a large number of new projects undertaken after their introduction of the design axioms [1].

The second axiom, the Information Axiom, deals with the variation of an FR (defined as the *system range*) with respect to the stated allowable variations of an FR (defined as the *design range*). According to the Information Axiom, *a system must minimize information content*. To minimize information content, the specific value of an FR, e.g., FRl, etc., must be within the specified *design range*, ΔFR1, etc.—i.e., the random variation of FRs, δFR1, must be $\Delta FR1 \ll \delta FR1$. The Information Axiom states that the information content should be zero, which occurs when the system range is inside the design range at all times. Robust design is the one that has zero information content, i.e., performs within the design range of FRs. Then, complexity is also zero.

Figure 3.2 illustrates the variation of FRs, and the system range versus the design range. The design range is the acceptable variation of FR, and the system range is the actual variation of FRs of the product.

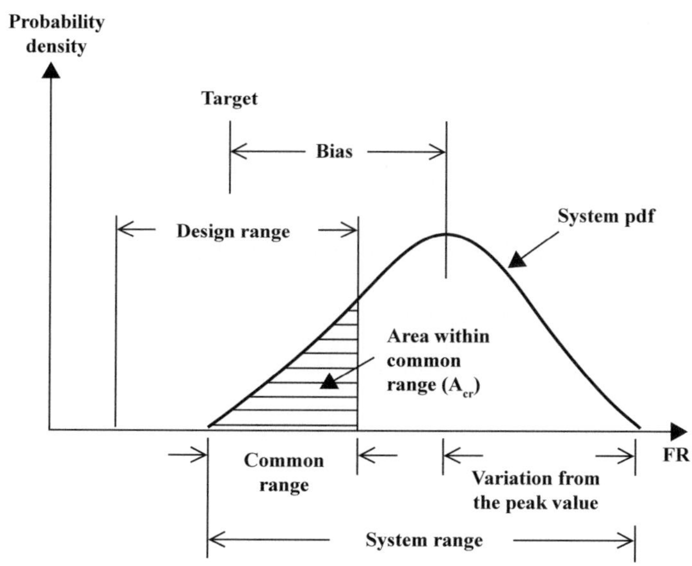

Fig. 3.2 Design range of FR versus System Range of FR

There are many theorems that follow from these two axioms [2]. For instance, Theorem 4 states that an ideal design has the same number of FRs and DPs, which means the design matrix is a square matrix.

When [D] is a diagonal matrix, the design is an *uncoupled* design, which satisfies the Independence Axiom. When it is a triangular matrix, the design is a *decoupled* design. In this case, DPs must be changed in a given order prescribed by the design matrix.

All other designs are *coupled* designs, which *violate the Independence Axiom*. For instance, suppose that one of the FRs of all battery-powered electric vehicle is to go as far as possible with electric power stored in the battery. Another FR is to reduce the weight of the vehicle for fast acceleration. If we choose the battery size as a DP to satisfy the first FR, it will negatively affect the second FR. Therefore, these two FRs are affected by the same DP, which makes it a coupled design that violates the Independence Axiom.

Even when there are many FRs and many DPs at a given level of design, an *uncoupled design* can be made to satisfy FRs, since the design can always be treated as a set of one-input/one-output problem regardless of the number of FRs involved. It does not matter whether the relationship between the FR and the DP is linear or nonlinear. On the other hand, when the design is a coupled design, it can only be made operational with a unique set of DPs and FRs, which makes it nearly impossible to satisfy the FRs at all times, because in the physical space, DP cannot be reconfigured easily.

The design for OLEV is a *decoupled* design. Therefore, our design is going to work if we set the values of DPs in the *order specified* by Table 3.1. It shows that FR3, FR4, FR7, and FR8 can be changed in any order, but DP2, DP5, and DP6 must be implemented to satisfy FR2, FR5, and FR6 before we set the value of DP1 to satisfy FR1. Similarly, SMFIR design must follow Table 2.2. Obviously, the design cannot violate the specified constraints.

3.4 System Architecture and the Role of the System Architect

To check for the functional coupling during the design process, system architecture should be constructed. There are many different ways to do this [7, 8]. One way is to create a design matrix for all FRs and DPs, including all decomposed FRs and DPs. For example, consider the design given by the following FRs and DPs with two levels of decomposition:

$$\begin{Bmatrix} FR1 \\ FR2 \end{Bmatrix} = \begin{bmatrix} A11 & 0 \\ A21 & A22 \end{bmatrix} \tag{3.2}$$

Table 3.1 System architecture for the illustrative decomposition example

		DP1			DP2	
		DP11	DP12	DP13	DP21	DP22
FR1	FR11	X	0	0	0	
	FR12	X	X	0		
	FR13	0	0	X		
FR2	FR21	A21			X	X
	FR22				0	X

FR1 and FR2 are decomposed as {FR11, FR12, FR13} and {FR21, FR22} as illustrated in Eqs. (2.3) and (2.4).

$$\begin{Bmatrix} FR11 \\ FR12 \\ FR13 \end{Bmatrix} = \begin{bmatrix} X00 \\ XX0 \\ 00X \end{bmatrix} \begin{Bmatrix} DP11 \\ DP12 \\ DP13 \end{Bmatrix} \tag{3.3}$$

$$\begin{Bmatrix} FR21 \\ FR22 \end{Bmatrix} = \begin{bmatrix} XX \\ 0X \end{bmatrix} \begin{Bmatrix} DP21 \\ DP22 \end{Bmatrix} \tag{3.4}$$

In the above examples, X indicates that there is a relationship between the FR and DP. The system architecture for the above design is shown in the following FR–DP design matrix table:

In the design sequence in Table 3.1, it should be noted that the decomposition process deals only with the diagonal elements. Therefore, we need to check whether the off-diagonal elements have unintended consequences on other FRs. For example, at the highest level, DP2 was assumed to have no influence on FR1. However, now that we have decomposed DP2 into DP21 and DP22 and the (SA) should check whether DP21 has any effect on FR11, FR12, and FR13 (and, similarly, DP22 on FR11, etc.). If so, the original assumption that FR1 and FR2 are independent of each other (i.e., a decoupled design) is not correct, and we need to go back and search for some other DP21 and DP22. If this is not done and a mistake is made, the project will be set back. The sooner the mistake is uncovered, the better. If we had not constructed the system architecture, we would not have discovered this potential design flaw until after the construction of OLEV and extensive testing, incurring additional cost and time delay.

The system architecture becomes more complicated with the number of levels of decomposition. It is best to use commercial software such as Acclaro (a software product to assist in axiomatic design) to keep track of all design decisions and consequences of changes on the independence of the highest level FRs. But all the decomposition in the world would not make a coupled system functional. Designers must avoid creating coupled designs by choosing DPs that will not couple FRs (i.e., the system must be either uncoupled or decoupled). If coupling of FRs is found, it is best to go back and change the design, i.e., select new DPs.

When the system is large with many FRs and DPs, and with many layers of decomposition, and involving a large number of designers working on different branches, we need a *system architect* to monitor the progress of the entire system design and identify coupling of FRs. A design decision made by a designer in a different group can create coupling of FRs in another group. When the number of FRs is very large, inadvertent coupling of FRs can occur, which will require major redesign and create havoc with scheduling and cost. The system architect serves as the custodian of the architecture of the entire system by constructing and maintaining FR and DP trees. When a DP at any branch causes coupling of FRs at any level, the system architect must stop the design process of that branch until a new DP is chosen to eliminate the coupling. Design changes can be made more easily and inexpensively when a coupled design is identified early in the project or design cycle.

The tool the design architect uses to check if the design decisions are acceptable is the design matrix. The design matrix for OLEV created by Hong and Park [7] is given in Table 3.2. The elements of the design matrix that could have caused coupling are circled.

Regardless of the actual dimensions or the number of FRs or the number of parts involved, the task of achieving FRs becomes more complex as the probability of satisfying a given FR decreases. The information content is defined as $I = -\log p$, where p is the probability of satisfying the given FR. Since in design, certain variations are allowed as long as the functional requirement, FR, is still satisfied to within its allowable tolerance (i.e., variations). When the system range does not overlap the design range at all, the information content is infinite and a new DP must be selected. Then the design is not acceptable, since the FR cannot be satisfied within the desired tolerance. In short, as previously mentioned, the complexity of a system increases as there is more functional coupling.[17]

3.5 Complexity Theory Based on Axiomatic Design

In developing a large system such as OLEV and SMFIR, it is useful to know how to deal with the notion of complexity, which we have defined as "*a measure of uncertainty in satisfying the FRs.*" In general, when the information content increases, the complexity increases, regardless of physical size or the number of components. This being the case, we can make use of complexity theory to improve the operation of the system.

[17]It should be noted that there are many diverse views on system architecture [9]. Computer programming, aerospace engineering, chemical engineering, and other fields have developed their specific systems engineering approaches. Many involve analysis and simulation of the designed systems to find root causes of system failure and devise means of improving systems. Some treat systems engineering as an optimization problem. Another view is that "system architecting" is what a smart and experienced system engineer do, using heuristic approaches and simple rules to develop system architecture [10]. In contrast to these approaches, in AD, the main function of the system architect is to monitor the creation of coupling by constructing the design matrix for the entire design [7, 8].

Table 3.2 Full design matrix of OLEV. A problem due to coupling of FRs is shown by an arrow (from [7])

							Three phase AC 370–440V	Converter circuit	Inverter circuit	Copper plate	Ferrite core set(s)	Circuit protection control	Location of return coil	Individual pick-up device	Number of individual pick-up device	Number of litz wire	Wire selecting circuit	Power converting circuit	Resonant circuit	Location of return coil	Rectifying circuit	Regulation circuit	Driving status sensor(s)	Battery status sensing circuit(s)	Device status sensing device(s)	Power status calculation algorithm	Motor power calculation algorithm	Battery control algorithm (1)	Battery control algorithm (2)	Drive battery power calculation algorithm	Device battery power calculation algorithm	Warning message control algorithm	Power delivery circuit(s)	Electric power managing circuit	Inverter circuit-drive motor	Motor control circuit	Battery control algorithm (3)	Battery control algorithm (4)	Device control circuit(s)	Mechanism(s) and mechanical part(s)	Thickness of asphalt	FRP plate	Plastic covering of electric wire	Exterior covering for waterproof	Mounting height from roadway	Bump of the cover	Welding to the vehicle body	Protection devices
1						Import electric power.	X	O	O	O	O	O	O	O	O	O	O	O	O	O	O	O	O	O	O	O	O	O	O	O	O	O	O	O	O	O	O	O	O	O	O	O	O	O	O	O	O	O
2	1	1	1			Transform 3Φ AC to DC.	X	X	O	O	O	O	O	O	O	O	O	O	O	O	O	O	O	O	O	O	O	O	O	O	O	O	O	O	O	O	O	O	O	O	O	O	O	O	O	O	O	O
			2			Transform DC to an AC electric power.	O	X	X	O	O	O	O	O	O	O	O	O	O	O	O	O	O	O	O	O	O	O	O	O	O	O	O	O	O	O	O	O	O	O	O	O	O	O	O	O	O	O
		2				Generate magnetic flux.	O	O	X	X	O	O	O	O	O	O	O	O	O	O	O	O	O	O	O	O	O	O	O	O	O	O	O	O	O	O	O	O	O	O	O	O	O	O	O	O	O	O
		3				Shape the magnetic flux distribution.	O	O	X	X	X	O	O	O	O	O	O	O	x	O	O	O	O	O	O	O	O	O	O	O	O	O	O	O	O	O	O	O	O	O	O	O	O	O	O	O	O	O
		4				Protect inverter device from counter electromotive force	O	O	O	X	O	X	O	O	O	O	O	O	O	O	O	O	O	O	O	O	O	O	O	O	O	O	O	O	O	O	O	O	O	O	O	O	O	O	O	O	O	O
		5				Neutralize magnetic flux generated within segments	O	O	O	X	X	O	X	O	O	O	O	O	O	O	O	O	O	O	O	O	O	O	O	O	O	O	O	O	O	O	O	O	O	O	O	O	O	O	O	O	O	O
	2	1	1			Pick-up magnetic flux over area about 2,000 X 1,200.	O	O	O	X	X	O	X	X	O	O	O	O	O	O	O	O	O	O	O	O	O	O	O	O	O	O	O	O	O	O	O	O	O	O	X	X	O	X	X	O	O	O
			2			Pick-up magnetic flux over length about 2,000 X 12,000.	O	O	O	O	O	O	O	X	X	O	O	O	O	O	O	O	O	O	O	O	O	O	O	O	O	O	O	O	O	O	O	O	O	O	O	O	O	O	O	O	O	O
		2	1			Convert same amount of electric power regardless of vehicle position	O	O	O	O	O	O	O	O	X	X	O	O	O	O	O	O	O	O	O	O	O	O	O	O	O	O	O	O	O	O	O	O	O	O	O	O	O	O	O	O	O	O
			2			Select litz wire set which can convert maximum electric power.	O	O	O	O	O	O	O	O	O	O	X	O	O	O	O	O	O	O	O	O	O	O	O	O	O	O	O	O	O	O	O	O	O	O	O	O	O	O	O	O	O	O
			3			Convert magnetic flux to electric power with selected litz wire.	O	O	O	O	O	O	O	O	O	O	O	X	x	O	O	O	O	O	O	O	O	O	O	O	O	O	O	O	O	O	O	O	O	O	O	O	O	O	O	O	O	O
	3					Resonate the inductive power transfer circuit.	O	X	X	x	O	O	X	X	O	X	X	O	X	O	O	O	O	O	O	O	O	O	O	O	O	O	O	O	O	O	O	O	O	O	O	O	O	O	O	O	O	O
	4					Neutralize magnetic flux.	O	O	O	X	X	O	O	O	O	O	O	O	O	X	O	O	O	O	O	O	O	O	O	O	O	O	O	O	O	O	O	O	O	O	O	O	O	O	O	O	O	O
3	1	1				Convert the AC electric power to DC electric power.	O	O	O	O	O	O	O	O	O	O	O	O	x	O	X	O	O	O	O	O	O	O	O	O	O	O	O	O	O	O	O	O	O	O	O	O	O	O	O	O	O	O
		2				Uniformize the maximum current to a 33[A].	O	O	O	O	O	O	O	O	O	O	O	O	O	O	X	X	O	O	O	O	O	O	O	O	O	O	O	O	O	O	O	O	O	O	O	O	O	O	O	O	O	O
	2	1	1			Check driving status of the vehicle such as velocity.	O	O	O	O	O	O	O	O	O	O	O	O	O	O	O	O	X	O	O	O	O	O	O	O	O	O	O	O	O	O	O	O	O	O	O	O	O	O	O	O	O	O
			2			Check power status of the vehicle.	O	O	O	O	O	O	O	O	O	O	O	O	O	O	O	X	O	X	O	O	O	O	O	O	O	O	O	O	O	O	O	O	O	O	O	O	O	O	O	O	O	O
			3			Check other status of vehicle such as light.	O	O	O	O	O	O	O	O	O	O	O	O	O	O	O	O	O	O	X	O	O	O	O	O	O	O	O	O	O	O	O	O	O	O	O	O	O	O	O	O	O	O
		2	1			Calculate required electric power and usable electric power of the vehicle.	O	O	O	O	O	O	O	O	O	O	O	O	O	O	O	O	X	O	X	X	O	O	O	O	O	O	O	O	O	O	O	O	O	O	O	O	O	O	O	O	O	O
			2	1		Decide electric power amount from pick-up device.	O	O	O	O	O	O	O	O	O	O	O	O	O	O	O	O	O	O	O	X	X	O	O	O	O	O	O	O	O	O	O	O	O	O	O	O	O	O	O	O	O	O
				2	1	Decide battery(s) to discharge.	O	O	O	O	O	O	O	O	O	O	O	O	O	O	O	O	O	O	O	X	X	X	O	O	O	O	O	O	O	O	O	O	O	O	O	O	O	O	O	O	O	O
					2	Decide discharging amount of each battery.	O	O	O	O	O	O	O	O	O	O	O	O	O	O	O	O	O	O	O	X	O	X	X	O	O	O	O	O	O	O	O	O	O	O	O	O	O	O	O	O	O	O
			3			Decide electric power amount to drive battery(s).	O	O	O	O	O	O	O	O	O	O	O	O	O	O	O	O	O	O	O	X	O	O	O	X	O	O	O	O	O	O	O	O	O	O	O	O	O	O	O	O	O	O
			4			Decide electric power amount to device battery(s).	O	O	O	O	O	O	O	O	O	O	O	O	O	O	O	O	O	O	O	X	O	O	O	O	X	O	O	O	O	O	O	O	O	O	O	O	O	O	O	O	O	O
			5			Send warning message to main control if there is insufficient electric power.	O	O	O	O	O	O	O	O	O	O	O	O	O	O	O	O	O	O	O	X	O	O	O	O	O	X	O	O	O	O	O	O	O	O	O	O	O	O	O	O	O	O
		3				Provide appropriate amount of electric power	O	O	O	O	O	O	O	O	O	O	O	O	O	O	O	O	O	O	O	O	X	X	X	X	X	O	X	O	O	O	O	O	O	O	O	O	O	O	O	O	O	O
	3	1	1			Merge electric power from pick-up and drive battery(s).	O	O	O	O	O	O	O	O	O	O	O	O	O	O	O	X	O	O	O	O	O	O	O	O	O	O	x	X	O	O	O	O	O	O	O	O	O	O	O	O	O	O
			2			Convert DC to AC (3Φ).	O	O	O	O	O	O	O	O	O	O	O	O	O	O	O	O	O	O	O	O	O	O	O	O	O	O	O	x	X	O	O	O	O	O	O	O	O	O	O	O	O	O
			3			Operate (rotate) drive motor.	O	O	O	O	O	O	O	O	O	O	O	O	O	O	O	O	O	O	O	O	O	O	O	O	O	O	O	O	x	X	O	O	O	O	O	O	O	O	O	O	O	O
		2	1			Decide battery(s) to charge.	O	O	O	O	O	O	O	O	O	O	O	O	O	O	O	O	O	O	O	X	O	O	O	X	O	O	O	O	O	O	X	O	O	O	O	O	O	O	O	O	O	O
			2			Decide charging amount of each battery.	O	O	O	O	O	O	O	O	O	O	O	O	O	O	O	O	O	O	O	X	O	O	O	X	O	O	O	O	O	O	x	X	O	O	O	O	O	O	O	O	O	O
		3				Operate other electric devices of the vehicle.	O	O	O	O	O	O	O	O	O	O	O	O	O	O	O	O	O	O	O	O	O	O	O	O	X	O	X	O	O	O	O	O	X	O	O	O	O	O	O	O	O	O
	4					Generate physical actions to drive the vehicle.	O	O	O	O	O	O	O	O	O	O	O	O	O	O	O	O	O	O	O	O	O	O	O	O	O	O	O	O	O	O	O	O	O	X	O	O	O	O	O	O	O	O
4	1	1				Support vertical loads to make vehicles pass through.	O	O	O	O	O	O	O	O	O	O	O	O	O	O	O	O	O	O	O	O	O	O	O	O	O	O	O	O	O	O	O	O	O	O	X	O	O	O	O	O	O	O
		2				Protect electric power supply coil.	O	O	O	X	O	O	O	O	O	O	O	O	O	O	O	O	O	O	O	O	O	O	O	O	O	O	O	O	O	O	O	O	O	O	x	X	O	O	O	O	O	O
		3				Protect roadway systems from water.	O	O	O	X	O	O	O	O	O	O	O	O	O	O	O	O	O	O	O	O	O	O	O	O	O	O	O	O	O	O	O	O	O	O	O	x	X	O	O	O	O	O
	2	1				Protect electric circuit(s) from water.	O	O	O	O	O	O	O	O	O	O	O	O	O	O	X	X	O	O	O	O	O	O	O	O	O	O	O	O	O	O	O	O	O	O	O	O	O	X	O	O	O	O
		2				Protect pick-up devices from external obstacles.	O	O	O	O	O	O	O	O	O	O	O	O	O	O	O	O	O	O	O	O	O	O	O	O	O	O	O	O	O	O	O	O	O	O	O	O	O	O	X	O	O	O
		3				Protect pick-up devices from external loads.	O	O	O	O	O	O	O	O	O	O	O	O	O	O	O	O	O	O	O	O	O	O	O	O	O	O	O	O	O	O	O	O	O	O	O	O	O	O	O	X	O	O
		4				Support pick-up devices from self-weight.	O	O	O	O	O	O	O	O	O	O	O	O	O	O	O	O	O	O	O	O	O	O	O	O	O	O	O	O	O	O	O	O	O	O	O	O	O	O	O	O	X	O
	3					Protect other electric devices from external environment.	O	O	O	O	O	O	O	O	O	O	O	O	O	O	O	O	X	X	X	O	O	O	O	O	O	O	O	O	O	O	O	O	O	O	O	O	O	O	O	O	O	X

There are many different kinds of complex systems depending on the relationship between the design range and the system range. The design range is the acceptable tolerance range of FRs, and the system range is what the designed system actually delivers in terms of its variation from the specified variation of the FRs. When the system range is inside the design range, the FR is always satisfied. When the system range is partly outside the design range, there is a finite probability that the system may not function as desired. These relationships are affected by the design matrix, which relates FRs to DPs. In the case of coupled designs, it is difficult to make the system range lie within the design range at all times. In some cases, the system range changes as a function of time; this occurs in many systems because of wear, tear, and variations of external factors.

To reiterate a point already made: *Designed systems, whether large or small, should not be complex.* However, some existing large systems are complex because it is difficult to satisfy the FRs of the system independently without affecting other FRs unless a deliberate effort is made. Often these highly coupled systems operate at a unique operating point: If any one of the FRs changes, all other FRs must change also. That is the reason the efficiency of IC engines, for example, cannot be increased from its low value of $\sim 17\%$. However, not all large systems with many components and physical parts are complex, if they are designed right. On the other hand, even small systems such as the ignition switch of an automobile can be complex if their FRs are coupled.

An example of a small system becoming complex is the GM ignition key. It is a very small component of an automobile, but it can be complex due to wrong design and create havoc. As shown in Fig. 3.3, the major part of the switch was small with a limited number of parts. Yet it was *complex*, because it could not satisfy its FRs at all times. It had two FRs: (1) Move the plunger across the ignition switch plate to enable spring loaded plunger to lock into the notch of the plate for either ignition or

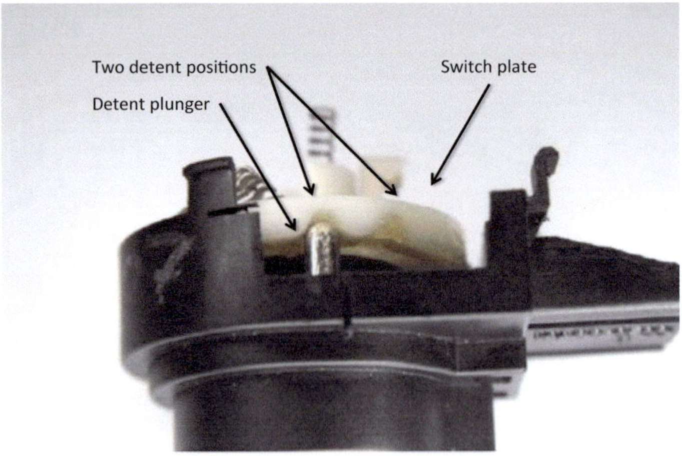

Fig. 3.3 An example of complex small system: the ignition key of an automobile

the accessory position. (2) Lock the plunger in the preset notched position. It is a coupled design, since there is only one design parameter (DP, the spring) for two FRs. Therefore, the DP required fine-tuning of the spring force to satisfy both of these FRs. Sometimes, when one of these FRs could not be satisfied, the vehicle would malfunction. That is the reason for the fatal accidents that bedeviled GM for so many years. General Motors Corporation spent a small fortune, in reputation and money, to solve the failure of their ignition key that resulted in the fatalities. They seemed to have made minor changes without eliminating the coupled FRs through fundamental design changes. As long as their replacement key design was a coupled design, it would have been difficult to make it perform its intended functions all the time without failure.[18] Lately, GM introduced a new design that has eliminated the coupled design.

At the highest level, there are two kinds of complexities: *time-independent complexity* and *time-dependent complexity*. As the name implies, the FRs of systems with time-independent complexity do not change as a function of time, whereas in systems with time-dependent complexity, the FRs of the system change as a function of time.

Time-independent complexity can further be divided into time-independent *real* complexity and time-independent *imaginary* complexity. Imaginary complexity is associated with decoupled designs. When the designer or engineer does not recognize that the design is decupled and start changing DPs in a random order, i.e., not following the design matrix, the design becomes unnecessarily complex, making it a factorial design problem. Decoupled designs are acceptable designs, but the sequence of varying DPs must follow the design matrix to avoid coupling of FRs. The real complexity is associated with coupled designs.

There are two kinds of time-dependent complexities: time-dependent *combinatorial* complexity and time-dependent *periodic* complexity. In time-dependent combinatorial complexity, the number of combinations continues to increase as a function of time. Many of these designs will ultimately fail.

Time-dependent periodic complexity has a definite period and, therefore, can be made to operate stably for a long time by reinitializing the system at the beginning of each period. (A system with functional periodicity can be made to operate stably for a long time as discussed just below.)

For a system to be reliable, it should not have *time-independent real complexity* or *time-dependent combinatorial complexity*. We can make systems with *time-independent imaginary complexity* and *time-dependent periodic complexity* to perform their intended functions. The most preferred design is the one with no time-independent real complexity, which is accomplished by eliminating the coupling of FRs and having no time-dependent complexities.

---

[18]In a letter to the CEO of GM, Mary Barra, dated May 18, 2014, I pointed out the coupling present in the company's ignition-key system design. A vice president replied, saying that GM had performed more tests on a new switch, but it appeared to me that this new switch was still a coupled design that would operate only at a narrow sweet spot.

3.6 Functional Periodicity

In many systems, FRs have functional periodicity. Such systems are designed to be stable over a long time. Familiar examples of systems that have functional periodicity are: circadian cycles of living beings, mitosis of cells, airline schedules, and printing machines. We can also introduce functional periodicity to many other systems, e.g., computers, vehicles, lithography machines. Machines that repeat the same functions throughout its life can accumulate errors, but can be *reinitialized* at the beginning of each cycle, as in the case of rebooting computers, to keep the system stable and capable of achieving its FRs indefinitely.

An OLEV bus system where the bus goes around a fixed route repeatedly should be designed to have a *functional period*. Each time the bus comes back to the same starting station, its system should be reinitialized so that it can complete the next cycle. This will ensure that the OLEV bus serves the route indefinitely, without any problems. In order to give functional periodicity to a system, we can, for example, design the electric power charging system such that the state of battery charge returns to the same state at the beginning of each period. Then the driver does not have to do anything other than drive the bus according to the established schedule.

If we can reinitialize the operation of OLEV after every cycle of operation, we can design the system to perform well during the next cycle, e.g., by properly scheduling the travel and stop times. Re-initialization of the system to the desirable beginning state is a good way of making any system functional and stable. Another possibility is simply measuring the state of the system at the beginning of each cycle (e.g., state of battery charge) and then setting the operating conditions for stable operation during the next period.

Chapter 4 presents a detailed explanation of how OLEV was designed at KAIST in 2009–2011 using the principles of axiomatic design. An axiomatic approach allows one to judge the quality of any design decision, no matter how novel, based on whether it conforms to time-tested axioms. In Appendix 1, two other examples of other large systems designed and implemented based on AD is given.

3.7 Conclusions

Chapter 3 presents theories for design of large systems that can guide the development of large systems. The Axiomatic Design theory and complexity theory outlined in this chapter provided the framework for making correct design decisions, which enabled the rapid design and implementation of On-Line Electric Vehicle (OLEV) and Shaped Magnetic Field in Resonance (SMFIR). A large number of well-known industrial firms engaged in developing similar large systems had encountered major cost overruns and missed schedules because of their reliance primarily on their experience and in-house legacy practices. It is shown that the root cause of cost overruns and project delays is due to the coupling of FRs during the design and implementation process. The solution to these problems is fairly simple and straight forward: follow the Axiomatic Design and the Complexity Theory in designing and implementing large systems in order to prevent the coupling of functional requirements (FRs). To prevent coupling of FRs in designing large systems, system architects should be assigned to prevent inadvertent introduction of coupled design.

Appendix 1

Other Large Technology Systems Created Based on the Design Axioms

To illustrate how Axiomatic Design theory can be applied in the design of any kind of large, complicated engineered system, [2, 3] the creation of two other products—MuCell, the trade name for microcellular plastics, and Mixalloy, a dispersion-strengthened Cu/TiB_2 alloy—along with descriptions of the important lessons learned in the course of their design, offer useful examples.

MuCell

At MIT in the 1980s, the graduate students and I working in the MIT-Industry Polymer Processing Program[19] developed microcellular plastics that were commercialized by Trexel, Inc. (a licensee of MIT), under the trade name MuCell. This product is used by most of the automotive companies in the world to increase fuel economy by reducing the weight of the vehicle and to increase productivity by shortening the production cycle. A number of Ph.D. theses have been written on processing of microcellular plastics at MIT and at other universities, and it is still active academic research field [2, 11, 12].

The idea for the microcellular plastics was created when Gordon Brown, Executive of the Eastman Kodak Company who represented the firm in the MIT-Industry Polymer Processing Program, stated in a conversation I had with him that if the MIT program could come up with an idea for reducing plastics consumption at Eastman Kodak it would be a major contribution to his firm, since many of their products were made of plastics. Based on his description of the project goal, I

[19]The MIT-Industry Polymer Program was established in 1973 with a major five-year grant from the National Science Foundation (NSF) to demonstrate how the university and industry could collaborate to promote innovation. Up to 14 industrial firms participated in the program, paying large membership fees to promote innovative research in polymer processing. Members shared the results of the research and cooperated in selecting research topics. NSF has created the Industry-University Cooperative Program at a large number of universities to replicate the MIT-Industry Polymer Processing Program.

came up with three FRs for microcellular plastics: introducing voids to reduce the consumption of plastics; controlling the size of the voids to maintain fracture toughness; and controlling the geometry to maintain the product shape. To satisfy these FRs, I proposed the idea of putting a large number of micro-voids (about 1 billion bubbles per cm^3) of the order of 10 microns in diameter into the plastic. The corresponding design parameters (DPs) were: a plethora of tiny voids in polymer matrix, the size of the voids, and molding them into the desired shape. In other words, we came up with an uncoupled system with three DPs that could fulfill the three FRs.

One of my graduate students demonstrated that we could make microcellular plastics in a batch process that consisted of dissolving gases (initially CO_2 and later also N_2) in polymers by increasing the pressure of the gas at a temperature higher than the glass transition temperature of the plastic, and then suddenly lowering the pressure to change the thermodynamic state of the polymer/gas solution quickly, creating a two-phase material consisting of the gas phase and the matrix phase (i.e., the polymer). However, this batch process was not amenable to mass production, which was required for industrial use. Four Ph.D. students and I therefore developed a continuous process of manufacturing microcellular plastics, and did so in about three years (which is a typical length of time for doctoral research), making sure that we did not violate the Independence Axiom. The resulting MIT technology was taken over by Trexel, which developed an industrial version of the MIT extrusion process and also extended it to injection molding. These plastic parts have tiny bubbles ($\sim$10 to 30 microns in diameter) in large numbers ($\sim$billion bubbles per cm^3). Many parts of automobiles are made of MuCell (trade name of Trexel). It offers materials saving ($\sim$5 to 15%), higher productivity because of the reduction of cycle time, better dimensional stability, and better properties due to the bubbles.

Mixalloy

Mixalloy is a dispersion-strengthened copper alloy made up of a pure copper matrix phase with a plethora of nanoscale TiB_2 particles dispersed throughout the matrix in order to strengthen the alloy without sacrificing other properties such as electrical conductivity, thermal conductivity, formability, ductility, and fracture toughness.[20] This alloy and the process of manufacturing it were invented to meet the industrial need for copper alloys with high strength, high ductility, and high conductivity at high temperatures—the three FRs in the design process, which resulted in a decoupled design for the alloy itself. The first application for the material was for spot welding tips used in assembling automobile bodies.

This project was remarkable in that we went directly from the concept for a new product, and for the process of making it, to production without the benefit of a preliminary laboratory demonstration to test the feasibility of the invention. We

[20] For more details, see Suh [2], p. 343, and patents such as Sanchez-Caldera, L.E., et al. U.S. Patents 4,706,730; 4,890,662; 4,999,050; and 5,071,618.

built a unique production-scale manufacturing system from scratch and made the alloy for the first time using the machine, the only one of its kind in the world. The project was finished in about three years, which included designing and building the production equipment, making samples of Mixalloy, testing them, building the 27,000-square-foot plant for commercial operation, working with potential customers (e.g., Chrysler, GM), hiring people, working with a hot-isostatic extrusion company, and starting to sell the product.

After conceiving the idea for the Cu/TiB_2 dispersion-strengthened alloy and the process of making it, we could not manufacture the alloy in a laboratory in order to test the viability of the basic concept. Only large-scale production equipment would enable us to reach a steady-state, isothermal flow of molten metal solutions at the required 1200 °C in order to verify our processing idea and make the alloy. Therefore, we built full-scale production equipment from the outset based on our theoretical reasoning and then make the Cu/TiB_2 alloy for the first time and determined the properties of Cu/TiB_2 dispersion-strengthened alloys. We designed and built the production equipment and all the auxiliary machines. Concurrently, we also built the factory that housed this production machine, which was about 20 feet high. In about three years, we were in production of Mixalloy with the first machine we designed and built without the benefit of laboratory-scale experiments. We sold the product to both domestic and international companies. We made only a minor change to the original equipment to go into production, which we built for the first time.

In designing Mixalloy and the machine for producing it, we did our best to make sure that we did not violate the Independence Axiom. However, we knowingly adopted a coupled design for the conduits through which the molten metal flows from the crucible to the mixing chamber because the majority of the team voted for the coupled design, thinking that it would be much easier to fabricate the conduit if we did not have to heat it to 1200 °C—a deceptively attractive idea. We could have gone into commercial production much sooner had we not violated the Independence Axiom. This one wrong decision diverted the development process by at least six months and wasted significant financial resources in a small startup company. Soon after switching to an uncoupled design for the conduit, we were able to manufacture Mixalloy commercially and shipped the product to companies that made spot welding tips for automakers in the USA, Japan, and Korea. Just to provide a perspective to the whole commercialization process, typical industrial firms might have taken ten to twelve years to do what we did in three years.

References

1. Suh NP (1990) The principles of design. Oxford University Press, New York
2. Suh NP (2001) Axiomatic design: advances and applications. Oxford University Press, New York
3. Suh NP (2005) Complexity: theory and applications. Oxford University Press, New York

4. Suh NP (2016) Challenges of designing and implementing large systems. In: Farid A, Suh NP (eds) Axiomatic design in large systems: complex products, buildings and manufacturing systems. Springer, London, ISBN:978-3-319-32387-9, doi:10.1007/978-3-319-32388-6 (Library of Congress Control Number: 2016936680, Springer International Publishing Switzerland 2016)
5. Hatamura Y, Abe S, Fuchigami S, Kasahara N (2015) The 2011 Fukushima nuclear power plant accident (How and why it happened). Woodhead Publishing, Elsevier. p 21
6. Nakao M, Kusaka K, Tsuchiya K, Iino K (2013) Axiomatic design aspect of the Fukushima-1 accident: electrical control interferes with all mechanical functions. In: Thompson MK (ed) Proceedings of the 7th International Conference on Axiomatic Design, ICAD-2013-17, p 113–118
7. Hong E, Park G (2010) Design information management of an on-line electric vehicle using axiomatic design. SAE Int J Mater Manuf 3(1):133–141. doi:10.4271/2010-01-0279
8. Lee T, Park G (2010) Managing system design process using axiomatic design: a case on KAIST mobile harbor project. SAE Int J Mater Manuf 3(1):125–132. doi:10.4271/2010-01-0278
9. Kossiakoff A, Sweet WN, Seymour SJ, Biemer SM (2011) Systems engineering—principles and practice, Wiley, New York
10. Rechtlin E (1990) System architecturing: creating and building complex systems. Prentice Hall, Englewood Cliffs
11. Okamoto KT (2003) Microcellular processing. Hanser, Cincinnati
12. Wong A, Guo H, Kumar V, Park CB, Suh NP (2016) Microcellular plastics. Encyclopedia of polymer science and technology

Part II
The Technology of OLEV and SMFIR

Chapter 4
Axiomatic Design in the Design of OLEV

Nam P. Suh and Dong Ho Cho

Abstract This chapter describes the design details of OLEV and SMFIR. It presents the details of the Axiomatic Design of OLEV and SMFIR, including the modeling of SMFIR. Also, the overall hardware system is described, including power level control of invertors, control of leakage of magnetic field, and the pickup of the transmitted electric power by the power pickup unit mounted on the bottom of the vehicle. It also presents the system architecture for overall software control.

SMFIR and OLEV were designed using Axiomatic Design (AD) and complexity theory as defined in AD, concepts that are presented in Chap. 3. To recapitulate, AD treats design tasks as mapping between four domains: the "customer" domain where the needs for design are described; the "functional" domain where the customer needs are translated into functional requirements (FRs) that must be satisfied by the design; the "physical" domain that consists of design parameters (DPs) chosen to satisfy FRs; and the "process" domain characterized by the process variable (PVs) that can satisfy chosen DPs. Each FR chosen to satisfy the "customer" must be independent of other FRs. The FRs at a given level of design hierarchy constitutes a vector. The DPs are also a vector. Because DPs are chosen to satisfy the FRs, these two vectors are related to each other by a design matrix [DM], i.e., {FRs} = [DM]{DPs} [1–3].

N.P. Suh (✉)
The Department of Mechanical Engineering, Massachusetts Insitute of Technology,
77 Masachusetts Avenue, Cambridge, MA 02139, USA
e-mail: npsuh@mit.edu

D.H. Cho
School of Electrical Engineering KAIST, 291 Daehak-ro, Yuseong-gu, Daejeon 34141,
South Korea
e-mail: dhcho@kaist.ac.kr

N.P. Suh and D.H. Cho (eds.), *The On-line Electric Vehicle*,
DOI 10.1007/978-3-319-51183-2_4

The foundation of AD consists of the two design axioms, the *Independence Axiom* and the *Information Axiom*. The Independence Axiom states that we must *"maintain the independence of FRs"* when we choose DPs. It can be satisfied only when the design matrix [DM] is either diagonal or triangular—all others are coupled designs, which should not be pursued. When [DM] is diagonal, the design is called an *uncoupled* design, which is an ideal design. When [DM] is triangular, the design is a *decoupled* design. Both of these designs satisfy the Independence Axiom. It should be noted that if the design is uncoupled, it can be analyzed as a set of a one-input and one-output problem regardless of the number of FRs and DPs, which is easy to model and optimize. Although it is a little bit more cumbersome, decoupled designs can also be solved as a set of one-output and one-input problems. However, unlike the case of uncoupled design, the analysis must follow the sequence given by the triangular matrix.

The Information Axiom states that among all the designs that satisfy the Independence Axiom, the better design is the one with the least *information content*. The information content, I, is defined in terms of the probability p of satisfying the functional requirement, FR. When the probability of satisfying an FR is equal to 1, the information is zero. When there are many FRs, the probability of satisfying each FR must be evaluated. The most robust design has zero information content. The Information Axiom states that the simpler is the design, the better is the design. (In general, the uncoupled design that satisfies the Independence Axiom has less information content than a coupled design that violates the Independence Axiom.)

There are a large number of theorems that have been derived based on these two axioms [2]. For example, Theorem 4 states that in an ideal design, the number of FRs and the number of DPs are the same. Another theorem states that if the number of FRs is greater than the number of DPs, the design is a coupled design. Also, if the number of FRs is smaller than the number of DPs, the design is either a redundant design or a coupled design, unless some of the extra DPs are fixed to be a constant.

4.1 Overall Design Framework of OLEV

We begin the AD design process by defining customer needs. In the case of OLEV and EGTS, the needs in the customer domain are the reduction of CO_2 in the atmosphere through electrification of ground transportation system (EGTS). Then, the functional requirements (FRs) OLEV must satisfy at the highest level are defined. Once FRs are selected in the functional domain, we identify design parameters, DPs, which can satisfy the FRs. The selection of FRs and DPs is subject to a set of constraints; the design must not violate. Once DPs are chosen, each one of these FRs and DPs may have to be decomposed, if a chosen DP is still conceptual and need further design details.

The FRs selected for OLEV at the highest level are the following set:

$$\begin{aligned}
&\text{FR1} = \text{Propel the vehicle with electric power}\\
&\text{FR2} = \text{Transfer electricity from an underground electric cable to the vehicle}\\
&\text{FR3} = \text{Steer the vehicle}\\
&\text{FR4} = \text{Brake the vehicle}\\
&\text{FR5} = \text{Reverse the direction of motion}\\
&\text{FR6} = \text{Change the vehicle speed}\\
&\text{FR7} = \text{Provide electric power when there is no external electric power supply}\\
&\text{FR8} = \text{Supply electric power to the underground cable}
\end{aligned} \tag{4.1}$$

It should be noted that FRs are expressed starting with a verb to distinguish them from DPs, which start out with a noun.

The constraints the final design should not violate are:

$$\begin{aligned}
&\text{C1} = \text{Safety regulations governing electric systems}\\
&\text{C2} = \text{Price of OLEV (should be competitive with cars with IC engines)}\\
&\text{C3} = \text{No emission of greenhouse gases}\\
&\text{C4} = \text{Long term durability and reliabilityof the system}\\
&\text{C5} = \text{Vehicle regulations for space clearance between the road and the vehicle}
\end{aligned} \tag{4.2}$$

The design parameters (DPs) chosen to satisfy the highest level FRs given by Eq. (4.1) are as follows:

$$\begin{aligned}
&\text{DP1} = \text{Electric motor}\\
&\text{DP2} = \text{Wireless power transfer system}\\
&\text{DP3} = \text{Mechanical steering system}\\
&\text{DP4} = \text{Hydraulic braking system}\\
&\text{DP5} = \text{Electric polarity}\\
&\text{DP6} = \text{Motor drive}\\
&\text{DP7} = \text{Rechargeable battery}\\
&\text{DP8} = \text{Electric power supply system}
\end{aligned} \tag{4.3}$$

The relationship between {FRs} and {DPs} is given by

$$\{\text{FRs}\} = [\text{DM}]\,\{\text{DPs}\} \tag{4.4}$$

Table 4.1 Design matrix [DM] for the OLEV

	DP1	DP2	DP3	DP4	DP5	DP6	DP7	DP8
FR1	X	X	0	0	0	X	X	X
FR2	0	X	0	0	0	0	0	X
FR3	0	0	X	0	0	0	0	0
FR4	0	0	0	X	0	0	0	0
FR5	0	0	0	0	X	0	0	0
FR6	0	0	0	0	0	X	X	0
FR7	0	0	0	0	0	0	X	0
FR8	0	0	0	0	0	0	0	X

For the eight FRs and DPs chosen for OLEV, the design matrix [DM] is given in Table 4.1 by checking whether or not a given DP affects the specific FR:

X indicates that an FR is affected by a DP. For example, DP2 affects FR1 in the design represented by Eq. (4.4).

The matrix [DM] given in Table 4.1 indicates that this design, at the highest level, is a decoupled design. This is an acceptable design, but the FRs must be satisfied by following a specific sequence given by the design matrix. For example, DP2, DP5, DP6, DP7, and DP8 must be set before setting the value of DP1 to satisfy FR1. If the design matrix had only diagonal elements, the DPs could be set without considering the specific order of changing DPs. In the absence of [DM], it would have taken repeated, time-consuming, and costly trial-and-error processes to come up with a combination that works! It is a n-factorial problem.

Once the design is finalized, the Xs can be determined through modeling and analysis or even experimentally. But during the early stages of design, it is more important that the design should be "*approximately correct* than *precisely wrong*". As designers proceed to the lower levels of design, the system becomes "precisely correct" in satisfying the highest level FRs.

In the OLEV project, tasks were assigned to technical teams headed by senior members under an organizational framework established by the project director. Engineering and management staff met weekly to review and discuss each team's progress and the progress of the project as a whole. An "architecture group" had the task of checking the system architecture to ensure that no design at any level would create a coupled system at the system level. [4, 5] A group headed by a civil engineering professor designed a safe and reliable structure for the underground power supply system that could withstand vehicle weights, heavy rain, and temperature fluctuations. Another group dealt with management of the power delivered to the vehicle between the motor and battery. And still another group designed the pickup system that captures the power delivered to the vehicle from magnetic field generated by 20 kHz high current in the embedded power cable. Another group worked on meeting government regulations on EMF radiated, safety, etc. The first vehicle the KAIST team converted to an OLEV was a hybrid bus (a bus with a diesel engine and an electric motor) designed by an industrial firm. Commercial OLEV buses now in operation in Korea in Gumi City, KAIST campus,

Seoul Grand Park, and a few others. The underground infrastructure were designed, manufactured, and installed by industrial firms in Korea.

4.1.1 SMFIR Using a Field Effect

The most challenging part of the OLEV project was the wireless transfer (FR2 and DP2) of large amounts of electric power over a large distance (e.g., 20–30 cm). From the beginning, the KAIST team thought that a field effect could be used to achieve this. The idea of using the field effect for wireless power transfer was a natural extension of the work that Professor Suh had performed with his students and colleagues at MIT,[1] and of prior work done by the editors' colleagues at KAIST, some of whom have authored chapters of this book. The central idea is to transmit electric or magnetic power by establishing either electric or magnetic oscillating fields above the ground from a set of electric or magnetic poles near the surface of the ground, and catch the power at the receiver mounted on the vehicle at resonant frequencies by tuning the impedance of the receiver unit. There are many advantages in using magnetic field above the ground rather than electric or electromagnetic field in terms of safety, penetration through ice and water, electromagnetic interference (EMI), and electromagnetic field (EMF). This is the basis for SMFIR (Shaped Magnetic Field in Resonance).

4.1.2 Decomposition of FR2 and DP2 (Design of SMFIR)

The KAIST team needed to decompose FR2 (transfer electricity from an underground electric cable to the vehicle) and DP2 (wireless power transfer system) in order to develop a detailed design for transmitting power from underground cables to the vehicle. Other FRs could be implemented using existing technologies, which are appropriate DPs that do not require new design and decomposition. The conceptual design that satisfies FR2 is shown in Fig. 4.1.

The idea is to create an alternating magnetic field above the ground that can reach the vehicle by creating two poles at a known distance apart. The magnetic field is created by sending alternating electric current around a magnetic (e.g., ferrite or other materials with high magnetic permeability and low loss) core, which sets up a north and a south pole at the two ends of the core. The distance between the two poles of the ferrite (or other) core determines the shape and the height of the magnetic field. The farther apart the poles are, the higher the magnetic field is. The frequency is chosen to maximize the power transfer to the pickup unit mounted at

[1]Mixing of powder [6], viscous fluid mixing [7], curing of composites [8], non-destructive evaluation of composites [9], and desalination [10].

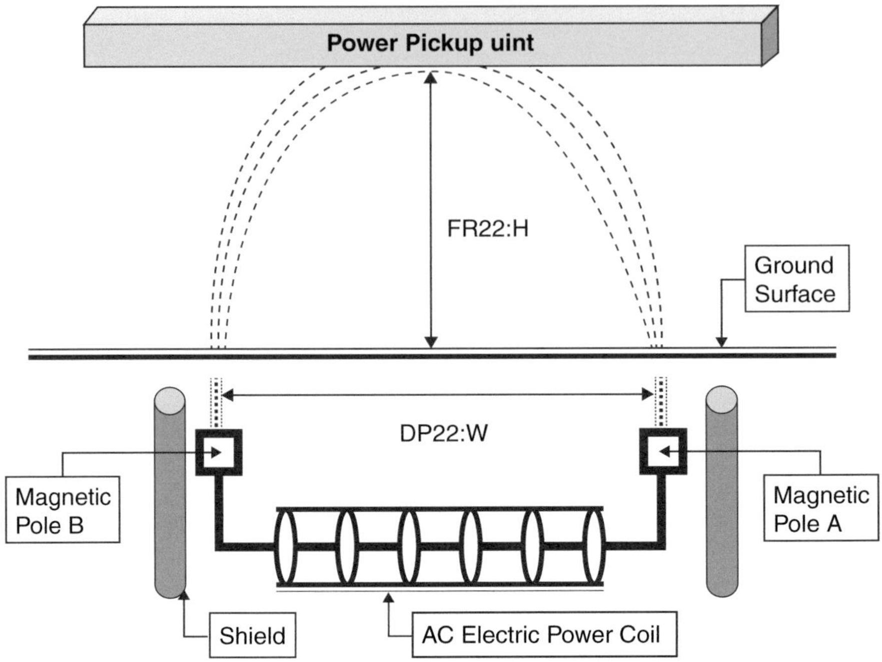

Fig. 4.1 Concept for sending power over a distance *H* wirelessly. First, a two-dimensional oscillating magnetic field is created above ground by sending oscillating electric current around a ferrite core of the underground power supply system. The magnetic field shape above the ground is determined by the distance between the magnetic poles W. The greater the distance is, the higher the reach of the magnetic field is. Once the field shape is determined, the field strength can be adjusted by controlling the electric power that generates the magnetic field. When we place a ferrite core above the ground (attached to the vehicle) opposite to the ferrite core below the surface, the shape of the magnetic field above the ground changes due to the high permeability of the ferrite core, i.e., more magnetic field permeates through the ferrite core above the ground. The oscillating magnetic field energy in the ferrite core is then captured by a set of electric conductors "wrapped around" the ferrite core of the power pickup system attached to the vehicle. Because the power pickup system is tuned to have the same natural frequency as the power supply system below the ground, the magnetic field forms a continuous loop from the power supply system to the power pickup system because they are in resonance. Note that the power transfer efficiency is maximum only when the power pickup system is in resonance with the underground power supply system, because only at resonance, a continuous magnetic loop is created from Magnetic Pole A to Magnetic Pole B through the top power pickup system without interference and a phase lag

the bottom of the vehicle at resonance with a minimum loss of power, i.e., high Q factor. Magnetic fields are preferred over electric fields, since magnetic fields are not affected by ice or many other materials that may be on the surface of the road. In designing the pickup and power cable modules, we chose ferrite due to its relatively low price and very high saturation magnetic flux density.

FR2 was decomposed into second-level FRs:

$$\begin{aligned}
&\text{FR21} = \text{Create an alternating magnetic field above the ground}\\
&\text{FR22} = \text{Control the shape of the magnetic field}\\
&\text{FR23} = \text{Control the power level of the magnetic field}\\
&\text{FR24} = \text{Pick up the energy of the magnetic field by the vehicle}\\
&\text{FR25} = \text{Confine the electromagnetic waves between the vehicle and}\\
&\qquad\quad \text{the underground power supply system (shielding of EMF)}\\
&\text{FR26} = \text{Deliver the electric power to the vehicle while it is in motion}
\end{aligned} \tag{4.5}$$

DP2 was decomposed into second-level DPs as follows:

$$\begin{aligned}
&\text{DP21} = \text{Electromagnet design-ferrite core inside electric field}\\
&\text{DP22} = \text{Distance between magnetic poles (W)}\\
&\text{DP23} = \text{Amplitude of the electric current that generates the magnetic}\\
&\qquad\quad \text{field around the underground ferrite core}\\
&\text{DP24} = \text{Resonating magnetic energy pickup unit on the vehicle}\\
&\text{DP25} = \text{(Passive or active) shield for stray electromagnetic field}\\
&\text{DP26} = \text{Two} - \text{dimensional magnetic field that does not vary along}\\
&\qquad\quad \text{the direction of vehicle motion}
\end{aligned} \tag{4.6}$$

The constraints that the design could not violate were:

$$\begin{aligned}
&\text{C1} = \text{Maximum allowable EMF level : 62.5 mG}\\
&\text{C2} = \text{Maximum weight of the pickup unit}\\
&\text{C3} = \text{Electric shock resistance of the system}\\
&\text{C4} = \text{Temperature rise should not exceed } 20\,^{\circ}\text{C}\\
&\text{C5} = \text{High magnetic permeability of the core material}\\
&\text{C6} = \text{Minimize power loss}
\end{aligned} \tag{4.7}$$

The design matrix for FR2 s and DP2 s is given in Table 4.2:

The design at the second level of FR2 and DP2 is a decoupled design. Therefore, DPs must be varied in the certain sequence specified in Table 4.2, i.e., DP23 should be set first to satisfy FR23 before setting the value of DP21 to satisfy FR21. Figure 4.2 shows the schematic arrangement of the SMFIR design that is given by the design matrix.

To satisfy FR21 (create an alternating magnetic field above the ground), DP21 (electromagnet design—ferrite core inside electric field) is the ratio W/H, which must be much larger than 1, i.e., $W/H \gg 1$. FR25, the shielding of EMF, can be satisfied by either reactive or passive shielding. Passive shielding would consist of placing a barrier in the ground and grounding the EMF picked up around the receiving unit; reactive shielding would consist of generating a signal that is

Table 4.2 Second-level design matrix for OLEV

	DP23	DP21	DP26	DP24	DP25	DP22
FR23	X	0	0	0	0	0
FR21	X	X	0	0	0	0
FR26	0	X	X	0	0	0
FR24	0	0	0	X	0	0
FR25	0	0	0	0	X	0
FR22	0	0	0	0	0	X

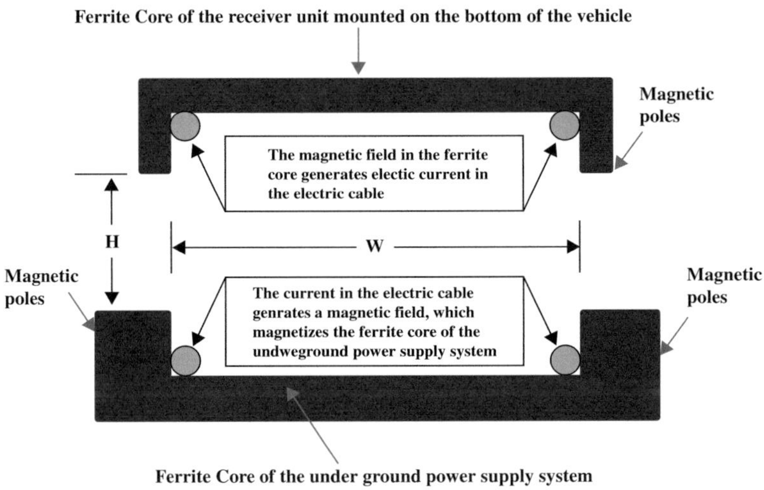

Fig. 4.2 Schematic design of SMFIR. To have the strongest field extend from the underground magnetic pole to the pole of the pickup unit on the vehicle (FR21), *L* must be much larger than *H*. Having the electric current flow perpendicularly to the ferrite cores (parallel to the direction of motion of the vehicle) satisfies FR26

opposite to the EMF emanating from the receiver unit. The design of DP25 is given in [11] and Chap. 14.

The current that flows in the electric cable loop generates a magnetic field in the ferrite core by aligning magnetic domains. Because the ferrite core has low magnetic permeability, the magnetic field is concentrated in the core. One end of the core becomes the S pole, and the other becomes the N pole. Since the electric current that flows in the cable is alternating current (AC), the magnetic field above the ground also oscillates, because the magnetic poles alternate between N and S. To create a magnetic field that is strongest above the ground, pointing toward the vehicle rather than directed toward the magnetic poles, the distance H from the magnetic pole in the ground to the poles of the pickup unit on the vehicle should be smaller than the distance W between the magnetic poles of the underground power supply system (see Fig. 4.2).

The frequency of the field must be carefully chosen. It should be high enough to reach the necessary volume and low enough to minimize loss of energy in the conductors and ferrite core. The initial frequency chosen was 20 kHz. The wire for electric current in the underground cable has to be Litz wire, which is a type of cable used in electronics to carry alternating current at frequencies up to about 1 MHz, in order to reduce the skin effect and proximity effect losses in conductors. For high-speed trains, the chosen frequency was 60 kHz. The advantages of using 60 kHz are as follows: If the current of embedded power line is maintained the same as for the 20 kHz case, the cost and weight of the power pickup system can be reduced by about 1/2 of the 20 kHz design, because the induced voltage will increase by three times. On the other hand, if the induced voltage of the power pickup system is the same as the 20 kHz case, the cost and weight of inverter and embedded power cable can be decreased by about 1/2 because the current of the inverter and embedded power cable decreases by 1/3.

Ahn et al. numerically determined the magnetic field between the underground power supply system and the pickup unit at the bottom of the OLEV, which is shown in Fig. 4.3 [12]. The ferrite core is a dual-rail unit (i.e., W-type), a superposition of two of the units shown in Fig. 4.2.

The idealized flow of the magnetic field is shown in Fig. 4.4 [13].

After determining the overall design, the KAIST group needed to determine the elements of the design matrix through modeling to develop optimum solutions. Obviously, the modeling is needed to be based on the magnetic field emanating from the poles of the ferrite buried underground and how it would induce the magnetic field in the ferrite core of the pickup unit mounted under the vehicle—but this would involve fairly complicated modeling that would require a numerical solution.

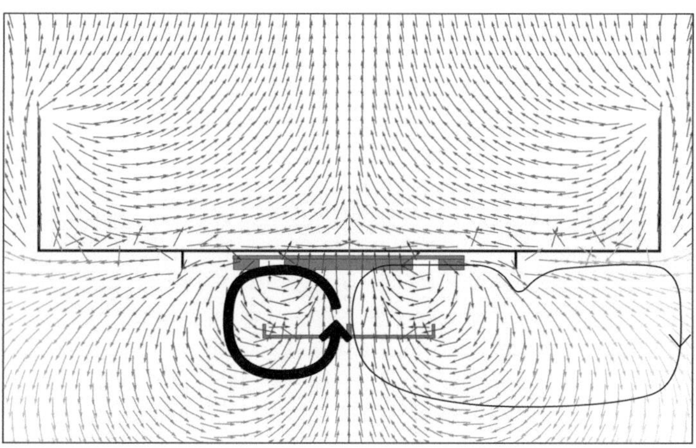

Fig. 4.3 Simulation of the magnetic field between the underground power supply system and the pickup unit mounted at the bottom of the vehicle

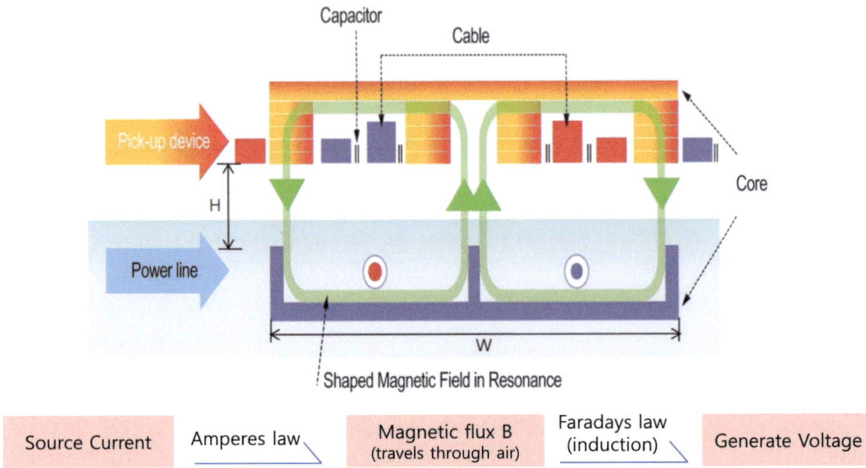

Fig. 4.4 An idealized magnetic field flow of the dual-rail system. The blue ferrite core is under the ground with an electric coil. The ferrite core in the pickup unit and power cable module controls the shape of the magnetic field

4.1.3 Electric Power Transfer to the Moving OLEV

SMFIR is unique in its transmission of electric power to moving vehicles, i.e., FR26. This feature is achieved by creating a two-dimensional magnetic field (DP26) that is not a function of the direction of vehicle motion, i.e., the magnetic field created is a two-dimensional planar field perpendicular to the vehicle's direction. Therefore, the pickup unit mounted on the vehicle sees the same magnetic field while the vehicle is moving, independent of the vehicle's position along the direction of motion, and the vehicle's motion does not affect power transfer and its efficiency. This is why OLEV receives electric power while in motion or stationary. Furthermore, the height of the magnetic field is controlled by placing magnetic poles at a predetermined distance apart.

The power transfer efficiency of SMFIR decreases when the magnetic poles of the receiver unit mounted on the bottom of the vehicle are not aligned with the poles of the underground power supply system. However, because these poles are far apart ($\sim$70 cm), the transmission efficiency is not too sensitive to slight misalignment, unlike the magnetic induction between two circular rings.

Many wireless power transfer systems use circular coils for the transmitter and the receiver, which requires that the centers of two coils should be well aligned and the coils should be close together for maximum power transfer. Such a device can be used only when a vehicle is stationary.

4.1.4 Design of the Power Pickup Unit Mounted on the Vehicle

FR24 (pickup the energy of the magnetic field by the vehicle) and DP24 (resonating magnetic energy pickup unit on the vehicle) must be decomposed to the next level to define the lower-level FR24s and DP24s. One way to decompose the FR24s is:

$$\begin{aligned}
&\text{FR241} = \text{Control the impedance of the pick-up coil to that of the incoming magnetic field}\\
&\text{FR242} = \text{Control the natural frequency of the pickup coil to match that of the magnetic field}\\
&\text{FR243} = \text{Control the temperature of the pickup coil within an acceptable range}\\
&\text{FR244} = \text{Control the flow path of the magnetic field}
\end{aligned} \tag{4.8}$$

DPs must be selected to satisfy these FRs. Since the pickup unit must be designed using a coil and ferrite core, it can be characterized as an RLC circuit built inside a ferrite core. We can choose capacitance and inductance to satisfy FR241, capacitance and inductance to satisfy FR242, a convective air channel for FR243, and a shaped ferrite core for FR244. Many physical configurations are possible, all variations on this basic idea. There are also many constraints at this level of design that cannot be violated. The details of these design issues are discussed in detail in subsequent chapters.

4.1.5 Shielding of Magnetic Radiation

In case of designing SMFIR, it is necessary to consider shielding of the magnetic radiation due to the wireless power transfer between the underground power supply system and the vehicle. The generated magnetic field may cause biological harm to the human body, and it may interfere with the electronic circuit and communication system, or it may cause malfunction of the logic system.

A combination of various shielding methods can be applied to reduce the leakage of the magnetic field from the wireless power transfer coils. *Magnetic shielding*, as shown in Fig. 4.5a, is also a common method to reduce the radiation. But this magnetic shield has following disadvantages such as heavy weight and limited shielding effectiveness. Also, the *reactive shield* in Fig. 4.5b can solve the problem of the active shield due to the usage of the leakage magnetic field to reduce the incident magnetic field. By adding capacitor in the shielding coil, the reactive magnetic shield generates the canceling magnetic field, thus no additional power source is needed [14].

The KAIST team tried both passive and reactive shields. The passive shield, which grounds the bus to the road, was too cumbersome, so the reactive shield was adopted. As shown in Fig. 4.6, we can see that the EMF emitted by OLEVsystem is much less than 6.25 μT [14, 13].

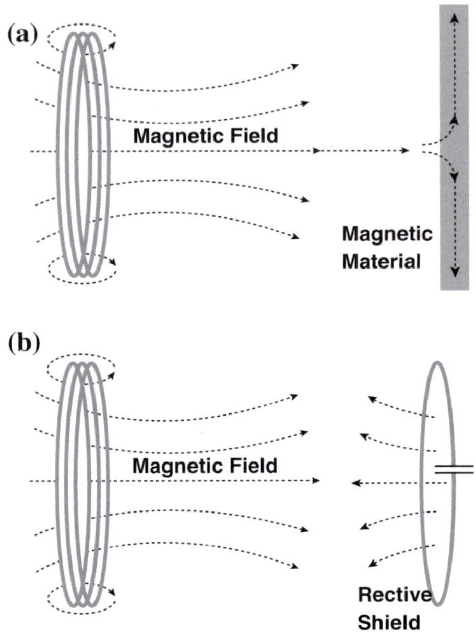

Fig. 4.5 Various shielding methods using: **a** magnetic shield; **b** reactive shield

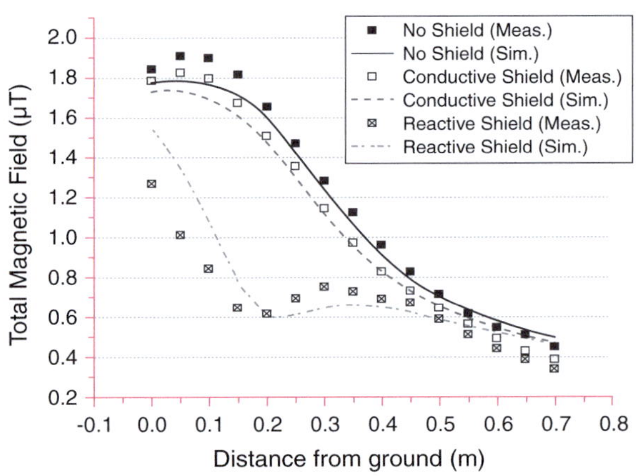

Fig. 4.6 Effect of reactive shield [14]

4.2 Modeling of SMFIR

OLEV is designed to receive electricity wirelessly from a road embedded power cable module (i.e., underground power supply system). When it is on the road with embedded power cable module, the power is used to propel the vehicle and/or charge a small battery on board which supplies the electric power to propel the vehicle on roads without the road embedded power cable module. The size of battery is determined to minimize the overall cost of the OLEV system, including the cost of the road embedded power cable module and the battery.[2] The road embedded power cable module is segmented to supply electric power only when the vehicle is on the top of the road. The total length of the power cable module imbedded under the roads is a function of the vehicle speed. About 2–20% of the road currently used by OLEV buses has the road embedded power cable module [15].

The power cable module imbedded underground to create the magnetic field can be one of the following three: W-type, I-type, and U-type. The W-type device is called the dual-rail type, with the common pole at the center and the other poles at the outside edge of the system to create two symmetric magnetic fields (Fig. 4.4). The three magnetic poles of the W-type are oriented perpendicular to the moving direction of the vehicle. A series of these power cable units are placed along the direction of the vehicle motion with the current supply cables extending along the moving direction of the automobiles to generate the magnetic field in the ferrite core. Therefore, the magnetic field generated is a 2-D field perpendicular to the direction of the vehicle motion. The field extends parallel to the direction of the vehicle motion for dynamic pickup of the power. In the U-type current supply device, electrodes are at two ends of the power supply system (Fig. 4.2). The magnetic field shape for the W-type is symmetric about the center of the power supply unit, whereas in the U-type, the field shape is a single loop, extending from one pole to the other pole. The height of the magnetic field is determined by the spacing between the electrodes. Given a fixed width for the underground power supply system, the U-type can reach twice gap between the ground and the bottom of the vehicle than the W-type. In contrast, in the I-type power supply device, electrodes are arranged parallel to the moving direction of OLEV.

In a case of the U-type current supply device, a distance, H, between the ground and the current collection device may be adjusted by changing a core width W. (Note that H is the FR22 and W is DP22.) This may be expressed by: H = (Constant) × f(W). Here, f(W) denotes a function of the core width W. Given a value of W, the current and voltage may be varied to control the power transferred. The design matrix given in Table 4.2 provides the sequence of calculating for FRs by varying DPs. In some of these designs, there may be more DPs than FRs. In this case, we should fix the DPs that cannot be varied and choose one that can easily vary FR. For example, the power level can be varied using either voltage or current, but normally voltage is fixed and cannot be varied. Therefore, we should use the current as the DP. Similarly, the impedance of an electric circuit can be changed by varying either capacitance or inductance, but it may be more practical to fix inductance at a constant value and choose capacitance as the DP. The DP should be chosen so that its random variation, i.e., δDP, will not adversely affect the FR.

[2]It should be noted that we can also use super capacitors rather than batteries to store electric charges when suitable super capacitors become available. The advantage of using super capacitor is that since the charging time of super capacitors is short relative to that of charging batteries, the length of the underground power supply unit can be short.

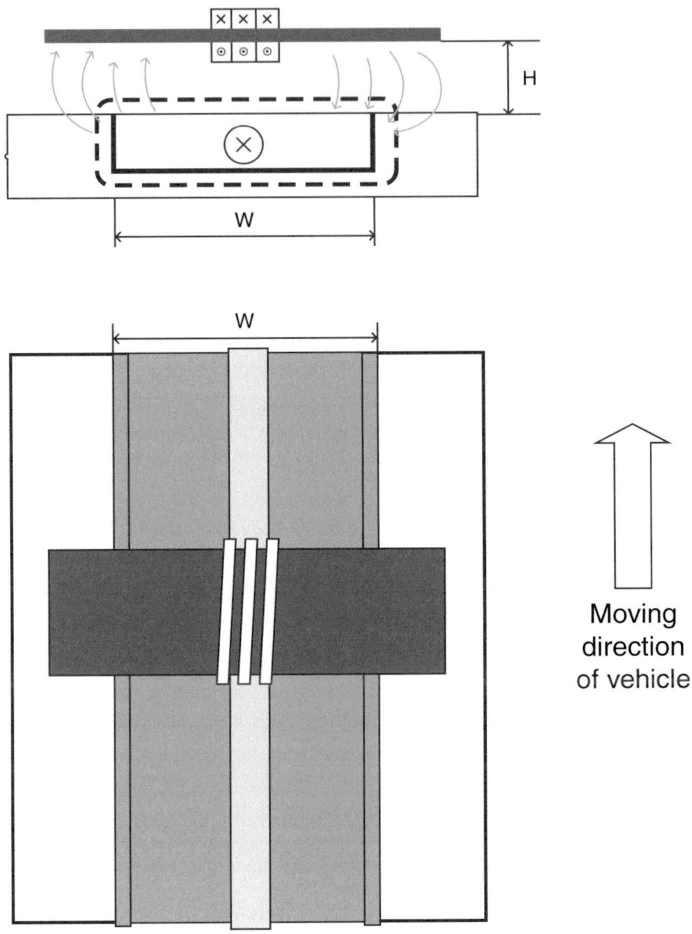

Fig. 4.7 Structure of U-type power supply system

The U-type current supply device may have a core structure as shown in Fig. 4.7, in which two magnetic poles (i.e., two top ends of the U shape) extend perpendicular to the surface and also perpendicular to a moving direction of the vehicle.

The W-type current supply device may have a structure similar to the U-type, except that two U shapes are joined at the center, making it an "E" shape that is turned 90 degrees counterclockwise. Therefore, it has three poles perpendicular to the road surface, the center pole being common to both the right-side pole and the left-side pole. Thus, two magnetic fields extend from the right-side pole to the center pole, and similarly, from the left-side pole toward the center pole. These three poles are aligned with each other and are perpendicular to the moving direction of the vehicle and to the road surface.

Once the gap distance H between the road embedded power cable module and the power pickup module (see Fig. 4.7) is known, we can determine the width between two poles of the U-type. In W-type, the distance between the left- and the right- pole should be two times the width of the U-type.

After the distance between the power cable module in ground and the power pickup module on the vehicle is fixed, the required magnetic field strength is determined. The magnitude of the magnetic field strength is determined based on the power needed to drive the vehicle and the power transmission efficiency. Then, the power of the power cable module may be determined based on H, and the strength of the required magnetic field. After the desired voltage level is established, the magnitude of the current can be determined. Finally, the electric cables that supply power to the power cable module underground is designed by specifying the diameter and a number of the electric cables.

Figure 4.8 is a flowchart illustrating a method for an optimal design of a power transfer systems in order to maximize the power collection. If we denote the power collected as P_c, and the power collecting voltage as V_c, P_c is proportional to the square of V_c. That is, $P_c \propto V_c^2$.

The power collecting voltage may be expressed as

$$\begin{aligned} V_c &= F(f_r, I_s, N_1, N_2, H, W, S_c, C_c). \\ V_c &\propto \frac{f_r I_s N_2 W}{N_1 H} \end{aligned} \tag{4.9}$$

where

f_r = resonant frequency
I_s = feed current
N_1 = number of winding coils of the primary side (the power cable module)
N_2 = number of the winding coils of the secondary side (the power pickup module)
H = distance between the power cable module and the power pickup module
W = width of a power transfer system, (i.e., a gap between magnetic poles in a case of the I-type)
S_c = power transfer core structure (of U-type, W-type, and the improved type)
C_c = material characteristic of a power transfer core, e.g., permeability and a frequency characteristic

The efficiency of current collected by the power pickup unit, E, may be expressed as

$$\begin{aligned} E &= \frac{\text{Regulator Output Power}}{\text{Current Supply Inverter Input Power}} = \frac{\text{Current Collecting Capacity}}{\text{Current Supply Line Capacity}} \\ &= \frac{P_c}{P_s} = \frac{\frac{V_c^2}{R_c}}{I_s^2 R_s} = \frac{V_c^2}{R_c I_s^2 R_s} \end{aligned} \tag{4.10}$$

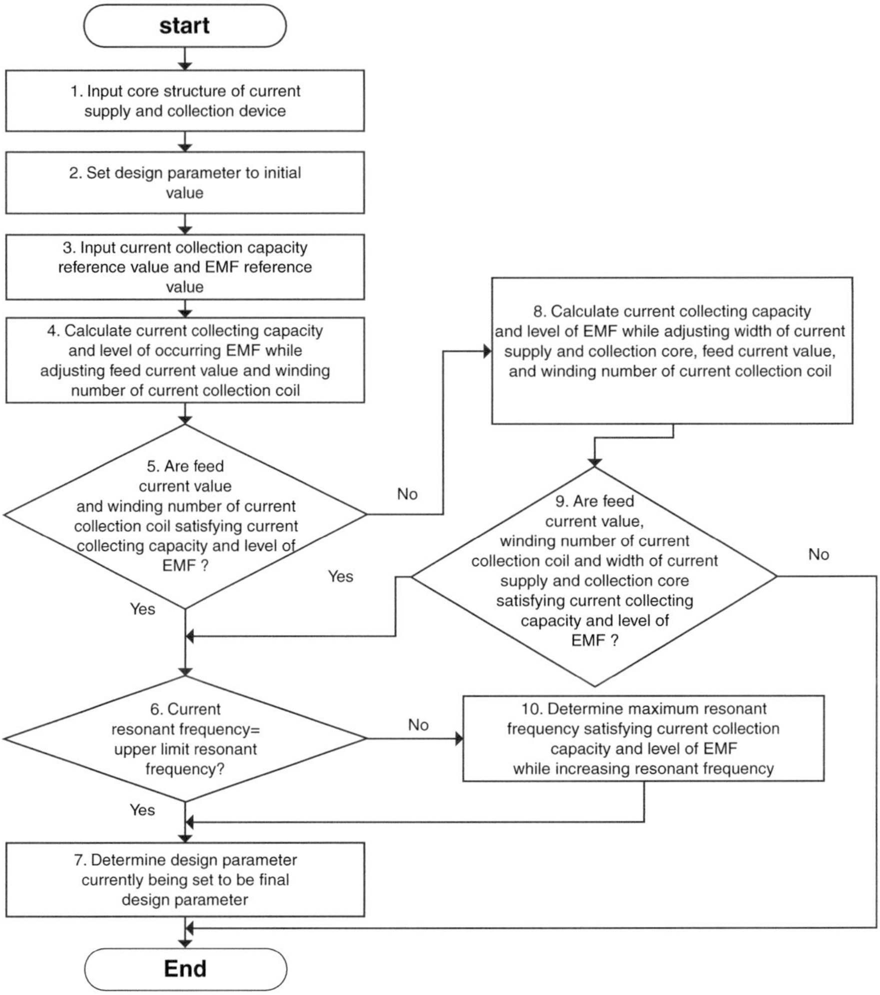

Fig. 4.8 Flow chart of optimal design method for power transfer system using a magnetic resonance scheme

where R_c is the resistance of the current collecting device, and R_s is the current supply line resistance.

In designing the current supply device, the distance, H, may be expressed as

$$H = F\left(f_r, I_s, \frac{N_2}{N_1}, W, S_c, C_c\right) \propto f_r I_s \frac{N_2}{N_1} W \tag{4.11}$$

For example, maintaining H to be greater than or equal to 12 cm may be necessary for a case of a sedan, and 20 cm for a case of a large-sized vehicle such as a bus. Then, given the value of H, W can be determined, fixing the value of all other variables.

The magnitude of EMF should be less than a predetermined value. In particular, when the magnitude of the EMF generated is represented as L_{emf}, it can be expressed as

$$L_{emf} = F\left(f_r, I_s, \frac{N_2}{N_1}, W, S_c, C_c, H\right) \propto f_r I_s \frac{N_2}{N_1} W H \tag{4.12}$$

For OLEV, the maximum allowable EMF is $L_{\mathrm{emf}} < 62.5$ mG, which must be satisfied in vicinities where people are likely to be present.

The core structure of the power transfer system can be determined using the following input parameters: the resonant frequency, feed current, a width of a power transfer core, and the number of windings of the current collection coil. The type of the power supply core structure may include an I-type, a U-type, a W-type, and modified I-type, U-type, and W-type.

The next step is to set the initial values for the design parameters: the minimum power collecting capacity value required by the power pickup module (i.e., "the power collecting capacity reference value"), a maximum EMF value allowed (i.e., "the EMF reference value"), and the minimum distance between the power cable module and the power pickup module.

Once the feed current and the number of windings of the power pickup coil are determined, other design parameters may be determined for the core structure. In addition, if the resonant frequency is less than a maximum allowable resonant frequency, the upper limit resonant frequency may be obtained by repeating the calculation of the power collecting capacity and the resulting EMF. Through this process, the upper limit resonant frequency, the feed current value, the winding number of the current collection coil, and the other design parameter can be finalized. When the input current and the number of windings of the pickup system need to be adjusted, the width of the power transfer core may be increased within the allowable range. The phrase "width of the power transfer core" is used to refer to the width W (see Figs. 4.4 and 4.7) of a W-type core and a U-type core, respectively.

Determination of the power collecting capacity and the EMF follows a similar process used for the power supply capacity, e.g., increasing the width of the power transfer cores, and adjusting the feed current value, the windings of current collection coil, etc. When the width of the power transfer core, a feed current value, and the number of windings of the current collection coil satisfy the design condition, the resonant frequency may then be determined. After determining the width of the power transfer core, the feed current value, and the number of windings of the current collection coil that satisfy the design condition, the upper limit of the resonant frequency can be obtained by repeating the calculation of the power collecting capacity and EMF while increasing the resonant frequency if it is less than the upper limit resonant frequency within an allowable range.

4.3 Overall Hardware System Design

4.3.1 Power Level Control of Invertor

Figure 4.9 shows the OLEV inverter with a two-segment controller and primary power tracks.

The main modules of inverter system are a three-phase full bridge silicon controlled rectifier (SCR) with a DC-link capacitor and a high-frequency pulse width modulation (PWM) module. SCR is used to convert three-phase AC power to DC voltage source. Using this voltage source, the inverter module converts DC voltage to AC current. For electrical isolation between two segments, the transformer is used, and the inverter output load of each segment consists of line inductance of the primary track segment and a resonance capacitor.

Track segmentation is done using additional half-bridge modules. For example, the first two half-bridge networks are used to supply current to the first segment. Then, the half-bridge network located on center is commonly used for both segmentations.

The inverter output current changes according to the load conditions, such as the state of charge (SOC) of the battery in the OLEV bus. However, it is necessary to maintain constant output current regardless of the amount of power consumed from the load for the purpose of generating uniform magnetic flux density, which is done by controlling the SCR rectifier as shown in Fig. 4.10. The first loop, $PI_1(z)$, controls the output current, and the second loop, $PI_2(z)$, controls the firing angle of the SCR to set the DC voltage for the purpose of obtaining the desired output current.

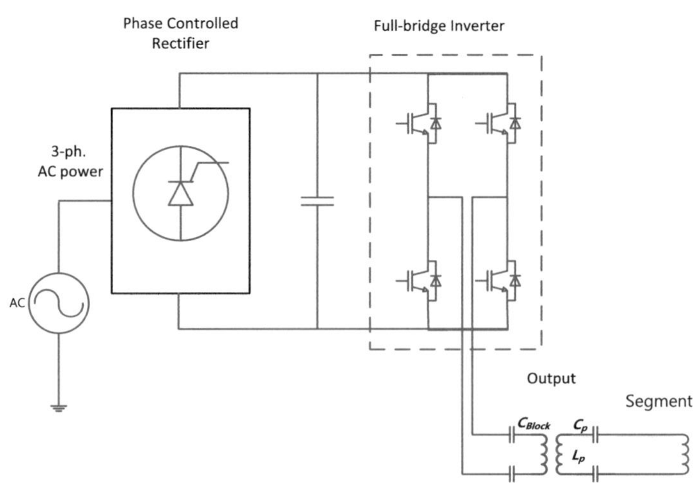

Fig. 4.9 OLEV inverter system

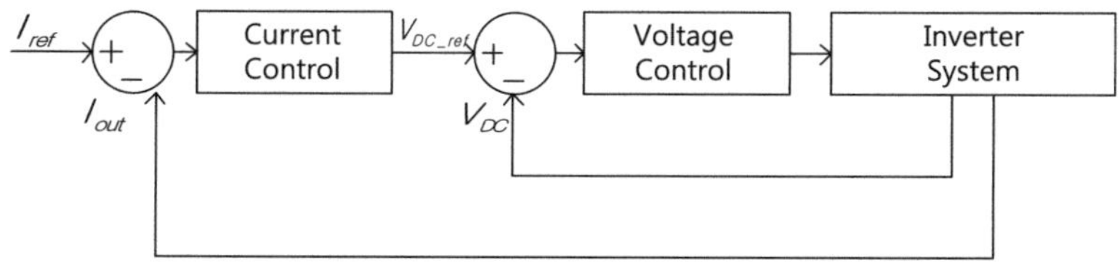

Fig. 4.10 Control algorithm block diagram

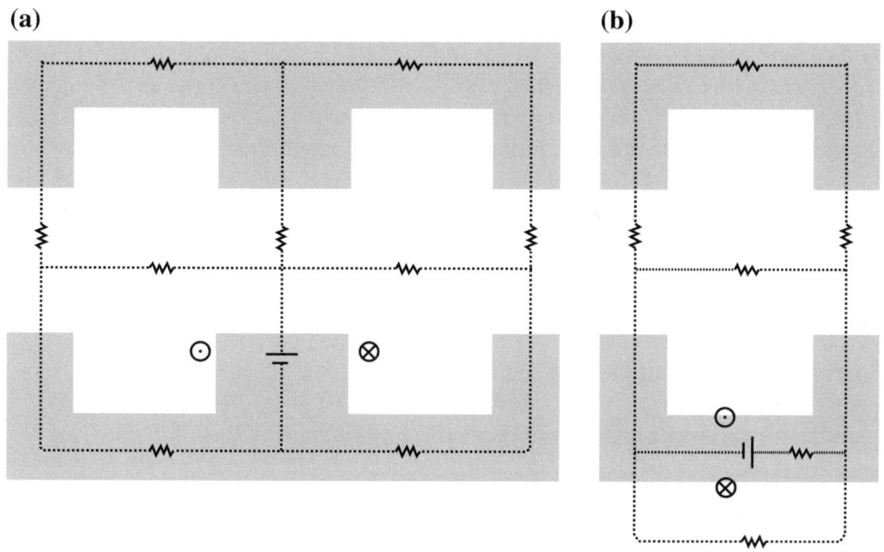

Fig. 4.11 Magnetic core structures of **a** EE-type and **b** UU-type

4.3.2 Control of Magnetic Field for Minimum Leakage

To maximize magnetic flux density, wireless power transfer systems use magnetic cores. The basic core structure can be categorized into two types such as EE-type and UU-type. As shown in Fig. 4.11, the EE-type consists of a pair of E-shaped cores, and the UU-type consists of a pair of U-shaped cores

In case of the EE-type structure, power line cables are wound around the center magnetic core pole, and two target magnetic loops and two main leakage magnetic loops are formed. However, in case of the UU-type structure, power line cables are wound around a pole or a horizontal bar of the primary magnetic core, and a target magnetic loop and two main leakage magnetic loops are formed.

To support the maximum power transfer, the coupling coefficient k should be maximized by adjusting the reluctances.

Here, the reluctance is proportional to the length l of the path and inversely proportional to the relative permeability μ and cross-sectional area A. Since relative permeability is the ratio of the permeability of specific media to that of free space, larger relative permeability of the core is more advantageous to reduce the core reluctance R_c. Thus, the ferrite having a relative permeability higher than 3000 was used in the pickup and power cable module.

4.3.3 Power Pickup from the Magnetic Field at Vehicle

The most important parameters to consider in designing a high power pickup module are the magnetomotive force and power loss in the pickup module. In order to increase the magnetomotive force, both the number of coil turns and the current value flowing in each coil turn need to be increased. Wires with bigger diameter have larger conductive area and higher allowable current. Thus, the use of Litz wire with larger number of wires allows us to flow higher current in the pickup coil while minimizing the power loss as well.

In Fig. 4.12, we can see the conventional circuit of the pickup module in the KAIST OLEV system.

In case of designing the core for a high power pickup, the most important key factor to be considered is how much magnetic flux the core is capable of receiving. To increase the amount of magnetic flux the core receives, the core with larger area has to be designed.

In the pickup module, the reactive voltage of about 10 kV occurs due to a very large L, which is difficult to insulate even if compensated. Hence, to reduce the peak voltage, the capacitance is not connected all at one point but is divided and distributed among the winding cables represented by the inductances.

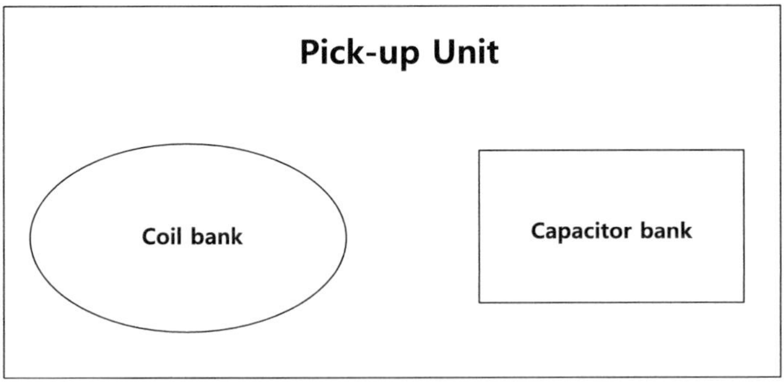

Fig. 4.12 Conventional resonance circuit of pickup module

4.4 Overall Software Control System Architecture

Figure 4.13 shows a schematic diagram of the power transfer control loop. The power pickup system determines the pickup output voltage reference and calculates pickup output voltage error from the pickup output voltage measured in the pickup device. This pickup output voltage error is sent to power supply system through communication transmitter.

The power supply system determines the power inverter current reference from the received pickup voltage error through communication receiver. Power inverter current error is calculated from the power inverter output current measured in the power inverter and power inverter controls the output voltage to make power current error to be almost zero.

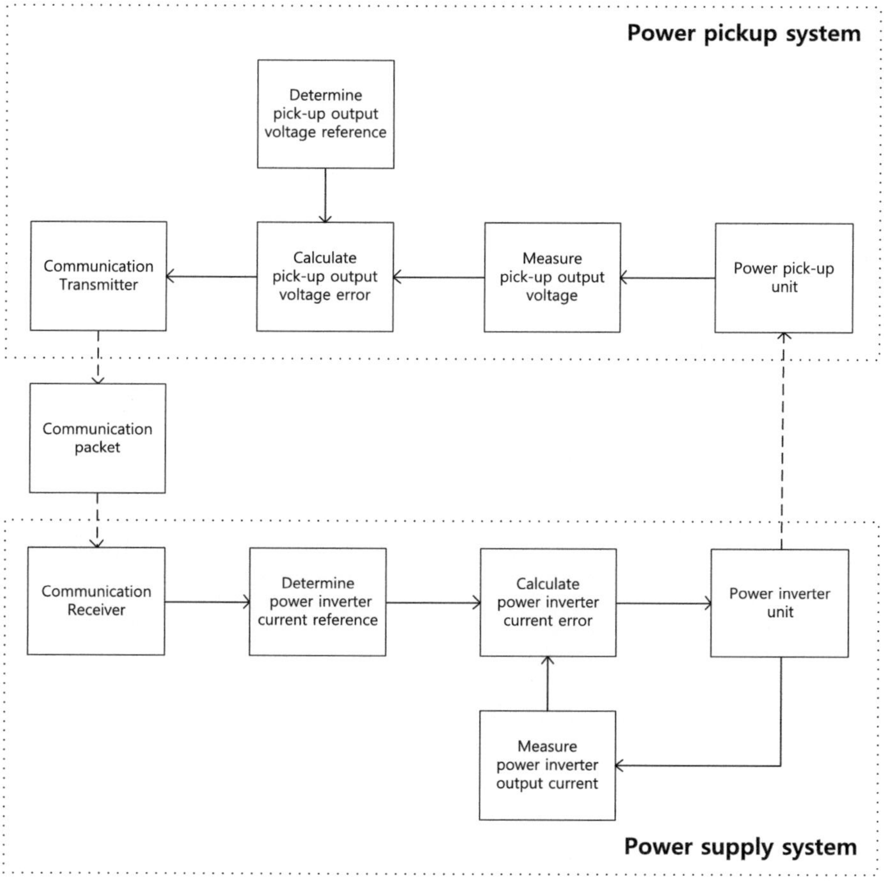

Fig. 4.13 Power transfer control loop

4.5 Conclusions

Chapter 4 presents the overall scheme of the design for OLEV and SMFIR, based on Axiomatic Design (AD) theory, including how SMFIR enables wireless power transfer over a large distance, how power is transmitted to a moving vehicle, and how EMF is controlled. Detailed procedure for determining key design parameters is also outlined, and the design of hardware for power supply, such as the inverter, is presented. The detailed design and operation of some of the major components of these systems are further elaborated in subsequent chapters.

References

1. Suh NP (1990) The principles of design. Oxford University Press, New York
2. Suh NP (2001) Axiomatic design: advances and applications. Oxford University Press, New York
3. Suh NP (2005) Complexity: theory and applications. Oxford University Press, New York
4. Hong EP, Park GJ (201) Collaborative design process of large-scale engineering systems using the axiomatic design approach. Proc Inst Mech Eng Part C: J Mech Eng Sci, 225(9): 2174–2188
5. Lee T, Park G (2010) Managing system design process using axiomatic design: a case on KAIST mobile harbor project. SAE Int J Mater Manuf 3(1):125–132. doi:10.4271/2010-01-0278
6. Suh NP, Tucker III CL (1977) Method and apparatus for mixing particles. US patent 4,034,966
7. Suh NP, Rotz C, Erwin L, Melcher JR, Hoburg JH (1979) Fluid mixing apparatus. US patent 4,174,907
8. Haven R, Suh NP (1983) High frequency electric field curing of polymeric composites. US patent 4,423,191
9. Suh NP, Tse M (1984) Method for non-destructive detection and characterization of flaws. US patent 4,443,764
10. Suh NP, Kim SG, et al (2007) Method and apparatus for permeating flow desalination, USSN 12/002,664, filed 19 Dec 2007
11. Kim JS, Kim JH et al (2013) Coil design and shielding methods for a magnetic resonant wireless power transfer system. IEEE 101(6):1332–1342
12. Ahn SY (2013) Overview of wireless power transfer technology. In: IEEE Conference on Intelligent Transportation Systems, 6–9 Oct. Hague, Netherlands
13. Cho DH (2013) WPT concept and applications. IEEE Conf Intell Transp Syst, 6–9 Oct. Hague, Netherlands
14. Ahn S et al (2014) Design and analysis of a resonant reactive shield for a wireless power electric vehicle. IEEE Trans Microw Theory Tech 62(4):1057–1066
15. Ahn SY, Suh NP, Cho DH (2013) Charging up the road. IEEE Spectr 50(4):48–54

Chapter 5
Magnetic Field Generation

Seungyoung Ahn

Abstract The fundamentals of electromagnetic field generation in wireless power transfer system are discussed, followed by explanation of the design of transmitting and receiving coils for electric vehicle application. Then, it presents the basic mechanism of magnetic field generation, the near- and far-field characteristics of the electromagnetic field, and the coil topologies for wireless power transfer system. Also, the coil structures of transmitter and receiver structure for OLEV application are discussed. To maximize the efficiency, the resonance frequency is carefully selected and the resonance circuit is constructed by using coil inductance and lumped capacitance. The use of the magnetic material for SMFIR design, the criteria for optimal selection of the operation frequency in OLEV system, and the role of skin effects are discussed.

5.1 Introduction

As mentioned in previous chapters, we generate magnetic fields around an *E*- or *U*-zshaped ferrite core by sending an alternating electric current through electric wire that "wraps" around the ferrite core of the power supply system under the ground. Then, the ends of the ferrite core (which are placed near the surface of the ground) become the magnetic poles, emanating the magnetic field into the space between these two poles. If there is nothing above the ground, the magnetic field above the ground would be simply a series of arc of the magnetic field, connecting from these two poles. However, if we place an identical magnetic ferrite core above the ground opposite to the electro magnet in the ground (either *E* or *U* type, facing

S. Ahn (✉)
The Cho Chun Shik Graduate School for Green Transportation KAIST,
Korea Advanced Institute of Science and Technology (KAIST),
291 Daehak-Ro, Yuseong-Gu Daejeon 3414, South Korea
e-mail: sahn@kaist.ac.kr

N.P. Suh and D.H. Cho (eds.), *The On-line Electric Vehicle*,
DOI 10.1007/978-3-319-51183-2_5

upside down), the magnetic field emanating from the pole of the underground ferrite core preferentially go through the ferrite core of above the ground because of the high permeability of the ferrite core relative to that of air. Although we cannot channel all of the magnetic field through the ferrite core above the ground because of the finite air gap between the opposing poles below and above the ground, we can capture a large fraction of it, depending on the design of ferrite cores. If we surround the ferrite core above the ground with a coil of electric wire, electric current is induced in the conductor. This electricity is used to either power the electric motor of the vehicle or charge electric batteries. For this to occur, we have to match the resonant frequency of the power pickup system with that of the underground power supply system by varying the impedance and reactance of the electric circuits so as to assure the "flow" of the magnetic field through the underground ferrite core and above ground ferrite core without interference and a phase lag. Because of the complex configuration of the underground power supply system and the power pickup system mounted on the vehicle, only numerical calculations can give precise power transfer efficiency. In this chapter, a much simpler standard analysis of magnetic field generation by a circular or dipole antenna is presented to show the general behavior of magnetic field around the resonant frequency.

The wireless power transfer system uses an electromagnetic field to transfer power wirelessly by using transmitting and receiving coils. In this chapter, the fundamentals of electromagnetic field generation are discussed and the design of transmitting and receiving coils is, respectively, explained for electric vehicle application. First, the basic mechanism of magnetic field generation and the characteristic of the electromagnetic field are explained. Also, the coil topologies are explained because the antenna structure generating a low frequency electromagnetic field plays an important role in control of the electromagnetic field. Recent studies on resonant wireless power transfer systems generally focus on the transfer of the power using a magnetic field rather than an electric field. The shape of the generated magnetic field thus determines the power efficiency and the power capacity in a wireless power transfer system. To enhance the electrical performance of the wireless power transfer system using a magnetic field, a high permeability material is frequently adopted to construct efficient paths for the magnetic field. In the OLEV system application, the use of the ferromagnetic material significantly improves the efficiency and capacity. In actual design of a wireless power transfer system, the frequency should be carefully chosen depending on the application and purpose of the system. Therefore, this chapter also deals with the dependency of electrical characteristics of a wireless power transfer system to explain how optimal selection of the operation frequency is achieved. Although an increase of the frequency improves the efficiency by increasing the induced voltage with the same amount of source current, the skin effect should be considered to design the coils of the wireless power transfer system with high efficiency by minimizing the ohmic loss at the coils.

5.2 Generation of Alternating Magnetic Field

5.2.1 *Electromagnetic Field Characteristics in Near- and Far-Field Regions*

To transfer power wirelessly without any physical connection where the transmitter and receiver are electrically separated, it is necessary to use an electric field or a magnetic field. The electromagnetic field can be the medium of transmission. According to Maxwell's equations, a time-varying electric field generates a magnetic field, and a time-varying magnetic field generates an electric field. Therefore, when the alternating current flows through a transmitter, both an electric field and a magnetic field are generated.

$$\begin{aligned} \nabla \times \boldsymbol{E} &= -\frac{\partial \boldsymbol{B}}{\partial t} \\ \nabla \times \boldsymbol{H} &= J + \frac{\partial \boldsymbol{D}}{\partial t} \\ \nabla \cdot \boldsymbol{D} &= \rho \\ \nabla \cdot \boldsymbol{B} &= 0 \end{aligned} \tag{5.1}$$

The ratio of the electric field to the magnetic field is defined as the wave impedance (Z_W).

$$Z_W = \frac{E}{H} \tag{5.2}$$

The wave impedance of a field is determined by the radiating antenna structure, and the media surrounding the field. However, the effect of the antenna structure becomes weaker if the distance between the source and the observation point increases. Therefore, the space surrounding a source of radiation can be broken into two regions. Far from the source, the properties of the field depend mainly on the medium through which the field is propagating. At a point close to the source, the field properties are determined primarily by the source characteristics. If we observe the magnitude of the electric field and the magnetic field in the far field region, a point far from the source, the ratio of the electric field to the magnetic field will converge to the intrinsic wave impedance. If we observe the electric field and magnetic field in the near field region, either the electric field or the magnetic field is predominant depending on the generating antenna structure of the transmitter.

If the observation point is close to the source, the field is in the near-field region. If the observation point is far from the source, the field is in the far-field region. In case of OLEV system, the operating frequency is in kHz range and the distance (or dimension) is in meter range, therefore, it satisfies the following near-field condition that the distance is much smaller than the wavelength normalized by 2π

$$\text{Distance} \ll \frac{\lambda}{2\pi} \tag{5.3}$$

while λ is the wavelength of the operating frequency [1].

As shown in Fig. 5.1, if the normalized distance is much larger than 1, it is in the far-field region and the wave impedance converges at around 377 Ω. If the normalized distance is much smaller than 1, it is in the near-field region; this region can be categorized into two conditions: electric field-predominant and magnetic field-predominant conditions. If a dipole antenna is used for the generation of the electromagnetic field, the electric field is predominant and the wave impedance is higher than 377 Ω. If a loop antenna is used, the magnetic field is predominant and the wave impedance is lower than 377 Ω. In most cases of wireless charging for an electric vehicle, the near-field region in the magnetic field-predominant condition is used.

Both the electric field and the magnetic field can be used for power transmission. However, most applications adopt magnetic field transmission because a magnetic field has the characteristics of less attenuation and reflection due to the high dielectric material, as a magnetic field is not as affected by permittivity. Also, a

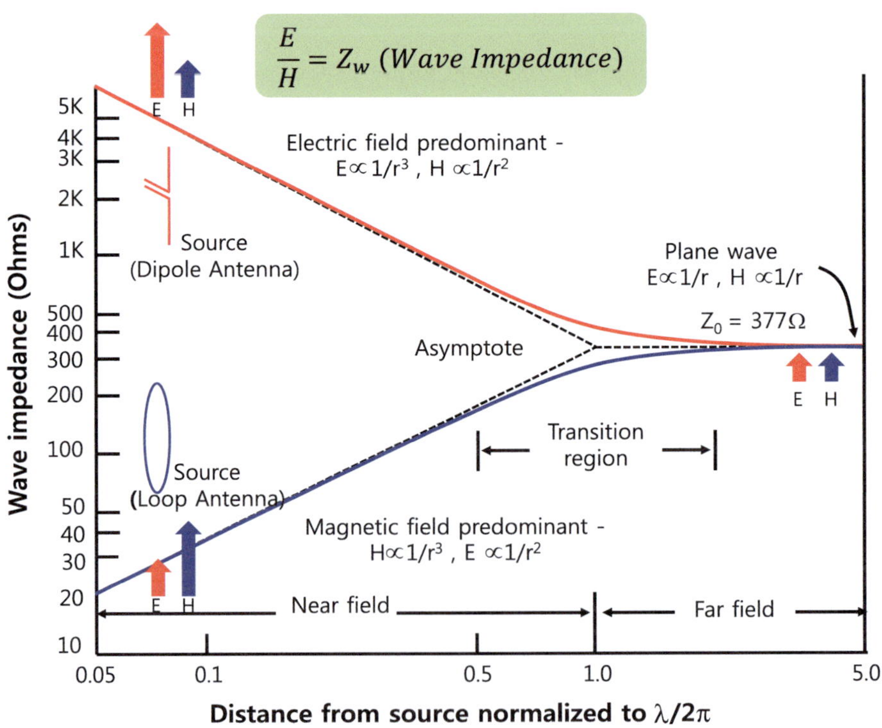

Fig. 5.1 Wave impedance depending on the distance from the source [1]

magnetic coupling structure is easier to implement than electric field coupling. The magnetic field can be generated and received by a loop antenna, and large wires can transmit and receive high power by means of a magnetic field. In electric field coupling, however, a large capacitance between the transmitter and the receiver is required. This requires large plates in the transmitter and receiver rather than a dipole antenna with wires.

5.2.2 *Coupling of the Generated Magnetic Field*

Wireless power transmission by induction has been used for a long time in the form of electric toothbrushes. These electric toothbrushes have to use inductive coupling technology as a means of recharging their batteries because they are constantly exposed to water. This type of wireless charging system can be categorized as a type of inductive coupling. Inductive coupling here generally implies wireless power transfer of a few millimeters from the transmitter coil to the receiver coil.

If the transmitting and receiving coils have short distance between them in a range of a few millimeters, the magnitude of magnetic field leakage could be very small and therefore the efficiency could be very high, in a range of 60–90%. The limitations of inductive coupling are that the distance cannot easily be increased and that the alignment of the transmitter and receiver should be accurate. Not fulfilling these conditions leads to degradation of the transfer power efficiency [2]. Therefore, the simple induction is not a proper method for OLEV applications because the air gap between transmitter and receiver should be more than 20 cm to transfer power wirelessly to moving electric vehicle. In the view of the coupling coefficient, the coupling coefficient of a traditional induction system generally is around 0.9 or higher, however, the coupling coefficient in OLEV system is around 0.1 or lower if the air gap is 20 cm.

The distance and alignment accuracy limitations can be resolved by applying a resonance circuit to the transmitter and receiver systems in a process also known as resonant magnetic coupling. This theory was clearly explained in earlier research [3] by means of coupled-mode theory. It explains that the power is efficiently transferred wirelessly although the coupling coefficient is very low. Other researchers later showed that this idea is also valid for different frequency ranges and power levels [4]. With the resonance effect, the OLEV system could have much higher efficiency and larger air gap than simple induction system.

5.2.3 *Topology Selection and Coil Design*

The core of a wireless power transfer system consists of a power source, a transmitter coil, a receiver coil, and the load. The design of each component is significant in the implementation of the wireless power transfer system because the coil is closely related to the dimensions, circuit design, transfer power, efficiency, magnetic field shape, and cost. Therefore, the source types, coil topologies, feeding types, and load types should be carefully determined at the beginning of the design.

Figure 5.2 shows the feeding types of transmitter coils and receiver coils. The left figure shows indirect-fed coils in which the source coil and transmitter coil are separated. In this case, the transmitter coil and the receiver coil have high Q-factor values because the resistance values of the source and load are not included; the wireless power transfer distance hence can be increased. However, as in other systems with a high Q-factor, this system can be very sensitive to design parameters such as the inductance value and the resonance frequency. Thus, a direct-fed wireless power transfer system can be useful when stability is more important than the transfer distance [5].

Figure 5.3 shows two types of transmitter and receiver coils with different resonance types. The left figure shows a self-resonant coil in which the spiral wire has its own resonance due to the inductance of the wire and capacitance between each turn of the wire. The right figure shows an LC resonant coil where the inductance of the coil and a lumped capacitor generates series resonance. The lumped capacitor component can create more loss than self-resonant fringing capacitance, but it is more controllable and easier to manufacture. Similar to Fig. 5.2, the selection of the

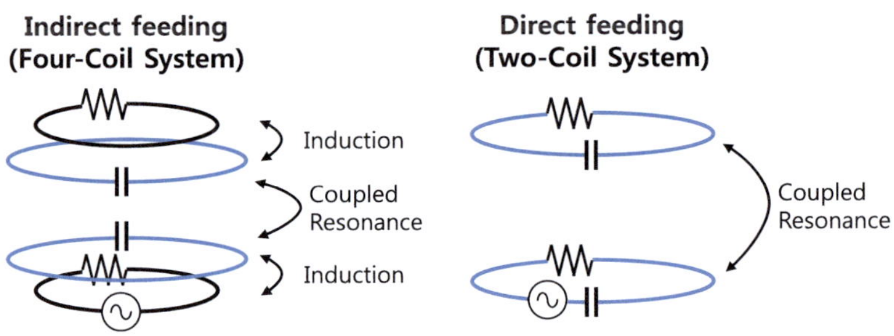

Fig. 5.2 Two types of feeding for transmitter and receiver coils: indirect-fed coil and direct-fed coil

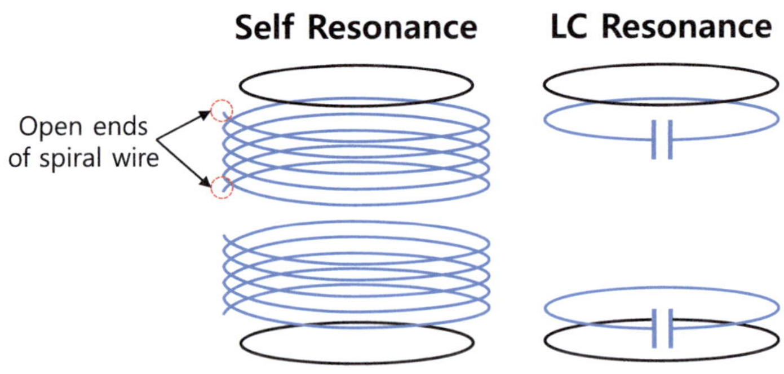

Fig. 5.3 Two resonance types for transmitter and receiver coils: self-resonance coil and LC resonance coil

resonance type determines the loss of the transmitter and the receiver coils and leads to a different coupling factor and altered distance.

To implement a wireless power transfer system with a longer distance but with higher efficiency, an indirect-fed self-resonant coil may be a better choice. However, when we focus on the design of a wireless power transfer system for electric vehicles, robustness and less sensitivity of the design parameters are important for direct-fed LC resonant coils. Figure 6.4 shows an example of basic topologies used with direct-fed LC resonant coils for a wireless power transfer system. There are 24 possible combinations of the system with two source types, two transmitter coil types, two receiver coil types, and three load types. In the selection of the resonance types between the series resonance and the in parallel resonance, the impedance of the load should be considered. For the maximum power transfer, the impedance of the receiver coil should be matched to the load. Therefore, if the load has higher impedance, a receiver coil with parallel resonance should be selected, whereas for lower load impedance, a receiver coil with series resonance should be selected [6] (Fig. 5.4).

To simplify the theory and to explain it with a simple equivalent circuit, two LC resonant circuits with mutual inductance are popularly used, as shown in Fig. 6.5. The transmitter coil and receiver coil have mutual inductance (M) for wireless power transmission, while they also have self-inductances. By adding the capacitances (C_S and C_L) in the circuits of the transmitter and receiver, the self-inductance of the inductance is canceled by the capacitance, and then the total impedance of each loop is decreased. To minimize the impedance maximally, the resonance frequency of the LC resonance should be the same as the frequency of the source (I_1). As a result, the imaginary part of the loop is minimized and the efficiency can be increased even if the distance is longer than inductive coupling. The calculation of the power transfer efficiency in the circuit is shown in Eq. (5.5). It is clear that

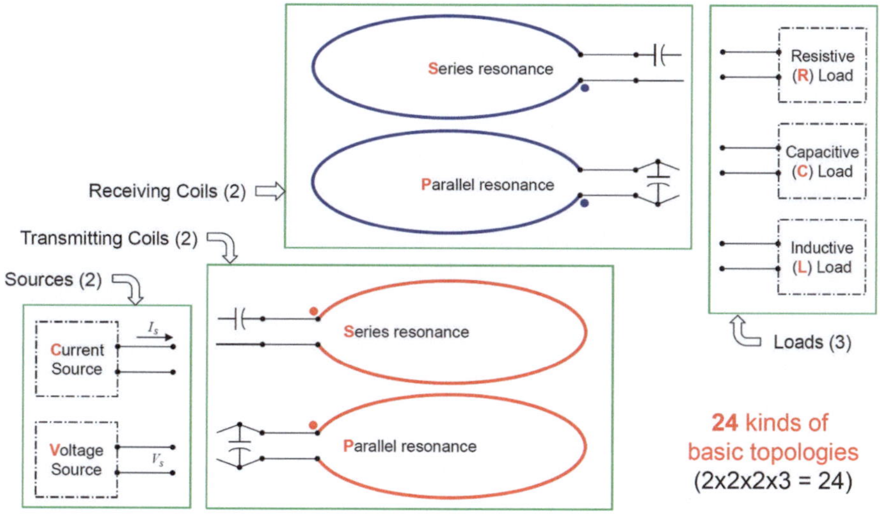

Fig. 5.4 Example topologies of direct-fed LC resonant coils for wireless power transfer system [7]

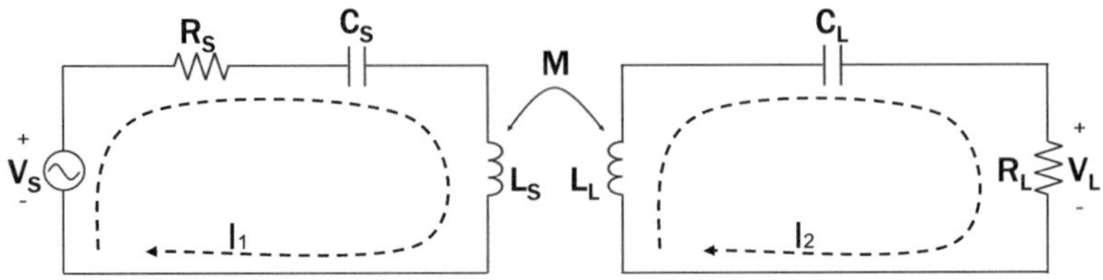

Fig. 5.5 Simplified equivalent circuit of voltage-source wireless power transfer system with resonant magnetic field

the frequency and coupling coefficient should be increased and the resistance should be decreased to improve the efficiency (Fig. 5.5).

$$\text{Induced Voltage:} \quad V_{\text{induced}} = j\omega M i_1 \tag{5.4}$$

$$\text{Efficiency:} \quad K = \frac{\omega^2 M^2}{R_S R_L + \omega^2 M^2} = \frac{1}{1 + \frac{R_S R_L}{\omega^2 M^2}} \tag{5.5}$$

5.3 Channeling of the Magnetic Field Using Materials with High Permeability

5.3.1 Advantage of Using High Permeability Material

Traditionally materials with high permeability have been used to control the path of the magnetic field in magnetic circuit applications such as actuators or transformers. In wireless power transfer systems, high permeability materials are useful to guide the magnetic field to an effective path. Guiding the magnetic field is very important in terms of both the wireless power transfer effectiveness and electromagnetic compatibility. The efficiency of the wireless power transfer system in OLEV is further enhanced by applying the channeling of the magnetic field using magnetic material.

5.3.2 Magnetic Field Guiding Using Ferrite in OLEV

Figure 5.6a shows the vertical magnetic flux type of the power lines and pickup coils [7]. There are two power lines with opposite current directions underneath the road surface, forming a current loop. Due to the current in the power lines, magnetic flux is induced around each power line. The magnetic fluxes from the two power lines are added at the location between the power lines, and the magnetic fluxes are canceled outside of the power lines. The pickup coil catches the vertical magnetic flux through copper coil loops around the ferrite core. The ferrite core is the most

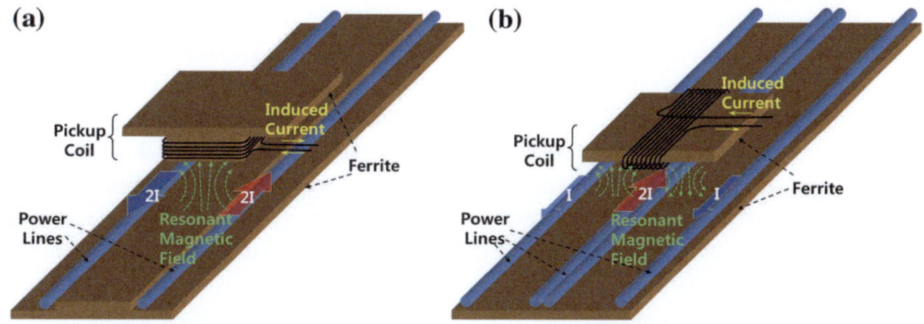

Fig. 5.6 Power lines and pickup coils for OLEV **a** Vertical magnetic flux type **b** Horizontal magnetic flux type

suitable material to apply a shaped magnetic field in resonance (SMFIR) technology, which is the key feature of wireless power transfer design in the OLEV to maximize the transfer power and to minimize the leakage electromagnetic field. This type of power line and pickup coil pair has the advantage of efficient power transfer because the direction of the magnetic flux from the power lines is the same as the direction of the flux to the pickup coil. As OLEV is designed to be powered both when the vehicle is stopped and when the vehicle is moving, the generated field is two-dimensional perpendicular to the vehicle motion, and there is no variation of magnetic field shape and coupling coefficient along the direction of vehicle motion. Therefore, the power is transferred wirelessly to a moving OLEV without any variation in the power and efficiency.

Figure 5.6b shows the horizontal flux type of power lines and pickup coil. In this type, there are four power lines forming two current loops. The copper pickup coils around the horizontal ferrite core receive the horizontal magnetic flux to generate power for vehicles. This type has a disadvantage in that the transfer power is less than the vertical flux type because the induced magnetic field directions of power lines and pickup coils are perpendicular. The difficulty of increasing the pickup coil loop size is another disadvantage because an increase of the pickup coil loop size entails increasing the total thickness of the pickup coil. However, this type of power line and pickup coil pair has a current distribution that generates a weaker magnetic field at the side of the vehicle.

5.4 Selection of the Operating Frequency of the Magnetic Field

5.4.1 Criteria of Optimal Operating Frequency

There are many factors that should be considered to determine the operating frequency of the wireless power transfer system: power transfer capacity, efficiency, coil size, electromagnetic field noise, and so on. The overall efficiency is one of the main factors, and it is affected by the losses at AC power generation, coils transmitting and receiving the magnetic field, and load including the rectifying circuit. If the frequency increases, the loss of the inverter and rectifier generally increases and the efficiency decreases. However, the magnetic coupling increases if the frequency increases because the induced voltage at the receiving coil is proportional to the operating frequency of the transmitting coil. Therefore, higher frequency is preferred from the viewpoint of the coil design, in a range of kHz or MHz. If the frequency increases to the GHz range, the resistive loss increases due to the small skin depth, which is in a range of a few μm, the radiation loss increases, and the loss at the magnetic material also increases. From the viewpoint of electromagnetic noise, the electromagnetic field (EMF) and electromagnetic interference (EMI) become serious if the operating frequency increases, as shown in Fig. 5.7.

5.4.2 *Resonance Frequency*

In the conventional WPT system with the simple passive elements in Fig. 5.5, the L_S and L_L are the inductances of the transmitting and receiving coils, respectively. When a transmitting coil is excited by the source, the transmitting and receiving coils are coupled magnetically (M is the mutual inductance and k is the magnetic coupling coefficient).

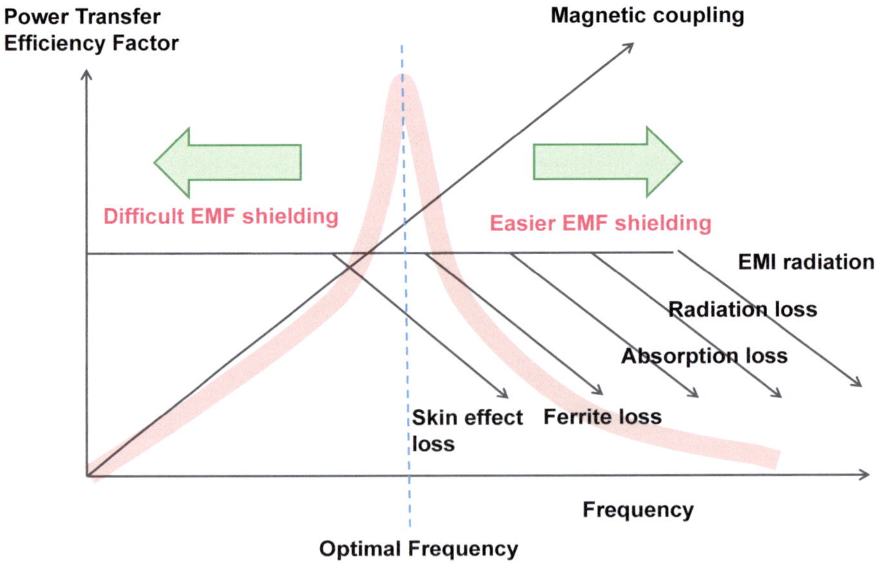

Fig. 5.7 Overview of determining the optimal frequency of wireless power transfer system

The equivalent circuit can be analyzed as expressed below by using Kirchhoff's voltage law.

$$\left(R_S + \frac{1}{j\omega C_S} + j\omega L_s\right) I_1 - j\omega M I_2 - V_S = 0 \tag{5.6}$$

$$\left(R_L + \frac{1}{j\omega C_L} + j\omega L_L\right) I_2 - j\omega M I_1 = 0 \tag{5.7}$$

For the maximum amount of power transfer, capacitors C_S and C_L are connected in series with each coil in the model. R_S is the internal resistance of the transmitting coil, and R_L is the load resistance. From the model, the loop impedance of transmitter Z_S and the loop impedance of receiver Z_R can be calculated as

$$Z_S = R_S + \frac{1}{j\omega C_L} + j\omega L_S \tag{5.8}$$

$$Z_L = R_L + \frac{1}{j\omega C_L} + j\omega L_L \tag{5.9}$$

The power transfer efficiency η and the maximum transferred power P_L can be calculated as

$$P_L = \mathrm{Re}\{V_L \cdot I_L^*\} = \frac{\omega^2 M^2 \cdot R_L \cdot I_S}{Z_L \cdot Z_L^*} \tag{5.10}$$

$$\eta = \frac{\mathrm{Re}\{V_L \cdot I_L^*\}}{\mathrm{Re}\{V_S \cdot I_S^*\}} = \frac{\omega^2 M^2 \cdot R_L}{\mathrm{Re}\{Z_S \cdot Z_L \cdot Z_L^* + \omega^2 M^2 \cdot Z_L^*)\}} \tag{5.11}$$

The power transfer efficiency is a function of the passive circuit parameters and the operating frequency, but with the fixed passive circuit parameters, the power transfer efficiency is generally maximized at the resonance frequency.

The conditions for the resonance frequency in the two loops are given below.

$$\frac{1}{j\omega C_S} + j\omega L_S = 0 \tag{5.12}$$

$$\frac{1}{j\omega C_L} + j\omega L_L = 0 \tag{5.13}$$

$$\omega_0 = \frac{1}{\sqrt{L_S C_S}} = \frac{1}{\sqrt{L_L C_L}} \tag{5.14}$$

$$f_0 = \frac{1}{2\pi\sqrt{L_S C_S}} = \frac{1}{2\pi\sqrt{L_L C_L}} \tag{5.15}$$

5.4.3 Q-Factor and Bandwidth

In the design of the coils of a wireless power transfer system, the Q-factor significantly affects the power transfer efficiency. The Q-factor is basically defined as the ratio of the bandwidth to the frequency.

$$Q = \frac{\Delta\omega}{\omega_0} \tag{5.16}$$

If the Q-factor increases, the impedance changes rapidly near the resonance frequency. As the wireless power transfer system utilizes the resonance effect, determining the proper Q-factor is very significant.

In a LC resonance system, the Q-factor can be expressed as a function of L, C, and R, as follows:

$$Q = \frac{\omega_0 L}{R} = \frac{1}{R}\sqrt{\frac{L}{C}} \tag{5.16}$$

Coils with a higher Q-factor can lead to higher power transfer efficiency if the frequency of the transmitted power and the resonance frequency of the system are identical. However, if there are some changes of the resonance frequency due to process variations, inaccuracy of lumped component values, or a dimension change caused by a temperature increase, efficiency can change. Therefore, for more stable operation of a wireless power transfer system considering manufacturing issues, the proper Q-factor should be selected first depending on the application requirements.

5.5 Conductor Design for Large Current Flow to Overcome the Skin Effect

5.5.1 Skin Effect

In Fig. 6.4, the transferred power P_L can be maximized when the reactive components of R_S is minimized because of the ohmic loss of the conductor. As some resistance inherently exists in the conductor, the use of conductors with lower loss is necessary to minimize the loss and consequently to maximize the efficiency. Selection of a cable with higher conductivity and thickness will help to reduce the loss. However, thicker cable does not significantly reduce the loss because the skin depth becomes less than the thickness of the cable. Therefore, consideration of the skin effect also should be a significant factor in minimizing the loss.

At low frequencies, the distribution of the current density inside a conductor is uniform. Looking at a cross section of a conducting wire, there is as much current flowing in the center as near the edge. At higher frequencies, the current density

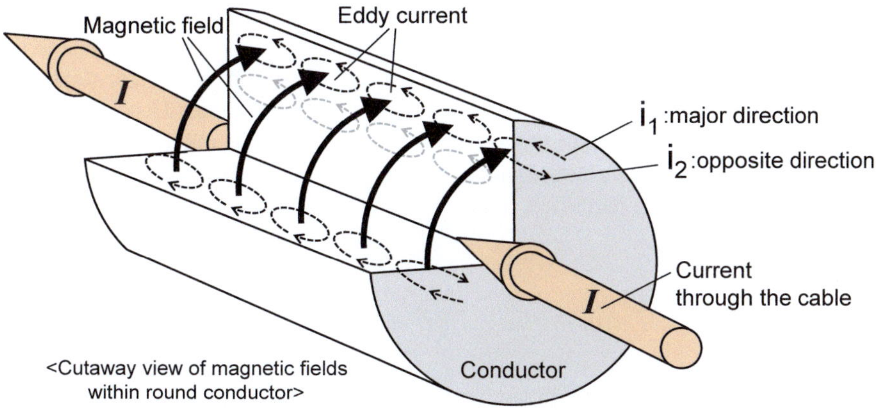

Fig. 5.8 Cutaway view of magnetic field within a round conductor

Fig. 5.9 Current density and magnetic field intensity along the line from the center of the conductor to outside

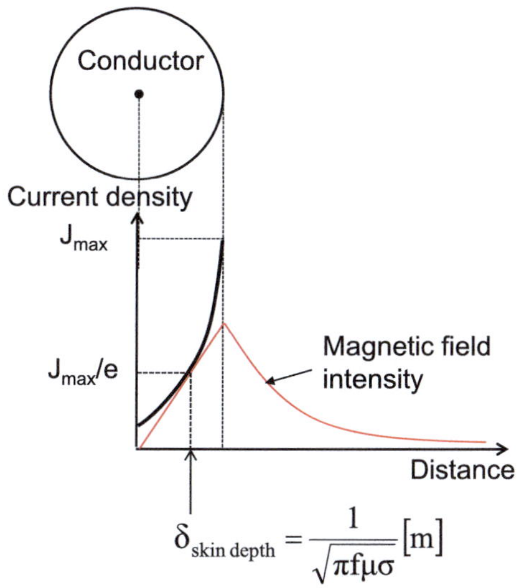

becomes larger near the surface of the wire and almost no current flows in the center [8].

Figure 5.8 shows the mechanism of the skin effect. Assuming the current is flowing through a conductor, the magnetic field is generated and the magnetic field generates eddy current. This eddy current then increases the effective current density around the periphery of the conductor while decreasing it in the middle. The skin depth is a measure of the depth at which the current density falls to 1/e of its value near the surface, as shown in Fig. 5.9.

As the skin depth is inversely proportional to the square root of the frequency, the skin depth decreases if the frequency increases. If the conductor is much thinner than the skin depth, the current flows through the entire cross section of the conductor, and then we can assume that the current density is almost uniform inside the conductor. In this case, the diameter of the conductor is the major factor to determine the resistance of the wireless power transfer system. However, if the operating frequency is so high that the conductor thickness is much larger than the skin depth, the current is concentrated to the skin depth from the surface, and therefore the resistance increases.

5.5.2 Cable for Transmitting Coil and Receiving Coil

As the resistance cable is significantly affected by the skin depth, litz wires with a bundle of thin wires are used if the total thickness of the cable is larger than the skin depth. The DC and AC resistances of cable with a solid wire and litz wires are calculated as follows:

$$R_{\text{dc,solid}} = \frac{1}{\sigma(\pi R^2)}\,[\Omega/\text{m}] \tag{5.17}$$

$$R_{\text{ac,solid}} = \frac{1}{\sigma(2\pi R \cdot \delta_{\text{skindepth},R})}\,[\Omega/\text{m}] \tag{5.18}$$

$$R_{\text{solid}} = \sqrt{\left((R_{\text{dc,solid}})^2 + (R_{\text{ac,solid}})^2\right)}\,[\Omega/\text{m}] \tag{5.19}$$

$$R_{\text{dc,litz}} = \frac{1}{N \times \sigma(\pi r^2)}\,[\Omega/\text{m}] \tag{5.20}$$

$$R_{\text{ac,litz}} = \frac{1}{N \times \sigma(2\pi r \cdot \delta_{\text{skindepth},r})}\,[\Omega/\text{m}] \tag{5.21}$$

$$R_{\text{litz}} = \sqrt{\left((R_{\text{dc,litz}})^2 + (R_{\text{ac,litz}})^2\right)}\,[\Omega/\text{m}] \tag{5.22}$$

Although the DC resistance is inversely proportional to the cross-sectional area of the metal, as shown in (5.17), the actual resistance is affected by the skin effect, as shown in (5.19). In the case of litz wire cable, the resistance is relatively simple in DC, as shown in (5.20); however, the actual resistance in AC becomes much more complex than (5.22) because of the proximity effect between multiple copper wires. To find the exact value of the resistance, a very precise 3-dimensional electromagnetic field simulation considering the structures of the coils or the measurements is required.

When solid cable and litz wire cables with a thickness of 1.25 mm are compared, solid cable has lower resistance at low frequency. However, the resistance of the solid cable increases rapidly if frequency higher than 500 Hz is used. Litz wire cables are necessary for the transmitting coil and the receiving coil in a wireless power transfer system, to minimize the loss and increase the efficiency because the wireless power transfer system operates in a kHz or MHz range where the skin effect significantly affects the total efficiency of the wireless power transfer system.

5.6 Conclusion

The fundamentals on the generation of magnetic field for high efficiency wireless power transfer system have been introduced in this chapter. As the magnitude and efficiency of transfer power are the most significant criteria for electrical performance of the system especially in electric vehicle application, the design parameters should be carefully determined to satisfy these requirements. The frequency of 20 kHz is a good choice for generation of the transfer power with high efficiency in the electric vehicle application which uses the power in kilowatt range. The loop antenna structure with low wave impedance is chosen as a transmitter coil for generation of magnetic field. The topology of current-source-type transmitter and series resonance is used for maximum power transfer with high efficiency for the low impedance loads in electric vehicle. The ferrimagnetic material also enhances the efficiency of the transfer power by shaping the magnetic field, and the skin effect is considered to minimize the ohmic loss in the cable by using litz wire. To satisfy the harsh requirement of wireless power transfer system in electric vehicles, it is necessary to combine each design technique and optimize as a system.

References

1. Ott HW (2008) Noise reduction techniques in electronic systems, 2nd edn. Wiley-Interscience
2. Kong S, Kim M, Koo K, Ahn S, Bae B, Kim J (2011) Analytical expressions for maximum transferred power in wireless power transfer systems. Paper presented at IEEE electromagnetic compatibility symposium, Long Beach, California
3. Kurs A, Karalis A, Moffatt R, Joannopoulos J, Fisher P, Soljačić M (2007) Wireless power transfer via strongly coupled magnetic resonances. Science 317(5834):83–86
4. Kim J, Son H, Kim D, Kim K, Park Y (2011) Efficiency of magnetic resonance WPT with two off-axis self-resonators. Paper presented at IMWS-IWPT, Kyoto, Japan
5. Hirayama H, Okuyama Y, Kikuma N, Sakakibara K (2011) Equivalent circuit of induction fed magnetic resonant WPT system. Paper presented at IMWS-IWPT, Kyoto, Japan
6. Kim J, Kim H, Kim I, Kim Y, Ahn S, Kim J, Kim J (2011) Comparison of series and parallel resonance circuit topologies of receiving coil for wireless power transfer. Paper presented at International forum on electric vehicle, Daejeon, Korea

7. Ahn S, Pak J, Song T, Lee H, Byun J, Kang D, Choi C-S, Chun Y, Rim C, Yim J, Cho D, Kim J (2010) Low frequency electromagnetic field reduction techniques for the on-line electric vehicle (OLEV). Paper presented at IEEE electromagnetic compatibility symposium, Port Lauderdale, Florida
8. Johnson H (1993) High-speed signal propagation—advanced black magic. Prentice hall

Chapter 6
Overview of Wireless Power Transfer System for Bus

Dong Ho Cho

Abstract The overall system architecture of wireless power transfer system for electric bus is presented. The inverter system converting 60 Hz power source into 20 kHz power source supplies 20 kHz current to the power cable module embedded under road. Many litz wire cables are used to eliminate skin effects in generating magnetic flux. The pickup module is attached at the bottom of bus to collect the magnetic flux. At magnetic resonance, induced electric power is delivered through regulator to motor or battery system. Using ferrite core structures such as EE type or UU type, magnetic flux is channeled to increase the power transfer efficiency as well as satisfy the EMF requirement. The power transfer process between the power cable module and the pickup module consists of four phases, i.e., selection, ping, identification, and configuration.

6.1 Introduction

In Chap. 4, we presented how the functional requirements (FRs) of the OLEV system are satisfied by generating a magnetic field from underground electric power sources through the use of magnetic poles of a ferrite core. This field emanating from the underground is captured above the ground by making a magnetic loop that goes through an inverted ferrite core attached to the vehicle. The distance between the underground magnetic poles determined the height of the magnetic field, which is a critical variable in wireless transfer of electric power, following axiomatic design theory. In Chap. 5, the principles of the magnetic field generation and transfer of the electromagnetic power using the resonance effect are presented. In this chapter, the system design for wireless power transfer is presented with detailed discussion of the physical components and their arrangements in the OLEV system for wireless power transfer from the underground cable to the vehicle.

D.H. Cho (✉)
School of Electrical Engineering KAIST, Korea Advanced Institute of Science and Technology (KAIST), 291 Daehak-Ro, Yuseong-Gu, Daejeon 34141, South Korea
e-mail: dhcho@kaista.ac.kr

N.P. Suh and D.H. Cho (eds.), *The On-line Electric Vehicle*,
DOI 10.1007/978-3-319-51183-2_6

This chapter presents the overall specifications of the OLEV wireless power transfer system consisting of the power supply system and the pickup system. The power supply system consists of an inverter and a power cable module, and the pickup system is made up of a pickup module as well as a rectifier module. The power cable module and the pickup module form the wireless power transfer system structure. With regard to the power cable module, the overall structure including the mechanical parameters is shown and the road-embedded power cable module design based on a single primary coil is described. Also, the design requirements for the coil of the pickup module, core, circuit, and mechanical components are presented. The requirements of the pickup system and regulator for the electric vehicle are also given, including the target development requirements and basic specifications of the vehicle system. Finally, power transfer phases from the power supply system to the pickup system are presented from the perspective of system control. They consist of four phases: selection, ping, identification and configuration, and power transfer.

6.2 System Overview

The OLEV wireless power transfer system consists of two major units: underground *power supply system* and the *power pickup system* mounted on the vehicle. The *power supply system* consists of the power cable module that consists of a primary coil and primary cores. The *power pickup system* consists of the pickup module comprising a secondary coil and secondary cores. Oscillating magnetic field is created in the underground power supply system by sending oscillating electric field around a ferrite core of the power supply system. This oscillating magnetic field of the underground power supply system is transmitted to the ferrite core of the power pickup unit mounted on the vehicle. The magnetic power transfer between the *power supply system* and the *power pickup system* is initiated from one of the poles of the underground electromagnet (i.e., the end of the underground ferrite core) and go through the loop created by the ferrite core of the *power pickup system* attached to the vehicle, which is then returned to the other pole of the underground electromagnet of the *power supply system*. The magnetic field in the power pickup system is converted to electric current by a coil of electric conductor wrapped around the ferrite core of the power pickup unit. Close spacing of the ferrite cores of the power supply system and the power pickup system ensures power transfer at an acceptable efficiency with a sufficient power transfer capacity. In addition, a magnetic shield minimizes the leakage of the magnetic field in order to maintain the leakage below the international and national standards [1–4].

Typically, a power cable module has a flat surface on the top, and a pickup module has a flat surface under the bottom of the vehicle. This minimizes the vertical spacing between primary cores and secondary cores. In addition, there is a positioning guide for horizontal alignment of the power cable module and the pickup module. The driver of the vehicle must actively align the pickup module under the vehicle to the power cable module embedded in the road, by steering the vehicle in close proximity above the surface on the power cable module, which is indicated on the surface of the road.

Figure 6.1 illustrates the basic system configuration. As shown, a power cable module comprises two main functional units: a power cable unit and a communications unit. The diagram schematically shows the primary coil as the magnetic field generating element of the power cable unit (note: physically it is under the ground and not visible). The communications unit exchanges data between the power supply system and the pickup system. Also shown in the diagram, the power supply system may contain multiple power cable modules of segmentations in order to serve running vehicles continuously. Finally, the inverter shown in the diagram comprises all other functionality of the power supply system, such as input power provisioning, control of multiple power cable modules, and user interfacing.

A pickup module comprises a pickup unit and a communications unit. Figure 6.1 explicitly shows the secondary coil as the magnetic field capturing element of the pickup unit, similar to the power cable unit of the power cable module. A vehicle contains a single or multiple pickup modules to ensure the required power. The communications unit sends and receives information data to and from the power supply system. These subsystems represent the main functionality of the pickup module. An important example subsystem is a battery that requires charging.

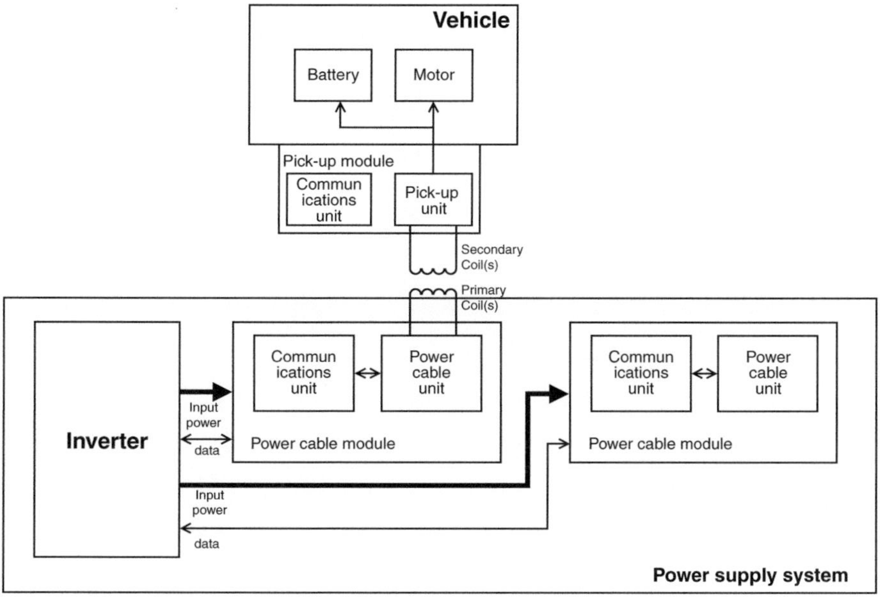

Fig. 6.1 Basic system overview

6.3 Power Cable Module

6.3.1 Introduction

The main function of the power cable module is the creation of magnetic fields that generate power in the vehicle. In a wireless power transfer system, a power cable makes source magnetic fields that travel to a pickup system. The power cable module as shown in Fig. 6.2 is located in the ground, and it consists of primary coils, ferrite cores, signal cables, steel reinforcements, EMF shielding, and common coils [5–7].

Primary coils create magnetic fields that reach the pickup unit. Each coil has a shield that is made of fiber-reinforced plastic (FRP). FRP is strong enough to withstand various conditions because of the strong fiber element. Primary coils are in the form of litz wire because AC current runs through primary coils. With AC current, the skin depth is usually very thin. To avoid this situation, primary coils are made with litz wire.

The main role of ferrite cores is directing the magnetic field to the ferrite cores of a pickup unit. Ferrite cores are composed of two vertical cores, 1 center core, and 1 flat-plated core. The center core has twofold greater thickness than that of the other cores. The saturation magnetic flux density of the ferrite cores is 490 mT at 25 °C and 390 mT at 100 °C. The Curie temperature of the ferrite cores is about 220 °C, and their magnetic permeability is 2400.

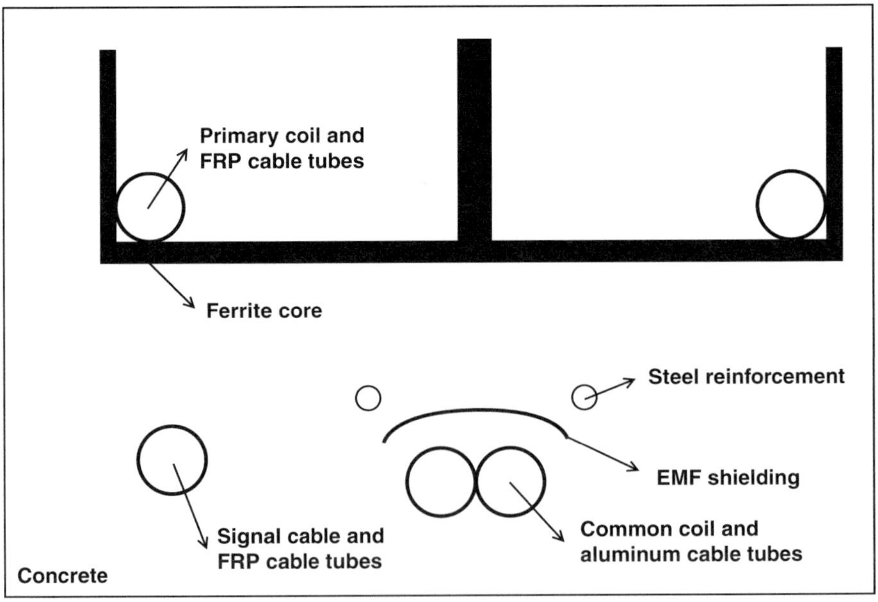

Fig. 6.2 Front view of power supply cable module

Signal cables are used for communication with the magnetic receiver, which operates when a vehicle enters the road-embedded power cable module.

Steel reinforcements make the power supply module much stronger in terms of structural mechanics. The steel reinforcement part can endure various structural situations such as heavy loads, shocks, and vibrations. The role of steel reinforcement is like the skeleton of the human body. Specifically, it retains the form of the power cable module.

EMF shielding shields the EMF from common coils. It is generally made of aluminum. Common coils are installed under ferrite cores in this power cable module and are shielded by using aluminum cable tubes.

6.3.2 Road-Embedded Power Cable Module Designs that Are Based an a Single Primary Coil

(1) Mechanical details

(a) Primary coil structure

The primary coil is one-turn winding litz wire having 95 sq. (4/0 AWG). The coil has a shield that is made of FRP.

(b) Ferrite core

The ferrite core has a E-shape. The center core has twofold greater thickness than that of the other cores. The ferrite core materials can be chosen among the following materials: PL-7 and PL-13.

(c) EMF shielding

The EMF shielding material is copper. It has an adjacent copper mesh with 30-mm spacing.

(d) Steel-reinforced concrete structure

The power supply system requires reinforced mechanical structures to keep its original shape for providing consistent AC current from the power cables. Steel reinforcement of rebar can create bigger tensile stress in the power supply system. When installing the steel reinforcement, the position of the steel reinforcement is very important because the shape of the magnetic field should not be affected by it.

(2) Electrical details

The power supply system shown in Fig. 6.3 uses a full-bridge resonance inverter to drive the primary coil. The inverter module consists of the rectifier, full-bridge inverter, blocking capacitor, output insulation transformer, and series resonance capacitor. The assembly of the inverter internal lines, primary coil, shielding, and ferrite core has a self-inductance $L_p = 55 \pm 10\%\ \mu\text{H}$, which varies with the length

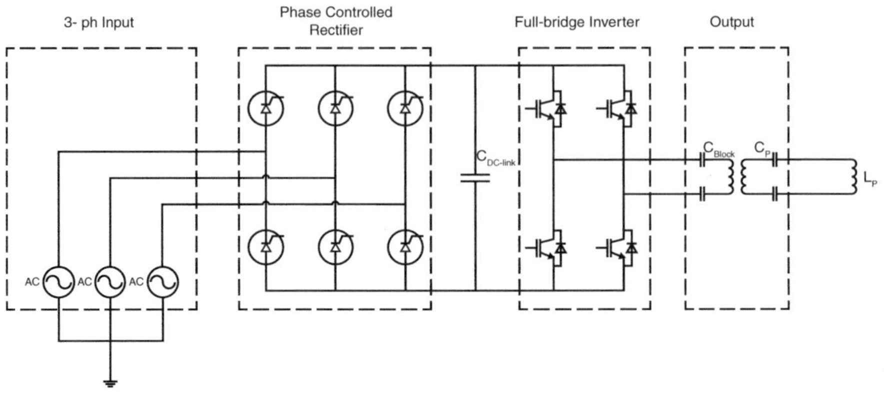

Fig. 6.3 Basic circuit of an inverter

of the primary coil. The inverter used 20.0 kHz switching frequency and 50% turn-on-duty. In order to maintain zero voltage switch (ZVS) operation of the inverter, the resonance frequency calculated with the series resonance inductance Lp and series resonance capacitance C_p should be slightly below the inverter switching frequency, on the basis of which $1.21 \pm 5\%$ μF was used as the capacitance for C_p. To maintain fixed output current, the power supply system uses a phase-controlled rectifier to control the DC-link voltage.

Control of the phase-controlled rectifier (PCR) is achieved using a PI controller. The DC-link voltage varies from 50 to 550 V, no load to full load, respectively.

The output module has blocking capacitors and an insulation transformer. It prevents current pass by ground coupling between primary coil(s).

6.4 Pickup Module

6.4.1 Introduction

The power pickup system consists of a E-shaped ferrite core and a secondary coil. As AC power is induced in the secondary coil, it is first rectified to DC and the regulator controls the flow of this induced power within the vehicle.

- The secondary coil wound around ferrite enhances the capability to receive more magnetic flux, and hence, more power is induced.
- The rectification circuit provides rectification of induced AC power to DC power.
- The regulator functions as a control unit of power flow and a communication module.

- Induced power in the pickup may be used to run the vehicle (load) and save any leftover energy in the battery. Otherwise, it can be simply wasted in the resistor box (active load) in case the vehicle does not need or have enough capacity to save energy.

6.4.2 Pickup Module Design Requirements

The common design requirements for the pickup can be classified according to each of its components: the coil, core, and mechanical components. Additionally, requirements in designing the circuit have to be considered.

(1) Coil design requirements

The primary design parameter in designing the coil for the pickup is the coil's impedance. Minimizing the coil's impedance leads to minimized power loss in the coil and thus results in high output with high power transfer efficiency. In order to minimize this impedance, the skin effect has to be taken into consideration and use of litz wire allows the coil to have smaller impedance.

Separate coating of each wire endows the coil with increased effective conductive area and thus results in minimized loss.

Table 6.1 lists the relevant parameters in designing the coil for the pickup.

Larger values of $\mathbf{d_w}$, N_w, and d_c lead to larger allowable current in the coil with less power loss. However, this might lead to increased pickup weight and cost.

(2) Core design requirements

Wireless power transfer systems use magnetic cores to maximize magnetic flux density. The basic core structure can be categorized into two types: EE type and UU type. The EE type is composed of a pair of E-shaped cores, and the UU type is composed of a pair of U-shaped cores. In the EE-type structure, power line cables are wound around the center magnetic core pole, and two target magnetic loops and two main leakage magnetic loops are formed. In the UU-type structure, power cables are wound around a pole or a horizontal bar of the primary magnetic core, and a target magnetic loop and two main leakage magnetic loops are formed.

The saturation of the core should also be considered. The saturation magnetic flux density of ferrite is about 0.4 T. As the magnetic flux density in the core becomes higher and closer to saturation, the rate of increment shrinks and the core loss rises. Heat generated due to the core loss increases the temperature of the

Table 6.1 Relevant parameters in designing coil for pickup

Parameter	Wire diameter	Number of wires in coil	Coil diameter (conductive area)	Allowable current
Symbol	$\mathbf{d_w}$	N_w	$\mathbf{d_c}$	$\mathbf{I_A}$

Table 6.2 Ferrite properties

Property	Symbol	Unit	Operating temperature
Initial permeability	$\mu_{\mathbf{iac}}$		
Core loss (100 kHz, 200 mT)	$P_{\mathbf{cv}}$	$\mathrm{kW}/\mathbf{m}^3$	25–120 °C
Saturation flux density (1194 A/m)	B_s	mT	25–100 °C
Remanence	B_r	mT	25 °C
Coercivity	H_c	A/m	25 °C
Curie temperature	T_c	°C	
Density	ρ	$\mathrm{Kg}/\mathbf{m}^3$	
Resistivity	d	Ω·m	25 °C

core, which may degrade the characteristics of the core such as core loss or permeability. Accordingly, if more magnetic flux is needed, the thickness of the core should be increased to reduce the magnetic flux loss, and this factor has to be considered with the pickup width constraint.

In designing the pickup module, we suggest the use of ferrite. Due to its relatively low price and very high saturation magnetic flux density, ferrite is the most appropriate material to be used for the core in a commercialized wireless power transfer system. Table 6.2 lists some important properties of this ferrite.

Since these key values may vary according to ambient temperature, pickups should be operated within the range where the properties listed in Table 6.2 do not vary severely.

(3) Mechanical components design requirements
As we have seen from the preceding section, properties of ferrite vary with temperature. This variance becomes clear, especially in terms of the saturation magnetic flux density, which is one of the most important properties for a core material. The saturation magnetic flux density value decreases by approximately 20% at 100 °C compared to the value at 25 °C. This phenomenon eventually leads to power loss. Therefore, designing mechanical components optimized for releasing heat properly is another key factor in designing high power pickup with high efficiency. In designing a heat radiation structure for the pickup, an aluminum heat sink can be used to implement the air cooling method. The air cooling method has advantages in simplicity, lightness, and safety. Additionally, vibration isolation and a waterproof structure are key factors that need to be considered in designing the mechanical components.

(4) Circuit design requirements
The pickup module is designed in accordance with common inductor design standards.

In this context, the mutual inductance (M) is a primary parameter in designing the pickup module. Figure 6.4 shows a number of wireless power transfer resonance circuits. Each circuit is presented with its mathematical model.

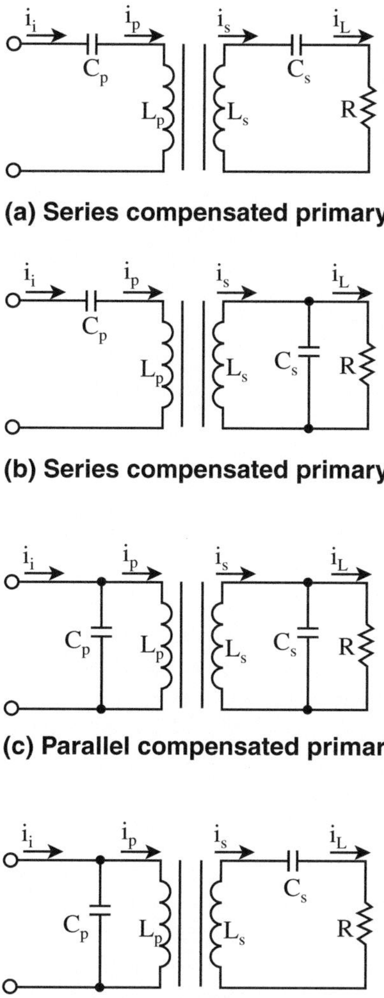

$$C_p = \frac{I}{W_o^2 L_p}$$

(a) Series compensated primary and secondary

$$C_p = \frac{I}{W_o^2 (L_p - M^2/L_s)}$$

(b) Series compensated primary and parallel compensated secondary

$$C_p = \frac{L_p - M^2/L_s}{\left(\frac{M^2R}{L_s^2}\right)^2 + W_o^2 (L_p - M^2/L_s)^2}$$

(c) Parallel compensated primary and secondary

$$C_p = \frac{L_p}{\left(\frac{W_o^2 M^2}{R}\right)^2 + W_o^2 L_p^2}$$

(d) Parallel compensated primary and series compensated secondary

Fig. 6.4 Wireless power transfer resonance circuit

Since the proposed pickup module is a device incorporated into vehicles, there are several additional parameters that affect the performance of the pickup module compared to common inverters. These include lateral displacement, number of vehicles, and so on. This indicates that the proposed pickup module needs to be designed such that its mutual inductance is not easily affected by external factors. In this sense, it is recommended to avoid the compensated primary and secondary circuits in parallel. The compensated primary in parallel and the compensated secondary circuit in series circuits in designing the pickup module for mutual inductances are easily affected by external factors. Change in mutual inductance leads to a change in the resonant capacitance, and hence, the circuit is no longer a resonance circuit. Moreover, these two types of circuits are easily affected by the load resistance as well. This brings problems related to both compensation and the PWM duty ratio value in controlling the power flow. On the other hand, the compensated circuits in series have no reflected reactance. Therefore, the use of compensated primary and secondary circuits in series is favorable in designing the pickup.

6.4.3 Pickup Module Design Guidelines

This section presents informative guidelines in designing the pickup. Each subsection describes guidelines in designing the coil, core, mechanical components, and circuits.

(1) Coil design
The most important parameters to consider in designing a high power pickup module for a wireless power transfer system are the magnetomotive force and power loss in the pickup module. In order to increase the magnetomotive force, both the number of coil turns and the current value flowing in each coil turn need to be increased. However, the increases in the coil turns and current value might lead to higher power loss, which is an obstacle in acquiring high power transfer efficiency. Wires with bigger diameter have larger conductive area and higher allowable current. Therefore, the use of litz wire with larger wires allows us to flow higher current in the pickup coil while minimizing the power loss as well.

(2) Core design
In designing the core for a high power pickup, the key factor to consider is how much magnetic flux, which is generated from the power cable line, the core is capable of receiving. In order to increase the amount of magnetic flux the core receives, the core has to be designed with larger area. Larger cross-sectional area leads to smaller resistance and thus smaller power loss. However, a core with larger area is more expensive and heavier. To solve this problem, core parts that are not effective in receiving magnetic flux are eliminated.

(3) Mechanical components design

(a) Heat release structure design
Due to electrical features of the components in the pickup module, the heat release structure must secure high power transfer output and efficiency. The heat release structure can be implemented with a water cooling method or an air cooling method. The use of the water cooling method may be effective. However, it causes many structural problems due to wires, the water tank, and so on. Therefore, the air cooling method is proposed in this design.

The air cooling module for the pickup is designed with the consideration of the pickup structure. The structure of the pickup module is very complicated and has many constraints, and therefore, the air cooling module for pickup module has to be designed to be as simple as possible. Additional heat transfer pads are adhered to the core and capacitors to implement the optimized heat release structure.

(b) Vibration isolation structure design
A vibration isolation structure design is implemented by reinforcing the expected vibrating parts of the pickup module with silicon rubber. Also, a grid pattern structure is used to disperse the vibration appropriately to mechanical parts.

(c) Water-proof structure design
A water-proof structure is designed according to IP 54 standards. Silicon rubber and PG 54 connectors are mainly used.

(d) Circuit design
Generally, in designing a high power pickup module, designs of multiple coils are necessary. In this case, however, multiple coils are placed in a narrow pickup core and the pickup takes on a form where multiple inductors coexist. The circuit design for the pickup module involves design of the inductor circuit, which is created by the inclusion of multiple coils. Since the design of the inductor circuit through multiple coils involves at least two circuits, the circuit design for the pickup must be focused on designing two pickup circuits as a single unit.

According to the described circuit design requirements, the basic circuit for the pickup is an LC series resonance circuit. Each circuit is designed such that the value of L is minimized. This is because a large value of L causes problems related to insulation, which would eventually lead to safety issues.

6.5 Requirements of Wireless Power Transfer System for Electric Vehicles

The maximum and rated power of the motor for an electric bus are 240 and 100 kW, respectively, which are very high compared to the case of a small electric car. Furthermore, the road to pickup gap, that is, the distance between the road and the bottom side of the pickup module, should be more than 200 mm to protect against collisions with obstacles on the road. From this, the target development requirements of our OLEV bus system were determined as presented in Table 6.3 [8–10].

Figure 6.5 shows an overall block diagram and power circuits for the wireless power transfer system. This system consists of four parts: the three-phase power inverter, road-embedded power cable module, pickup module, and regulator. In this case, the power inverter plays a role of converting three-phase AC voltage to

Table 6.3 Target development requirements and basic specifications of the system

Items	Values	Comments
Gap between road and pickup	200 mm	Distance between road and bottom side of pickup module
Rated three-phase AC voltage	480 Vac ± 10%	
Inverter switching frequency	20 kHz	
Rated power of regulator	100 kW	
Total power transfer efficiency	80%	Power efficiency from three-phase input of power inverter to regulator DC output

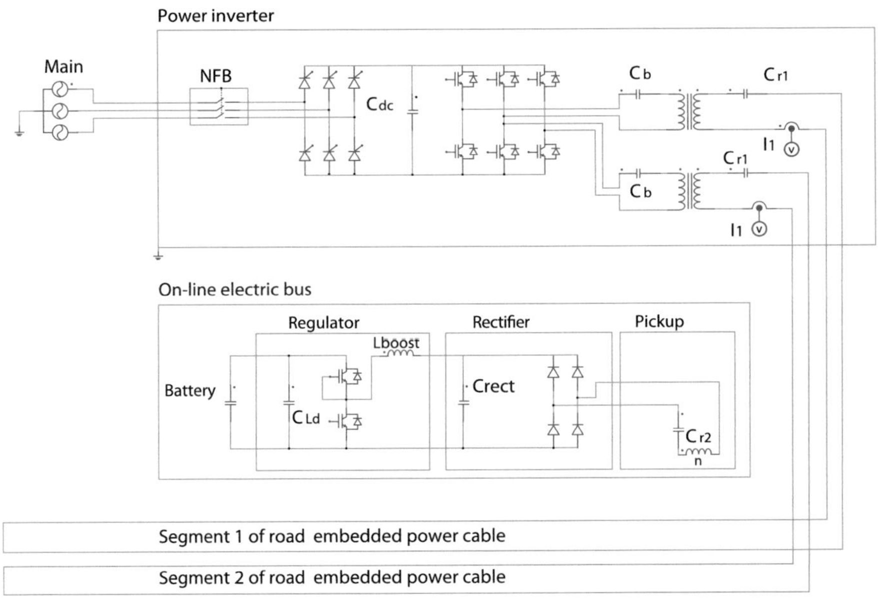

Fig. 6.5 Overall block diagram and power circuits for wireless power transfer system

constant single-phase AC current by controlling the DC voltage through the phase angle of the three-phase phase-controlled rectifier.

As illustrated in the internal circuit of the power inverter in Fig. 6.5, three-phase AC voltage is rectified and controlled as a variable DC-link voltage by the phase-controlled rectifier. This DC-link voltage is converted to an isolated single-phase voltage source by a single-phase inverter and high-frequency transformer, where a DC blocking capacitor C_b is incorporated to prevent saturation of the transformer, and the turn ratio for this transformer is 1:1. The controller in the power inverter has a dual-loop structure, where the outer loop is used for controlling the inverter output current to be constant and the inner loop is used for the DC-link voltage control loop to follow the voltage reference obtained from the outer current control loop. By this dual-loop structure of the controller, it is possible to obtain constant inverter output current.

As shown in the secondary part of the transformer, the equivalent inductance L_{r1} is measured by the power cable and the core in the embedded road power cable, whose value varies with the length of the power cable and is about 20 uH in the case of a 5-m power cable and is higher with a longer power cable. In the case of the embedded power cable shown in Fig. 6.5, the power cable is divided into two segments (or tracks), that is, #1, #2. Because the shown road-embedded power cable has two segments and the power inverter can supply current to each segment, the common cable is necessary on the bottom side of the power cable. In our case, each segment of the road consists of 1-turn of high-frequency power cable with internal litz wires and a dual-rail type (E-type) ferrite core.

The resonant capacitance C_{r1} existing at the secondary part of the transformer makes the resonant frequency between L_{r1} and C_{r1} be slightly below 20 kHz, that is, the inverter switching frequency. From this, the equivalent impedance of the embedded rail including C_{r1} operates as an inductance at a no-load condition and its current has a 90° lagging phase compared to the inverter output voltage. As the output load increases, the lagging phase decreases. Finally, its equivalent impedance becomes almost a resistor at a full-load condition and its current has the same phase, that is, 0° phase, compared to the inverter voltage. A full-load condition happens when the electric bus is situated on the road-embedded power cable and the regulator supplies 100 kW rated power from the power cable to the battery installed in the bus.

The road-embedded power cable and pickup module operate as primary and secondary sides of an E-type transformer. As mentioned before, the road-embedded power cable module in the primary side has 1-turn of power cable and the pickup module has n-turns of cables. Several mH of inductance L_{r2} is measured in the pickup module due to n-turns of wiring in the pickup module, and additional resonant capacitance C_{r2} is necessary to tune the resonant frequency between L_{r2} and C_{r2} to 20 kHz, that is, the inverter switching frequency. AC voltage at the output of the pickup module is rectified to low DC voltage by single-phase diode bridge. The applied regulator is a boost-type DC/DC converter, and it converts the low rectified DC voltage to high DC voltage to charge the battery.

In the case of the proposed OLEV system, five pickup modules are incorporated between the front and rear wheels, and the rated power of each pickup module becomes 20 kW considering the total 100 kW rated power of the regulator.

6.6 System Control

6.6.1 Introduction

In Fig. 6.6, wireless power transfer control between a power supply system and a power pickup system is shown. Power supply system can be expanded to serve multiple independent power pickup systems. The power supply system is composed of three main functional units, which are resonator, power inverter, and a signaling and control unit. The power pickup system also comprises three main functional units, which are resonator, power conversion unit, such as rectifier and regulator and a signaling and control unit [8–10].

The control and communication protocol for the WPT system is designed as the bidirectional architecture and is used to deliver power pickup system characteristics to the power supply system as well as to provide feedback to enable efficiency optimization.

In the power transfer process, the power supply system and power pickup system control the amount of power that is transferred. Figure 6.7 presents a schematic diagram of the power transfer control loop, which basically operates as follows: The power pickup system selects a desired control point such as a desired output current and/or voltage and temperature measured somewhere in the pickup device. In addition, the power pickup system determines its actual control point. Note that the power pickup system may use any approach to determine a control point. Moreover, the power pickup system may change this approach at any time during

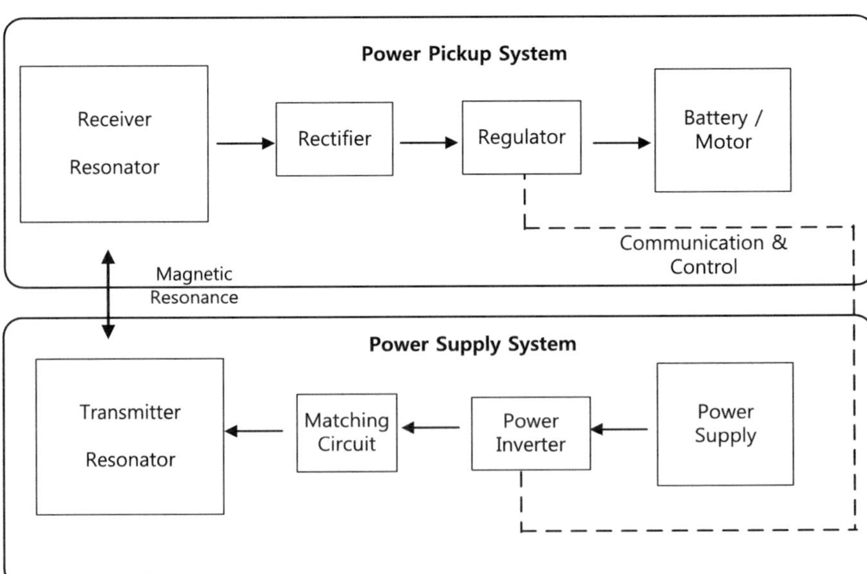

Fig. 6.6 Conceptual architecture of power transfer control

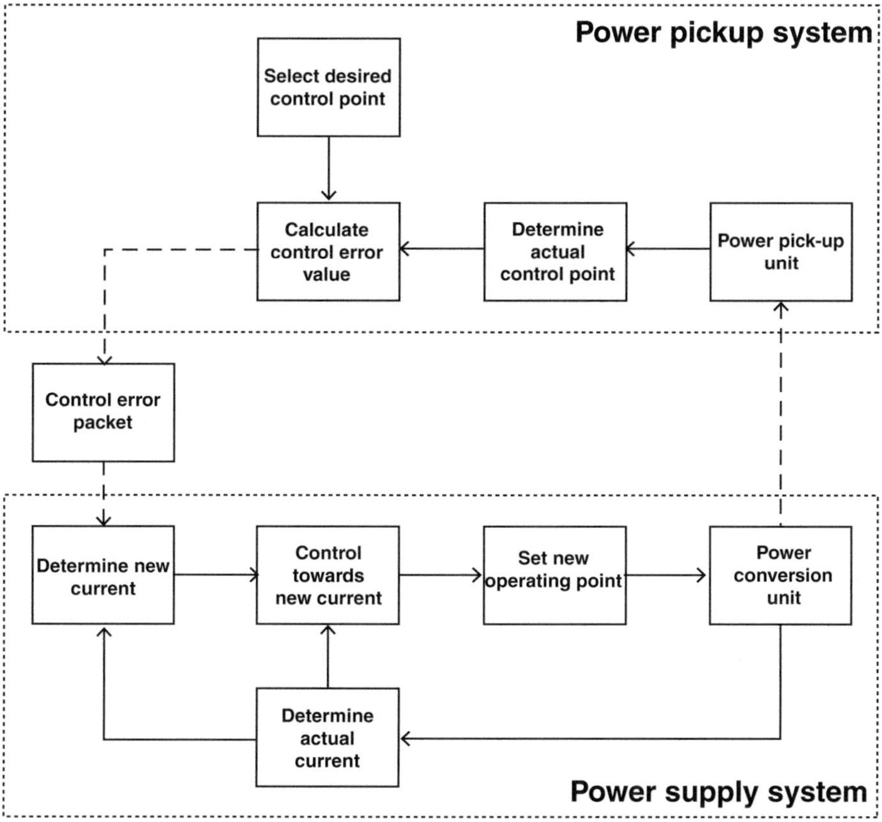

Fig. 6.7 Power transfer control loop

the power transfer phase. Using the desired control point and actual control point, the power pickup system calculates a control error value, for example, by simply taking the (relative) difference of the two output voltages or currents. The result is negative if the power pickup system requires less power in order to reach its desired control point, and positive if the power pickup system requires more power in order to reach its desired control point. Subsequently, the power pickup system transmits this control error value to the power supply system.

The power supply system uses the control error value and the actual current to determine a new current. After the system stabilizes from the communication of the control error packet, the power supply system has a short time window to control its actual primary cell current toward the new current. Subsequently, the power supply system keeps its operating point fixed in order to enable the power pickup system to communicate additional control and status information.

6.6.2 Power Supply System Perspective

(1) Sensing part
There are several sensing parts, that is, current, voltage, temperature, and so on. The current sensor is used to obtain the output current of the inverter, and the voltage sensor is used for sensing the DC-link voltage.

(2) Communication part
In this transmitter, CAN communication is used between the power inverter and the vehicle recognition sensor to provide power on the road when the vehicle enters the road.

(3) Cooling part
The cooling sensor operates when heat in the heat sink increases above the fixed temperature threshold and stops the power inverter in the transmitter.

(4) Power control part
The power control part controls the output current of the power inverter and the DC-link voltage by controlling the PWM pulse in the buck converter. That is, the outer loop controls the output current of the power inverter and the inner loop controls the DC-link voltage.

6.6.3 Power Pickup System Perspective

(1) Sensing part
The current sensing part is composed of a current sensor, I–V convert circuit, and extra peripheral circuit. The measured inductor currents processed through digital signal processing (DSP) are used to control the desired output current. The voltage sensing part measures all input port voltages and one output voltage. The measured input port voltage is used to balance each input. The measured output voltage processed through DSP is used to control the desired output voltage.

(2) Communication part
The communication part handles diagnosis, system monitoring, and parameter modification. The CAN communication part is in charge of diagnosis such as overheating, over current, and voltage by transmitting the status of the system to other parts. The SCI communication part is in charge of system monitoring and parameter modification by receiving major parameters such as current, voltage temperature, and control parameters from the sensing part. It can manage parameters by modifying control parameters.

(3) Cooling part
The cooling part measures the temperature of each semiconductor within the operating range. To reduce power consumption of the main controller, the cooling fan motor is operated when the temperature exceeds the operating range considering the semiconductor specifications. If malfunction occurs during operation, it is reported through the SPI channel and the main controller stops power control.

(4) Power control part
The power control part controls the output voltage and current using a PWM pulse generated by the PWM module. It receives the reference voltage, which is the desired voltage we wish to maintain from another system. It also calculates the reference current continuously. It then generates a PWM signal in proportion to the reference current.

6.7 Conclusion

In this chapter, in view of standard, first, a basic system configuration consisting of road-embedded power cable module and pickup module was suggested. Then, the communication unit was defined to exchange data between power cable module and pickup module. The structure of power cable module based on primary coil, capacitor, ferrite core, signal cable, and EMF shielding was proposed to generate shaped magnetic flux efficiently. Also, the structure of pickup module was suggested to maximize power capacity and efficiency in the consideration of secondary coil, capacitor, ferrite core, and mechanical structure related to heat release, vibration release and water proof. In addition, the requirements of power cable module and pickup module for large wireless power capacity were defined to support compatibility between different pickup module and common power cable module. Finally, to support the compatible operation of different pickup module for common power cable module, the system control and operation of both modules including state definition, state transition and action were suggested.

References

1. Suh NP, Cho DH, Rim CT (2010) Design of on-line electric vehicle (OLEV). Proceedings of the 20th CIRP design conference, Nantes, 19–21 Apr. 2010
2. Ahn SY, Suh NP, Cho DH (2013) Charging up the road. IEEE Spectr 50(4):48–54
3. Ahn SY, Cho DH (2013) Future wireless power transportation system. 2013 Asia-Pacific microwave conference proceedings (APMC), Seoul, Korea, 5–8 Nov 2013
4. Shin JG, Shin SY, Kim YS, Ahn SY, Lee SH, Jung JH, Jeon SJ, Cho DH (2014) Design and implementation of shaped magnetic-resonance-based wireless power transfer system for roadway-powered moving electric vehicles. IEEE Trans Industr Electron 61(3):1179–1192

5. Ahn SY, Pak JS, Song TG, Lee HJ, Byun J-G, Kang DS, Choi CS, Kim EK, Ryu JY, Kim MJ, Cha YM, Chun YB, Rim CT, Yim JH, Cho DH, Kim JH (2010) Low frequency electromagnetic field reduction techniques for the on-line electric vehicle (OLEV). IEEE international symposium on electromagnetic compatibility (EMC), Fort Lauderdale, FL, USA, 25–30 Jul. 2010
6. Ahn SY, Lee HJ, Byun JG, Kang DS, Song TG, Chun YB, Yim JH, Cho DH, Kim JH (2010) Design of electromagnetic field (EMF) for a novel on-line electric vehicle (OLEV). BEMS (Bioelectromagnetics), Seoul, South Korea, 14–18 June 2010
7. Ahn SY, Lee JY, Cho DH, Kim J (2011) Magnetic field design for low EMF and high efficiency wireless power transfer system in on-line electric vehicle. Proceedings of the 21th CIRP design conference, KAIST, Korea, Mar. 2011
8. Kim JS, Kim JH, Kong SK, Kim HS, Suh I-S, Suh NP, Cho DH, Kim JH, Ahn SY (2013) Coil design and shielding methods for a magnetic resonant wireless power transfer system. Proc IEEE 101(6):1332–1342
9. Ko YD, Jang YJ (2013) The optimal system design of the online electric vehicle utilizing wireless power transmission technology. IEEE Trans Intell Transp Syst 14(3):1255–1265
10. Jung GH, Song BY, Shin SY, Lee SH, Shin JG, Kim YS, Jeon SJ (2012) High efficient inductive power supply and pickup system for on-line electric bus. Electric vehicle conference (IEVC), Greenville, South Carolina USA, 4–8 Mar. 2012

Chapter 7
Magnetic Energy Pickup Using Resonance

Uooyeol Yoon

Abstract The concept of wireless power transfer technology is explained based on the principle of magnetic induction and magnetic resonance. Two technologies, magnetic induction and resonance, for wireless power transfer are compared in terms of expense, equivalent circuit, physics, loss, and performance. In magnetic induction, the magnetic field of one coil passes through the other coil. Two coils may be physically contained in a single unit, such as the primary and secondary sides of a transformer, or may be separated. In magnetic resonance, high-quality factor resonators enable efficient energy transfer at lower coupling rates, i.e., at greater distances and/or with more positional freedom than is otherwise possible.

7.1 Introduction

This chapter explores the concept of wireless power transfer technology with magnetic induction and magnetic resonance. The materials presented in this chapter provide the basic foundation for designing and understanding the pickup module in a wireless power transfer system. First, we review the conventional resonance in a wireless power transfer system in order to explain the principle of magnetic induction and magnetic resonance. Then, two technologies, magnetic induction and magnetic resonance, for wireless power transfer are compared in terms of expense, equivalent circuit, physics, loss, and performance. This chapter then reviews the application for magnetic induction and magnetic resonance. The strategy for LC tuning is presented, using an equivalent circuit of a wireless power transfer system. The real power, apparent power, and reactive power are described using a compensation capacitor and power factor. Finally, the tuning of the resonant frequency with transmitting and receiving part of the circuits of a wireless power transfer

U. Yoon (✉)
Korea Advanced Institute of Science and Technology (KAIST),
Wireless Power Transfer Research Center KAIST, 303 Truth Hall 193 Munji-Ro,
Yuseong-Gu Daejeon 305-732, South Korea
e-mail: uyoon@kaist.ac.kr

N.P. Suh and D.H. Cho (eds.), *The On-line Electric Vehicle*,
DOI 10.1007/978-3-319-51183-2_7

system is presented, followed by the description of automatic tuning in order to prevent fluctuation of the resonance frequency of the pickup module. The design decisions made at these lower-level FRs and DPs were checked by the system architect to be sure that they do not in advertently introduce coupling of FRs as described in Chap. 3 [1].

7.2 Concept of Resonance in Physical Science and Engineering

7.2.1 Magnetic Induction in Wireless Power Transfer System

Typically, two coils in close proximity to each other undergo mutual inductive coupling when they are configured such that time-varying current flow through one wire induces a voltage across the ends of the other wire through electromagnetic induction. The amount of inductive coupling between two conductors is measured by their mutual inductance. The concept of inductive coupling is illustrated in Fig. 7.1. The coupling between two wires can be increased by winding them into coils and placing them close together on a common axis. Therefore, the magnetic field of one coil passes through the other coil. Two coils may be physically contained in a single unit, such as the primary and secondary sides of a transformer, or may be separated. Coupling may be intentional or unintentional.

The main idea of inductive coupling wireless power transfer is presented in Fig. 7.2. There is a transmitter side (left side) and receiver side (right side). The transmitter side consists of an inverter circuit (or power amplifier) and a transmitting coil L_1. The transmitting coil is tuned to resonance with the tuning capacitor C_1. The receiver side consists of a pickup module coil (or receiving coil) L_2, a tuning capacitor C_2, a rectifier, and a regulator part.

Electric current flowing through a conductor, such as a wire, carries electrical energy. When electric current passes through a circuit, an electric field is formed in

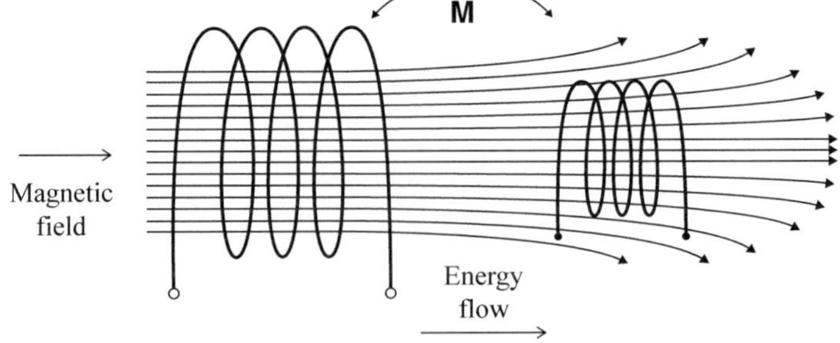

Fig. 7.1 Concept of inductive coupling in wireless power transfer system

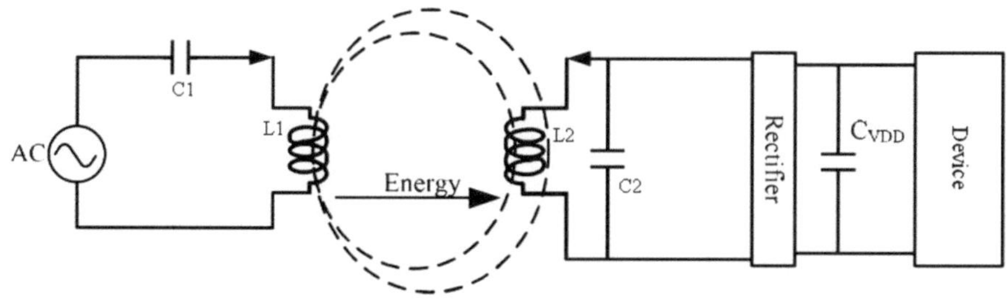

Fig. 7.2 Basic concept of inductive coupling in wireless power transfer system for lossless

the dielectric surrounding the conductor; magnetic field lines around the conductor and lines of electric force extend radially about the conductor.

In a direct current circuit, if the current is continuous, the fields are constant; there is a condition of stress in the space surrounding the conductor, which represents stored electric and magnetic energy, just as a compressed spring or a moving mass represents stored energy. In an alternating current circuit, the fields also alternate; that is, with every half wave of current and of voltage, the magnetic field and the electric field start at the conductor and run outward into space with the speed of light. Where these alternating fields impinge on another conductor, voltage and current are induced.

Any change in the electrical conditions of the circuit, whether internal or external, involves a readjustment of the stored magnetic and electric field energy of the circuit, that is, a so-called transient. A transient is the general character of a condenser discharge through an inductive circuit. The phenomenon of the condenser discharge through an inductive circuit therefore is of the greatest importance to the engineer, as it is the foremost cause of high-voltage and high-frequency problems in electric circuits. Electromagnetic induction is proportional to the intensity of the current and voltage in the conductor, which produces the fields, and to the frequency. The higher the frequency is, the more intense the inductive effect is. Energy is transferred from a conductor that produces the fields (the primary) to any conductor on which the fields impinge (the secondary). Part of the energy from the primary conductor passes inductively into the secondary conductor, and the energy decreases rapidly along the primary conductor. A high-frequency current does not pass for long distance along a conductor but rapidly transfers its energy by induction to adjacent conductors. Higher induction resulting from higher frequency accounts for the apparent difference in the propagation of high-frequency disturbances, unlike the propagation of the low-frequency power of alternating current systems. As the frequency becomes higher, the inductive effects that transfer energy from circuit to circuit across space become more preponderant. Furthermore, the energy decreases more rapidly and the current decreases along the circuit. This action of an electrical transformer is the simplest form of wireless power transfer system. The primary and secondary circuits of a transformer are not directly connected. Energy transfer takes place through a process known as mutual induction.

Principal functions are stepping the primary voltage either up or down and electrical isolation. Mobile phones, electric toothbrush battery chargers, and electrical power distribution transformers are examples of devices that exploit this principle. Induction cookers also use this method. The main drawback of this basic form of wireless transmission is short range.

In electric power transmission and distribution, the phenomena inside the conductor are of main importance, and the electric field of the conductor is usually observed only incidentally. Inversely, in the use of electric power for radio telecommunications, only the electric and magnetic fields outside of the conductor, that is, far-field electromagnetic radiation, are of importance in transmitting a message. The phenomenon in the conductor, the current in the launching structure, is not exploited.

7.2.2 Magnetic Resonance in Wireless Power Transfer System

A MIT research group explored many techniques for transmitting power over “mid-range” distances and came up with a non-radiative approach that uses resonance to enhance the efficiency of the energy transfer. High-quality factor resonators enable efficient energy transfer at lower coupling rates, i.e., at greater distances and/or with more positional freedom than is otherwise possible. The MIT research group demonstrated the high-resonant technique using a magnetic field to transfer energy over a mid-range distance of 2 m. In some instances, this technology is also referred to as “magnetic resonance,” and it is often contrasted to “induction” for its ability to efficiently transfer power over a range of distances with positional and directional offsets. Since the initial demonstration, the use of MR (magnetic resonance)-wireless power transfer system, or magnetic resonance, has enabled efficient wireless energy transfer in a wide range of applications that were not possible before [2].

In a previous work at MIT, the impedance matching was accomplished by inductive coupling into the source resonator and out of the device resonator [2]. This approach provides a way to tune the input coupling and therefore the input impedance, by adjusting the alignment between the source input coupling coil and the source resonator. Similarly, this approach provides a way to tune the output coupling and therefore the output impedance, by adjusting the alignment between the device output coupling coil and the device resonator.

Figure 7.3 shows the concept of a shaped magnetic field in the resonance technology of the KAIST OLEV system. The pickup module is composed of ferrite core blocks, pickup cables, compensation capacitors, and an FRP case. The case protects the inside physically and insulates the outside from electrocution. The road-embedded power cable module cables and pickup module cables are litz wires. The road-embedded power cable module and pickup module cables are wound around the ferrite cores, as shown in Fig. 6.3. In order to add up the induced voltages, the center winding is clockwise, and the other windings are counter-clockwise in the OLEV system [3].

Figure 7.4 shows the equivalent circuit model for the inductive coupling wireless power transfer system. In this figure, we used the subscript 1 to represent the power line part (transmitting part) and the subscript 2 to represent the pickup part (receiving part). The input voltage v_1 of the power line and induced voltage v_2 of the pickup are indicated in (7.1) and (7.2).

$$v_1 = r_1 i_1 + L_1 \frac{di_1}{dt} + M \frac{di_2}{dt} \tag{7.1}$$

$$v_2 = M \frac{di_1}{dt} + L_2 \frac{di_2}{dt} + r_2 i_2 \tag{7.2}$$

where M is the mutual inductance, L is the self-inductance, and r is the loss of the transmission line. Here, only the terms including the mutual inductance contribute

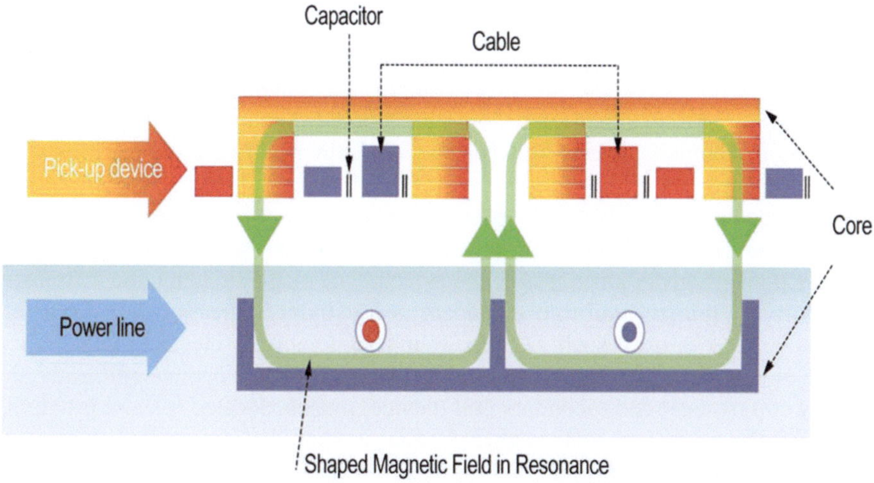

Fig. 7.3 Shaped magnetic field in resonance technology of KAIST

Fig. 7.4 Equivalent circuit model of wireless power transfer system

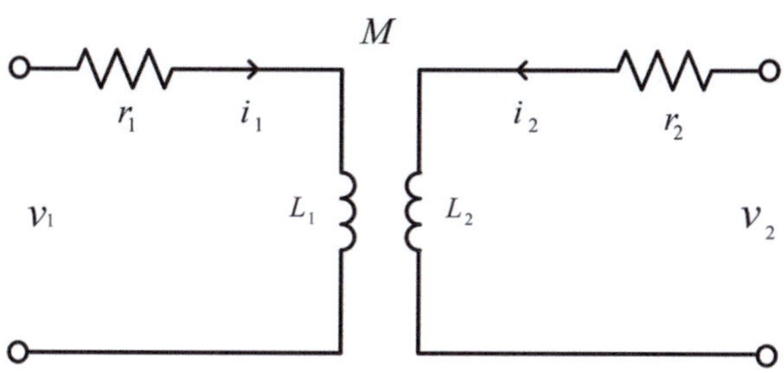

to the power transfer phenomenon, while the terms including the self-inductance have no contribution. In a typical power system, series compensation requires a resonance capacitor to reduce the reverse voltage generated by self-inductance. Without this compensation, reactive power occurs from coil inductance of the road-embedded power cable module and pickup module cable. This reactive power results in the degradation of performance in the wireless power transfer system, and hence, resonance capacitors have been used in the wireless power transfer system drawbacks of heaviness and increased expense.

7.2.3 Magnetic Induction Versus Magnetic Resonance in Wireless Power Transfer System

The idea of transmitting power through the air has been around for more than a century, with Nikola Tesla's pioneering ideas and experiments perhaps being the most well-known early attempts to do so. He had a vision of wirelessly distributing power over large distances using the earth's ionosphere. Most approaches to wireless power transfer use an electromagnetic (EM) field of some frequency as the means by which the energy is sent. Efficient transmission over large distances is possible with this approach; however, complicated pointing and tracking mechanisms are needed to maintain proper alignment between moving transmitters and/or receivers. In addition, objects between the transmitter and receiver can block the beam, interrupting the power transmission and, depending on the power level, possibly causing harm. It is also possible to transmit power using non-radiative fields. As an example, the operation of a transformer can be considered a form of wireless power transfer since it uses the principle of magnetic induction to transfer energy from a primary coil to a secondary coil without a direct electrical connection. Inductive chargers, such as those found commonly in electric toothbrushes, operate on this same principle. However, for these systems to operate efficiently, the primary coil (source) and secondary coil (device) must be located in close proximity and carefully positioned with respect to one another. From a technical point of view, this means the magnetic coupling between the source and device coils must be large for proper operation [3].

7.3 Fine-Tuning Using Capacitance

7.3.1 LC Tuning in Wireless Power Transfer System

Figure 7.5 shows the equivalent resonance circuit model of the KAIST OLEV system. In this figure, capacitors are connected to the cables to compensate the inductance of the winding cables. To make the impedance zero at the operating frequency, the capacitance is selected by the following Eq. (7.3).

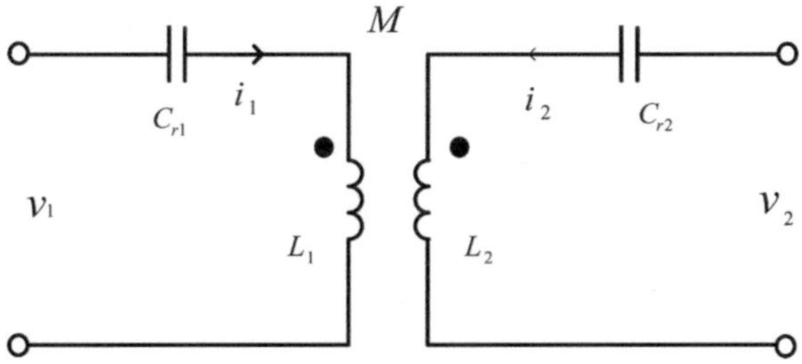

Fig. 7.5 Equivalent resonance circuit of KAIST OLEV system

$$C_{r2} = \frac{1}{4\pi^2 f^2 L_2} \tag{7.3}$$

In the case of the OLEV system pickup module, the reactive voltage due to a very large L is near 10 kV. This high voltage is difficult to insulate even if compensated. Hence, the capacitance is not connected all at one point but instead is divided and distributed among the winding cables to reduce the peak voltage. This connection method drastically reduces the peak voltage caused by the inductances [4].

The range of induction for wireless power transmission is near field distances up to the small wave length used. Near-field energy is non-radiative. Some radiative losses do, however, occur. In addition, there are usually resistive losses. With electrodynamic induction, electric current flowing through a primary coil creates a magnetic field that acts on a secondary coil producing a current within it. Coupling must be tight in order to achieve high efficiency. As the distance from the primary coil is increased, more and more of the magnetic field misses the secondary coil. Even over a relatively short range, inductive coupling is grossly inefficient, wasting much of the transmitted energy. The receiver must be directly adjacent to the transmitter or the induction unit in order to efficiently couple with it. The application of resonance increases the transmission range somewhat. When resonant coupling is used, the transmitter and receiver inductors are tuned to the same natural frequency. Performance can be further improved by modifying the drive current from a sinusoidal to a non-sinusoidal transient waveform. In this way, significant power may be transmitted between two mutually attuned LC circuits having a relatively low coefficient of coupling. Transmitting and receiving coils are usually single-layer solenoids or flat spirals with parallel capacitors, which, in combination, allow the receiving element to be tuned to the transmitter frequency. Resonance is used in both the wireless power transfer system pad (the transmitter circuit) and the receiver module (embedded in the load) to maximize energy transfer efficiency.

7.3.2 Feasibility of Automatic Tuning Using Capacitance

The resonance frequency of the LC resonance circuit in the pickup module may be affected by operating conditions such as temperature. Fluctuation of the resonance frequency in the pickup module circuit can degrade the transfer efficiency of the wireless power transfer system. Figure 7.6 shows the conventional circuit of the pickup module in the KAIST OLEV system.

For automatic tuning of the resonance circuit in the pickup module, the following two methods, as shown in Figs. 7.7 and 7.8, can be considered.

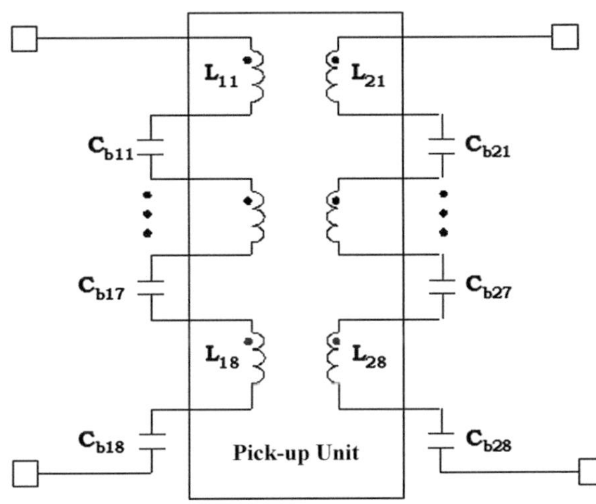

Fig. 7.6 Conventional resonance circuit of pickup module

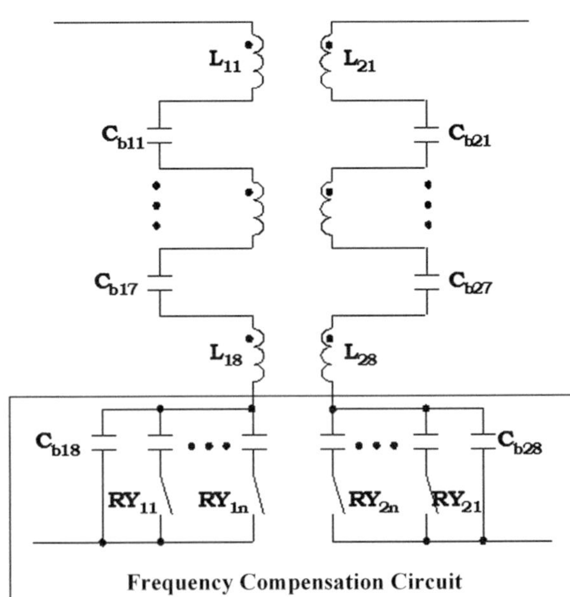

Fig. 7.7 Method 1 of resonance frequency compensation circuit

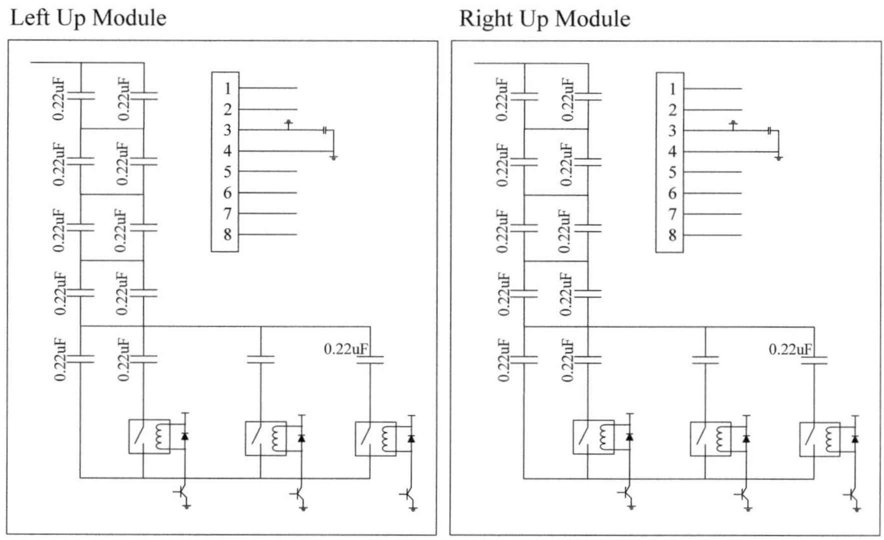

Fig. 7.8 Method 2 of resonance frequency compensation circuit

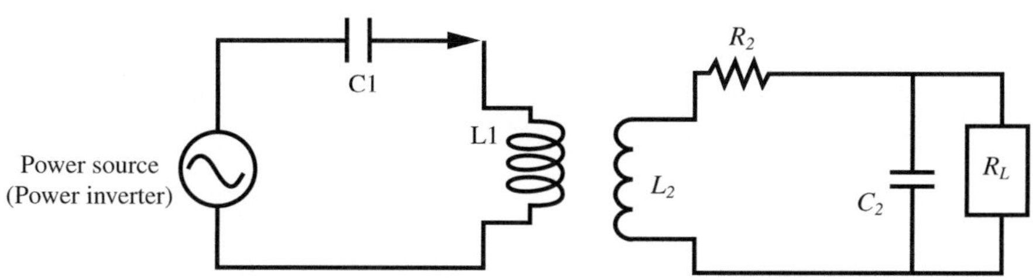

Fig. 7.9 Transmitting and receiving part circuit of wireless power transfer system

7.3.3 Resonance Frequency in Tuning

Figure 7.9 shows the simplified transmitting and receiving part circuit in the wireless power transfer system. In Sect. 7.1, we presented the resonance capacitor for the compensation of reverse-drop voltage. The tuning capacitor C_1 of the transmitting part can be calculated by

$$C_1 = \frac{1}{\omega^2 L_1} \tag{7.4}$$

where C_1 is the tuning capacitor, L_1 is the inductance of transmitting coil, and ω is the angle velocity $2\pi f$. Similarly, the receiving part tuning capacitor C_2 can be calculated by

$$C_2 = \frac{1}{\omega^2 L_2} \tag{7.5}$$

where C_2 is the tuning capacitor of the receiving part, and L_2 is the inductance of the receiving coil.

7.4 Magnetic Energy

7.4.1 Power Loss

The electric charge displacement in the conductor produces a magnetic field and resultant lines of electric force. The magnetic field is maximal in the direction concentric, or approximately so, to the conductor. That is, a ferromagnetic body tends to set itself in a direction at right angles to the conductor. The electric field is maximal in a direction radial, or approximately so, to the conductor. The electric field component tends in a direction radial to the conductor, and dielectric bodies may be attracted or repelled radially to the conductor. Where the electric circuit consists of several conductors, the electric fields of the conductors superimpose upon each other, and the resultant magnetic field lines and lines of electric force are not concentric and radial, respectively, except approximately in the immediate neighborhood of the conductor. Between parallel conductors, they are conjugates of circles. Neither the power consumption in the conductor nor that in the magnetic field or the electric field is proportional to the flow of energy through the circuit. However, the product of the intensity of the magnetic field and the intensity of the electric field is proportional to the flow of energy or the power, and the power is therefore resolved into a product of two components, current and voltage, which are, respectively, chosen proportional to the intensity of the magnetic field and of the electric field. The component known as the current is defined as that factor of the electric power which is proportional to the magnetic field, and the other component, the voltage, is defined as that factor of the electric power which is proportional to the electric field [4].

The power loss factor λ_{loss} of magnetic induction technology can be expressed as

$$\lambda_{\text{loss}} = \frac{P_{\text{out}}}{P_{\text{Loss}}} \tag{7.6}$$

This loss factor is the sum of all losses related to the transferred power. A detailed analysis results in a minimum loss factor, which can be achieved by a given wireless power transfer system, if the generator and load are properly matched [1].

$$\lambda_{\min} = \frac{1}{(kQ)^2}\left[2 + \sqrt{(kQ)^2 + 1} + \frac{1}{\sqrt{(kQ)^2 + 1}}\right] + \frac{1}{\sqrt{(kQ)^2 + 1}} \tag{7.7}$$

where k is the coupling factor between the primary coil and the secondary coil, and Q is the quality factor of the system.

In electrical engineering, the power factor of an AC electrical power system is defined as the ratio of the real power flowing to the load, to the apparent power in the circuit, and is a dimensionless number between −1 and 1. Real power is the capacity of the circuit for performing work in a particular time. Apparent power is the product of the current and voltage of the circuit. Due to energy stored in the load and returned to the source, or due to a nonlinear load that distorts the wave shape of the current drawn from the source, the apparent power will be greater than the real power. A negative power factor occurs when the device that is normally the load generates power which then flows back toward the device that is normally considered the generator. In an electric power system, a load with a low power factor draws more current than a load with a high power factor for the same amount of useful power transferred. Higher current increases the energy lost in the distribution system and requires larger wires and other equipment. Because of the costs of more equipment and wasted energy, electrical utilities will usually charge a higher cost to industrial or commercial customers where there is a low power factor. Linear loads with a low power factor (such as induction motors) can be corrected with a passive network of capacitors or inductors. Nonlinear loads, such as rectifiers, distort the current drawn from the system. In such cases, active or passive power factor correction may be used to counteract the distortion and raise the power factor. The devices for correction of the power factor may be at a central substation, spread out over a distribution system, or built into power-consuming equipment. The power factor is defined as

$$\text{Power factor} = \frac{P_{\text{real}}}{P_{\text{apparent}}} \tag{7.8}$$

where P_{real} is the real power (as known as active power), and P_{apparent} is the apparent power. In the case of a sinusoidal waveform of source power, the apparent power P_{apparent} can be expressed as

$$P_{\text{apparent}} = \sqrt{P_{\text{real}}^2 + P_{\text{reactive}}^2} \tag{7.9}$$

where P_{reactive} is the reactive power. The power factor triangle below illustrates how the real power, reactive power, and apparent power relate to each other to give the power factor angle [5].

A high power factor is generally desirable in a transmission system to reduce transmission losses and improve voltage regulation at the load. It is often desirable to adjust the power factor of a system to a value near 1.0. When reactive elements supply or absorb reactive power near the load, the apparent power is reduced. Power factor correction may be applied by an electric power transmission utility to improve the stability and efficiency of the transmission network. Individual electrical customers who are charged by their utility for a low power factor may install correction equipment to reduce those costs. Power factor correction brings the power factor of an AC power circuit closer to 1 by supplying reactive power of opposite sign, adding capacitors or inductors that act to cancel the inductive or capacitive effects of the load, respectively. For example, the inductive effect of motor loads may be offset by locally connected capacitors. If a load has a capacitive value, inductors (also known as reactors in this context) are connected to correct the power factor. In the electricity industry, inductors are said to consume reactive power and capacitors are said to supply it, even though the energy is just moving back and forth on each AC cycle. The reactive elements can create voltage fluctuations and harmonic noise when switched on or off. They will supply or sink reactive power regardless of whether there is a corresponding load operating nearby, increasing the system's no-load losses. In the worst case, reactive elements can interact with the system and with each other to create resonant conditions, resulting in system instability and severe overvoltage fluctuations. As such, reactive elements cannot simply be applied without an engineering analysis.

7.4.2 *Resonance Energy*

At resonance, the current and charge density profiles are π/2 out of phase from each other, meaning that the real part of one is maximal when the real part of the other is zero. Equivalently, the energy contained in the coil is at certain points in time completely due to the current, and at other points, it is completely due to the charge. In other words, we can define an effective inductance L and an effective capacitance C for each coil as

$$L = \frac{\mu_0}{4\pi|I_0|^2} \iint d\underline{r} d\underline{r}' \frac{\underline{J}(\underline{r}) \cdot \underline{J}(\underline{r}')}{|\underline{r} - \underline{r}'|} \tag{7.10}$$

$$\frac{1}{C} = \frac{\mu_0}{4\pi\varepsilon_0|q_0|^2} \iint d\underline{r} d\underline{r}' \frac{\rho(\underline{r}) \cdot \rho(\underline{r}')}{|\underline{r} - \underline{r}'|} \tag{7.11}$$

where the spatial current $\underline{J}(\underline{r})$ and charge density $\rho(\underline{r})$ are, respectively, obtained from the current and charge densities along the isolated coil, in conjunction with the geometry of the wireless power transfer system. As defined, L and C have the property that the energy U contained in the coil is given by

$$U = \frac{1}{2} L |I_0|^2 = \frac{1}{2C} |q_0|^2 \tag{7.12}$$

Given this relation and the equation of continuity, the resulting resonant frequency is

$$f_0 = \frac{1}{2\pi\sqrt{LC}} \tag{7.13}$$

A coil as a standard oscillator in coupled-mode theory can then be defined by

$$a(t) = \sqrt{\frac{L}{2}} I_0(t) \tag{7.14}$$

The power dissipated by noting the sinusoidal profile of the current distribution can be estimated as follows:

$$I_{avg} = |I_0|^2 / 2 \tag{7.15}$$

which implies the spatial average of the peak current squared. For a coil with n turns and with material of conductivity σ, the modified standard formulas for ohmic (R_0) and radiation (R_r) resistance are described as

$$R_0 = \sqrt{\frac{\mu_0 \omega}{2\sigma} \frac{1}{4\pi a}} \tag{7.16}$$

$$R_r = \sqrt{\frac{\mu_0}{\varepsilon_0}} \left[\frac{\pi}{12} n^2 \left(\frac{\omega r}{c} \right)^4 + \frac{2}{3\pi^3} \left(\frac{\omega h}{c} \right)^2 \right] \tag{7.17}$$

The first term in (7.17) is a magnetic dipole radiation term (assuming $r << 2\pi c/\omega$, where c is the speed of light); the second term is due to the electric dipole of the coil and is smaller than the first term for our experimental parameters. The coupled-mode theory decay constant for the coil and its quality factor are expressed as [2]. In these equations, Γ is the decay rate, and Q is the quality factor.

$$\Gamma = \frac{R_0 + R_r}{2L} \tag{7.18}$$

$$Q = \frac{\omega}{2\Gamma} \tag{7.19}$$

In physics, the quality factor Q is defined in terms of the ratio of the energy stored in the oscillating resonator to the energy dissipated per cycle by damping processes. For an electrically resonant system, the Q factor represents the effect of electrical resistance.

7.5 Conclusions

We compared with two concepts of wireless power transfer technology with magnetic induction and magnetic resonance. Two technologies, magnetic induction and magnetic resonance, for wireless power transfer are compared in terms of expense, equivalent circuit, physics, loss, and performance. Furthermore, we physically explained the principle of magnetic induction and resonance by the resonance phenomenon. The wireless power transfer system needs the LC tuning as shown in the equivalent circuit, and its strategy was presented. Also, we discussed the compensation capacitor and power factor to explain the real power, apparent power, and reactive power. This chapter might be useful for designing and understanding the wireless power transfer system.

References

1. Waffenschmidt E (2010) Wireless power for mobile devices *VDE-Kongress 2010*, Leipzig, Germany, pp 8–9. Nov 2010
2. Kurs A, Karalis A, Moffatt R, Joannopoulos JD, Fisher P, Solajcic M (2007) Wireless power transfer via strongly coupled magnetic resonances. Science 317(5834):83–86
3. Shin J, Shin S, Kim Y, Ahn S, Lee S, Jung G, Jeon S, Cho D (2014) Design and implementation of shaped magnetic-resonance-based wireless power transfer system for roadway-powered moving electric vehicles. IEEE Trans Ind Electron 61(3):1179 1192
4. Ansys NFC seminar, "Wireless power transfer" http://www.Ansys.com
5. Orange & Rockland, "Energy & safety, power factor" http://www.oru.com

Chapter 8
Selection of Optimum Frequency and Optimization

Uooyeol Yoon

Abstract The issues related to the selection of the optimum frequency are presented, including the optimization of power transfer and transfer capacity in a wireless power transfer system. Power loss due to Joule heating and material loss is an important factor to consider, including skin effects. At high frequencies, electric current in conductor is confined to the skin of the conductor, which limits the conduction of electric current and leads to current loss. Low magnetic reluctance of the core induces high magnetic flux density, which increases the performance of the wireless power transfer system. Low core loss and resistance are required in a typical system that uses a ferrite core. Switching devices and components for power electronics as well as the international standard and recent trend in industry for wireless power transfer technology are also discussed.

8.1 Introduction

This chapter presents the issues that arise in selecting the optimum frequency for wireless power transmission and in optimization of the overall system performance in a wireless power transfer system. The main focus is on the power loss due to the Joule heating (i.e., ohmic resistive heating) and "material" loss due to the use of the coil and ferrite core. Because of the high frequency used in OLEV, the skin effect in the coil that transmits electricity requires special considerations. Litz wire is used to minimize the skin effect in the OLEV system. To concentrate the magnetic flux, the OLEV system uses ferrite core both in the power supply system and the power pickup system. In addition to these issues, this chapter describes the switching devices and the components used in power electronics (e.g., thyristor, IGBT, and

U. Yoon (✉)
Korea Advanced Institute of Science and Technology (KAIST),
Wireless Power Transfer Research Center KAIST, 303 Truth Hall 193 Munji-Ro,
Yuseong-Gu Daejeon 305-732, South Korea
e-mail: uyoon@kaist.ac.kr

N.P. Suh and D.H. Cho (eds.), *The On-line Electric Vehicle*,
DOI 10.1007/978-3-319-51183-2_8

the diode of the rectifier). Finally, recent trends in wireless power transfer technology for vehicles and international standards are discussed.

8.2 Issues: Heating Versus Power Transfer, Transfer Capacity

The power transfer efficiency is a central issue in a wireless power transfer system. The power loss occurs due to Joule heating, also known as ohmic heating and resistive heating. The amount of heat released is proportional to the square of the current as

$$P_{\mathrm{Loss}} \propto I^2 \cdot R \tag{8.1}$$

This relationship is known as Joule's first law. The SI (Standard International) unit of energy was subsequently named the joule and given the symbol J. The commonly known unit of power, the watt, is equivalent to one joule per second. Joule heating is independent of the direction of current, unlike heating due to the Peltier effect [1].

To obtain more detailed results, a numerical analysis is generally carried out. An example is shown in Fig. 8.1. The simulated results were computed by Icepack of ANSYS. In Fig. 8.1, the region designated with red color is for high temperature and the blue region for low temperature after an elapse of one hour of heating due to ohmic heating.

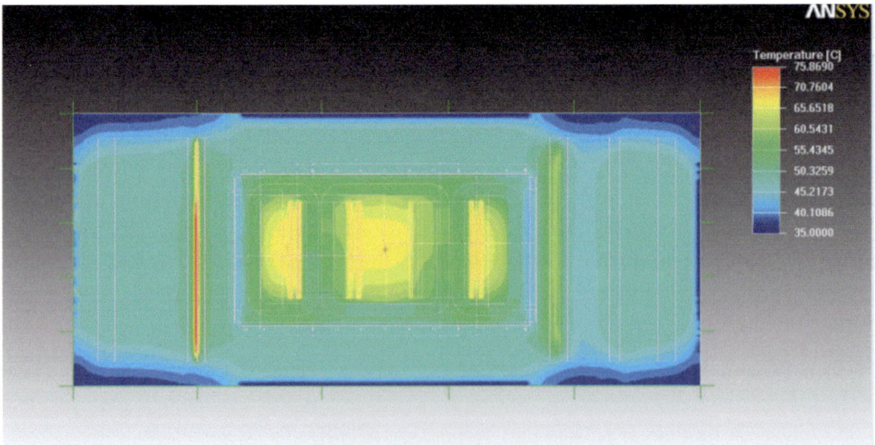

Fig. 8.1 Example picture of heating analysis simulation

8.3 Materials Limitation

8.3.1 Coil Limitation Due to Skin Effect

At high frequencies, electric current in conductor is confined to the skin of the conductor, which limits the conduction of electric current. The skin effect can cause the loss of current flowing in the coil. The skin depth δ is given by [2]

$$\delta = \sqrt{\frac{2}{\omega\mu\sigma}} \tag{8.2}$$

where ω is the angular velocity $2\pi f$, μ is the permeability, and σ is the electric conductivity. The intensity of the current in the conductor decays at an exponential rate toward the core of the conductor. Figure 8.2 shows the current distribution on a cross section of the conductor at low and high frequencies. The light color indicated the high current distribution and the dark color the low distribution. The area of current flowing at low frequency is wider than that at high frequency, indicating that the conducting resistance or loss in the coil is greater at high frequency than at low frequency.

To solve this problem, a typical wireless power transfer system with inductive coupling uses a coil consisting of Litz wire. As shown in Fig. 8.3, by using the Litz wire, the loss problem due to the skin effect can be solved [3].

8.3.2 Core Limitation

The inductive coupling wireless power transfer system uses a ferromagnetic material core to maximize the magnetic flux density. The magnetic reluctance R_m is given by the following equation:

$$R_m = \frac{\ell}{\mu_0 \mu_r A} \tag{8.3}$$

where ℓ is the length of the magnetic flux path, μ_0 is the permeability of a vacuum, μ_r is the relative permeability, and A is the cross section of the magnetic flux path. The low magnetic reluctance induces high magnetic flux density, which increases the performance of the wireless power transfer system. Furthermore, low values of the core loss and resistance are always required in a typical system using a ferrite core. For example, ferrite PL-13 is one of the most efficient ferrite core materials. Figure 8.4 shows the complex permeability of PL-13.

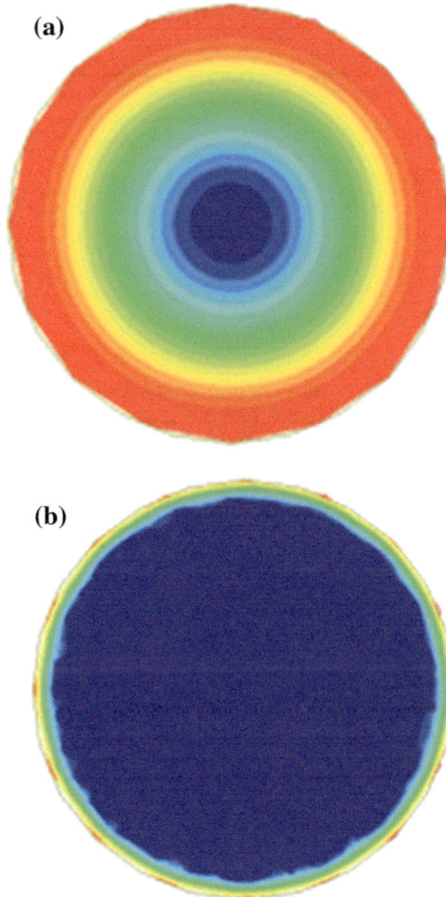

Fig. 8.2 Current distribution of cross section of conductor. **a** Low frequency, **b** high frequency

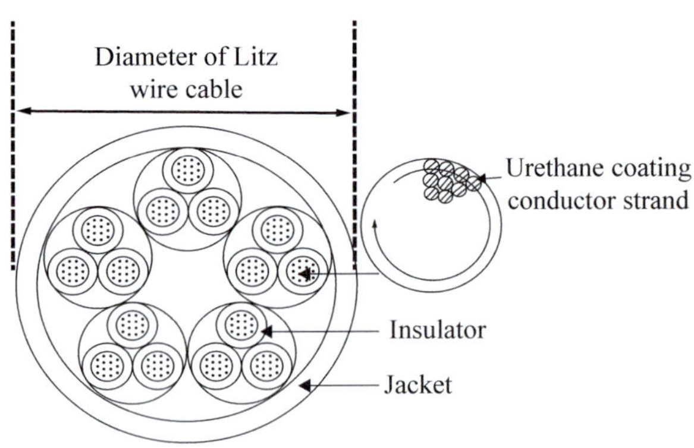

Fig. 8.3 Example of Litz wire for a wireless power transfer system

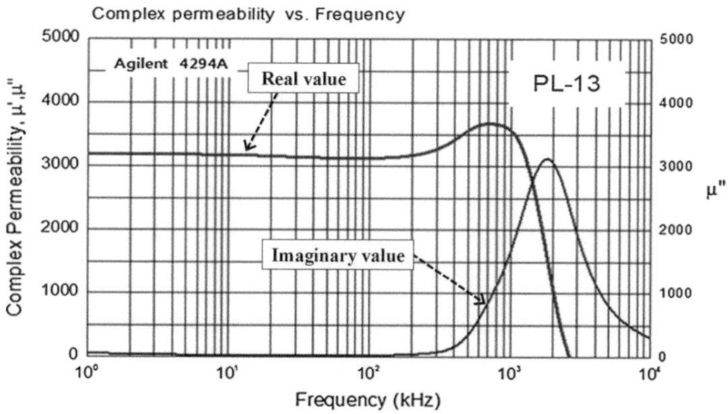

Fig. 8.4 Complex permeability of PL-13 in terms of frequency

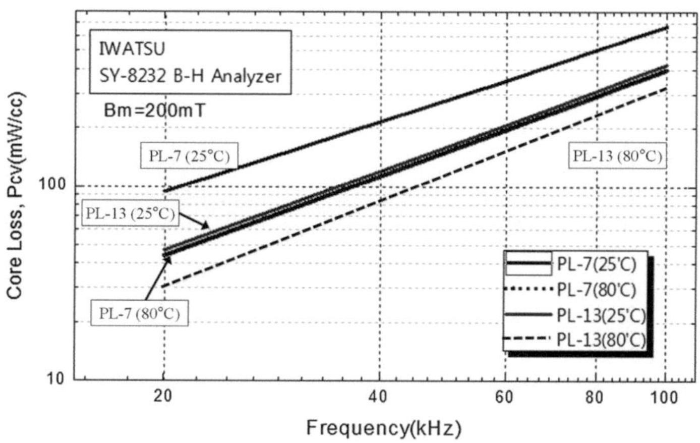

Fig. 8.5 Core loss of ferrite core in terms of frequency

To understand the frequency characteristic of a ferrite material, the characteristic curve of complex permeability in terms of frequency has been used. The complex permeability is given by

$$\bar{\mu} = \mu' - j\mu'' \tag{8.4}$$

where μ' is the real value of permeability and μ'' is an imaginary value of permeability. Figure 8.5 presents the core loss of ferrite core PL-7 and PL-13. From Fig. 8.5, we can see that the PL-13 ferrite core is appropriate for the wireless power transfer system.

8.4 Switching Devices

Switching devices in wireless power transfer system have been used in inverter and rectifier. In this application, switching devices played a role in controlling the inverter operation and converting the power between of AC and DC. This section describes some of switching devices used for this purpose.

8.4.1 *Power Electronics Switching Devices with Different Resonant Frequencies*

(1) Thyristor

Thyristor is a typical switching device for high power application, which is also called the silicon controlled rectifier (SCR). The thyristor typically consists of 3 terminals: an anode, cathode, and gate. If a signal is excited at the gate of the thyristor, the reverse current is excited in the main circuit without supplying gate current. Furthermore, current flow is maintained until it drops below the preset holding current. In summary, the thyristor is a power electric element that can withstand the high voltage with high power capacity without a turnoff switch function. This element is used mainly in low electrical power systems with a low operating frequency. Series and parallel connections with several elements are generally used in high power transformers requiring high power capacity [4].

Figure 8.6 shows the voltage–current characteristics of the thyristor. In a conventional thyristor, once it has been switched on by the gate terminal, the device remains latched in the on-state, providing the anode current exceeds the latching current (I_L). As long as the anode remains positively biased, it cannot be switched off until the anode current falls below the holding current (I_H). A thyristor can be switched off if the external circuit causes the anode to become negatively biased (a method known as natural, or line, commutation). In some applications, this is done by switching a second thyristor to discharge a capacitor into the cathode of the first thyristor. This method is called forced commutation. After the current in the thyristor has extinguished, a finite time delay must elapse before the anode can again be positively biased and retain the thyristor in the off-state. This minimum delay is called the circuit commutated turnoff time (t_Q). Attempting to positively bias the anode within this time causes the thyristor to be self-triggered by the remaining charge carriers (holes and electrons) that have not yet recombined. For applications with frequencies higher than the domestic AC main supply (e.g., 50 or 60 Hz), thyristors with lower values of t_Q are required. Such fast thyristors can be made by diffusing heavy metal ions such as gold or platinum, which act as charge combination centers, into the silicon. Today, fast thyristors are more commonly made by electron or proton irradiation of silicon, or by ion implantation. Irradiation

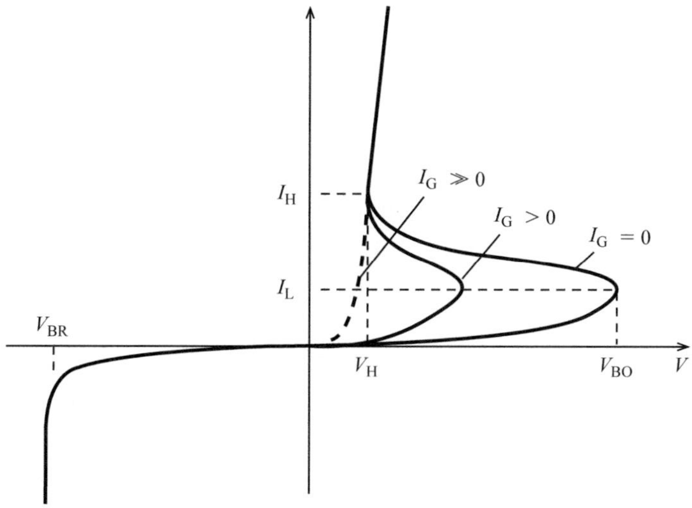

Fig. 8.6 Voltage–current characteristics of thyristor [4]

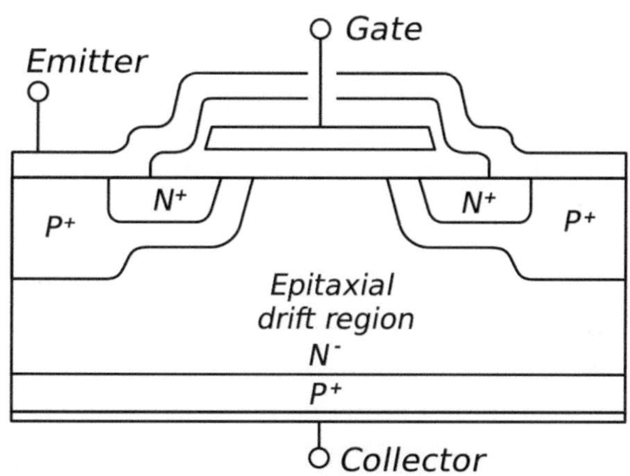

Fig. 8.7 Cross section of a typical IGBT [5]

is more versatile than heavy metal doping because it permits the dosage to be adjusted in fine steps, even at a later stage in the processing of the silicon.

(2) IGBT

An insulated-gate bipolar transistor (IGBT) combines the simple gate-drive characteristics of the metal oxide semiconductor field-effect transistor (MOSFET) with the high current and low saturation voltage capability of bipolar transistors [5]. The IGBT combines an isolated gate FET for the control input and a bipolar power transistor as a switch in a single device. The IGBT is used in medium- to high-power applications such as switched-mode power supplies, traction motor

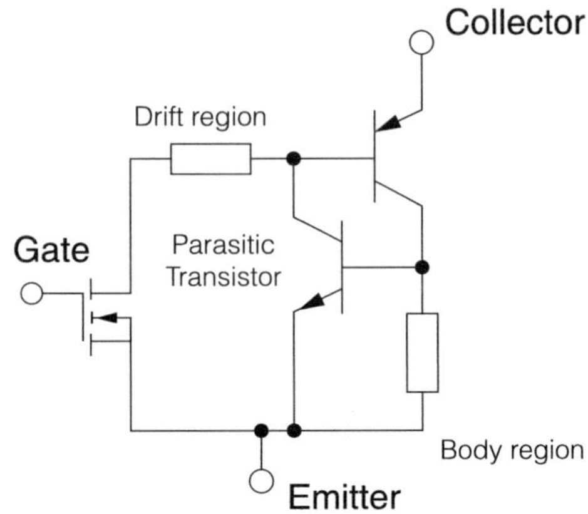

Fig. 8.8 Equivalent circuit model of IGBT

control, and induction heating. Large IGBT modules typically consist of many devices in parallel and can have very high current handling capabilities on the order of hundreds of amperes with blocking voltages of 6000 V, equating to hundreds of kilowatts. In a wireless power transfer system, this has mainly been used in high-speed inverters below the 50 kHz operating frequency. The IGBT is the fruit of the most advanced metal oxide silicon (MOS) and bipolar technology, and it has characteristics of low saturation voltage and high switching speed with high voltage. This semiconductor can be a next-generation power element with high efficiency and high speed in a high power system. Figure 8.7 shows the cross section and equivalent circuit model of a typical IGBT. Figure 8.8 shows the equivalent circuit model of an IGBT.

8.4.2 *Diode for Rectifier*

The diode is a two-terminal electronic component with asymmetric conductance, that is, it has low (ideally zero) resistance to current flow in one direction and high (ideally infinite) resistance in the other. A semiconductor diode, the most common type today, is a crystalline piece of semiconductor material with a p–n junction connected to two electrical terminals.

8.5 Trends in Resonance Frequency in Wireless Power Transfer System for Vehicles

The power that can be taken from a homogeneous magnetic field B is dependent on the induced voltage V_{ind} in the used receiver coil. Considering it as a loop, for a sinusoidal signal shape, this is expressed as

$$V_{\text{ind}} = 2\pi f \cdot B \cdot A \quad (8.5)$$

where f is the frequency and A is the loop area. With the same flux density, higher power can be transferred at higher frequencies. This means that the product of the maximum magnetic flux density times the frequency is relevant for the power transmission.

There are ongoing international standardization activities such as IEC (International Electro-technical Commission) TC69 PT61980 and SAE (Society of Automotive Engineers) J2954 for wireless charging in electric vehicles. In these communities, it is announced that 85 or 140 kHz will be used as the resonance frequency for wireless charging in electric vehicles. This is used for wireless charging in small-sized passenger cars under 30KW capacity. For the case of an electric bus with system capacity up to 200 KW, 20 kHz resonance frequency is being used. Standardization of the resonance frequency for wireless power transfer system in heavy-duty electric vehicles is under discussion. In SAE J2954, 85 kHz was recently selected as the resonant frequency for wireless power transfer system in the area of passenger cars up 30 KW system capacity. For the high power area up to 200 KW, including electric buses, discussions are ongoing. In IEC PT 61980, the operation frequency is not decided yet, and 85 and 140 kHz are being discussed.

As described previously, as the operating frequency becomes higher, the system capacity for power transfer increases. It results in a reduction in weight in the pickup system, which is installed in the electric vehicle. This is one of the reasons why EV manufacturers prefer higher operation frequency. But for wireless power transfer system, systems with high capacity such as an electric bus, the safety issues including electric safety and EMF (electromagnetic field) at high frequency above 20 kHz have not yet been sufficiently verified. Also, power electronics devices including IGBT described previously are not sufficiently available for high power systems at high frequencies.

In summary, the selection of optimum frequency should be made in consideration of system capacity, the availability of power electronics devices, and safety (electric safety and EMF).

8.6 Conclusion

This chapter presents the optimization and selection of the optimum frequency in a wireless power transfer system. Power loss in the wireless power transfer system consists of the ohmic heating loss and the material loss. Two kinds of material loss were presented with the coil and ferrite core. At a high frequency, power electronics system must deal with the skin effect in conducting wires. The use of ferrite material as a core to channel magnetic field is affected by the core loss and its magnetic permeability. Finally, the switching devices and components for power electronics such as the thyristor, IGBT, and the diode of the rectifier was introduced.

References

1. Butterworth S (1926) Effective resistance of inductance coils at radio frequency—part i. Wirel Eng 3:203–210
2. Fink DG, Beaty HW (1999) Standard handbook for electrical engineering, 14th edn. McGraw Hill, Boston
3. Hayt WH, Buck JA (2001) Engineering electromagnetics, 8th edn. McGraw Hill, New York
4. Meier AV (2006) Electric power system: a conceptual introduction. Wiley, New York
5. Nakagawa A (1987) Safe operating area for 1200-V nonlatchup bipolar-mode MOSFET's. IEEE Trans Electron Devices 34(2):351–355

Chapter 9
Optimum Design of Wireless Power Transfer System

Gu Ho Jung

Abstract This chapter presents an optimum design of the power circuit and core structure for the OLEV wireless power transfer system. The design requirements such as a road-to-pickup gap are given first, followed by the description of the overall block diagram and power circuit. The equivalent equations in terms of time and phase domains are obtained from the equivalent circuit. The cores structure is designed for the EE and UU types, using the equivalent equations and the properties of the magnetic core.

9.1 Introduction

This chapter introduces the design method for the wireless power transfer system for OLEV in terms of the power circuit and core structure, including the related equivalent circuits and equations necessary for the design. The OLEV system consists of four components, an inverter, a road embedded power cable module, a pickup module with a rectifier, and a regulator [1, 2]. The operation of these components is explained briefly. It also discusses the design requirements and basic specifications such as the gap between the road and the pickup unit and the power transfer efficiency of the total system.

The overall block diagram and power circuit are given, and related parameters are suggested, from which the equivalent circuits and equations are obtained in time and phase domains. Also, the waveform of the line-to-line inverter output voltage and inverter phase current at the low- and full-load conditions is explained [3, 4].

We design the core structure based on the EE and UU types. The equivalent equations for the reluctances of the target loop and the leakage loop are, respectively, expressed. The equation of the coupling coefficient k is then given, and the

G.H. Jung (✉)
Korea Advanced Institute of Science and Technology (KAIST), Wireless Power Transfer Research Center KAIST, 303 Truth Hall 193 Munji-Ro, Yuseong-Gu Daejeon 305-732, South Korea
e-mail: ghjung9595@kaist.ac.kr

N.P. Suh and D.H. Cho (eds.), *The On-line Electric Vehicle*,
DOI 10.1007/978-3-319-51183-2_9

properties of the magnetic core are explained. Finally, the applied core design for weight reduction is given.

As discussed in the preceding chapters, the OLEV system consists of four components, a three-phase inverter, a road embedded power supply system, a power pickup system with rectifier, and a regulator. The inverter supplies high-frequency current to the rail, which consists of high-frequency cable and a magnetic core. When the inverter supplies high-frequency current to the rail, the rail generates a high-frequency magnetic field through the power cable and core, and the pickup module with a rectifier converts this magnetic field into low DC voltage. Using this low DC voltage, the regulator outputs high DC voltage appropriate for a battery.

9.2 Design Requirements

Before we start the design of the OLEV electric bus, it is necessary to determine several design constraints, as listed in Table 9.1. As indicated in the first row of Table 9.1, the gap between the roads to the pickup module is set at 200 mm to prevent the contact with obstacles on the road. The rated three-phase AC voltage is 480 Vac ± 10%, which can be changed depending on the main voltage of the location where the inverter is installed.

An important constraint is the switching frequency of the inverter. The Korean government allotted two frequency bands, 20 ± 1 and 60 ± 1 kHz for wireless power transfer systems. As the switching frequency becomes higher, we can obtain a smaller road embedded power module and pickup module, respectively, thereby lowering the implementation cost of the total system. However, 20 kHz switching frequency for the inverter was chosen to increase the power transfer efficiency with

Table 9.1 Design constraints for OLEV bus system

Items	Values	Comments
Road to pickup gap	200 mm	Distance between road and bottom side of pickup module
Rated three-phase AC voltage	480 Vac ± 10%	This value can be changed depending on the main voltage of the place where inverter is installed
Inverter switching frequency	20 kHz	20 ± 1 kHz
EMF	Below 62.5 mG	Government regulation in Korea
Rated output power of regulator	100 kW	
Total power transfer efficiency	80%	Total power efficiency calculated from three-phase input of inverter to regulator DC output

lower switching loss of the available IGBT (insulated-gate bipolar transistor) power device while avoiding the range of audible frequency below 20 kHz.

The EMF (Electro-Magnetic Field) measured around the OLEV bus should be below 62.5 mG, as specified by the Korean government. Furthermore, the maximum and rated power of driving motor for the electric bus is 240 and 100 kW, respectively, which are very high compared to a small electric car. The regulator output is connected to a battery, and the target rated output power of the regulator in our case was chosen to be the same with the rated power of an electric driving motor.

Finally, the minimum efficiency of power transfer from three-phase input of the inverter to the regulator DC output connected to the battery is specified as 80%. The efficiency is the ratio of the active power measured at the regulator DC output divided by the active power measured at three-phase input of the inverter.

9.3 Design of Optimum Parameters for the Overall System

Figure 9.1 is an overall block diagram and the power circuits for the OLEV system, which consists of an inverter, road embedded power cable module, pickup module, and a regulator. Three-phase AC voltage is rectified as DC voltage, which is controlled as a variable DC-link voltage by the phase-controlled rectifier. This DC-link voltage is converted to an isolated single-phase voltage source by a single-phase inverter and a high-frequency transformer.

There are two segments (or tracks): #1, #2. Because the road embedded power cable module has two segments and the inverter can supply current to each segment, common cable is necessary on the bottom side of the rail. Resonant capacitance C_{r1} existing at the secondary part of the transformer makes the resonant frequency between L_{r1} and C_{r1} be slightly below 20 kH (Fig. 9.1).

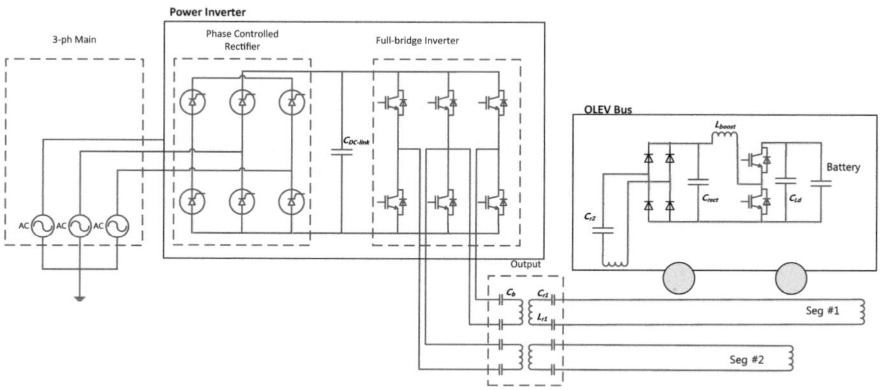

Fig. 9.1 Overall block diagram and power circuits for OLEV system

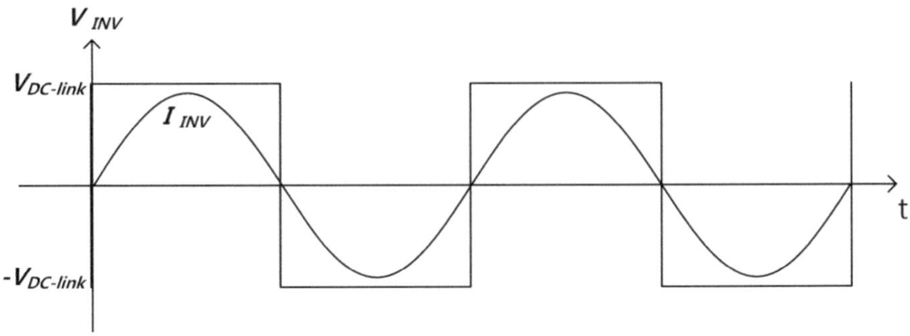

Fig. 9.2 Waveform of line-to-line inverter output voltage V_{inv} and inverter phase current I_{inv} at full-load condition; x-axis: time, y-axis: voltage (V_{inv}) and current (I_{inv})

As the output load increases, the lagging phase decreases. Finally, its equivalent impedance almost becomes like resistance at a full road condition, and its current has the same phase, that is, 0° phase, compared to the inverter voltage, as illustrated in Fig. 9.2. A full-load condition happens when the electric bus is situated over the embedded rail and the regulator supplies 100 kW rated power from the rail to the battery installed in the bus.

The embedded road rail and pickup module operate as primary and secondary sides of an E-type "transformer." As mentioned before, the embedded rail in the primary side has 1-turn of power cable, and the pickup module has n-turns of cables. Inductance L_{r2} on a scale of several mH is measured in the pickup module due to n-turns of wiring in the pickup module, and additional resonant capacitance C_{r2} is necessary to tune the resonant frequency between L_{r2} and C_{r2} to 20 kHz, that is, the inverter switching frequency. AC voltage at the output of the pickup module is rectified to DC voltage around 400–500 V by a single-phase diode bridge. The boost-type DC/DC regulator then converts the low rectified DC voltage to high DC voltage to charge the battery.

In the case of the proposed OLEV system, 5 pieces of pickup modules are incorporated between the front and rear wheels, and the rated power of each pickup module is 20 kW considering the total 100 kW rated power of the regulator.

Figure 9.3a shows the equivalent circuit of the proposed wireless power transfer system for the OLEV system in the time domain. Here, L_1 and R_1 at the primary side of the high-frequency transformer are the inductance and resistance caused by the embedded road rail and transformer. C_{r1} is the resonant capacitance to make the resonant frequency between L_1 and C_{r1} be the same as the inverter switching frequency, from which the impedance created by C_{r1} and L_1 becomes almost zero. In the meantime, L_2 and R_2 at the secondary side of the transformer are the inductance and resistance created by high-frequency litz wires with several tens of turns in each pickup module. Similar to operation of C_{r1}, C_{r2} is the resonant capacitance that makes the resonant frequency between L_2 and C_{r2} be the same as the inverter switching frequency f_{inv}, and the impedance created by C_{r2} and L_2

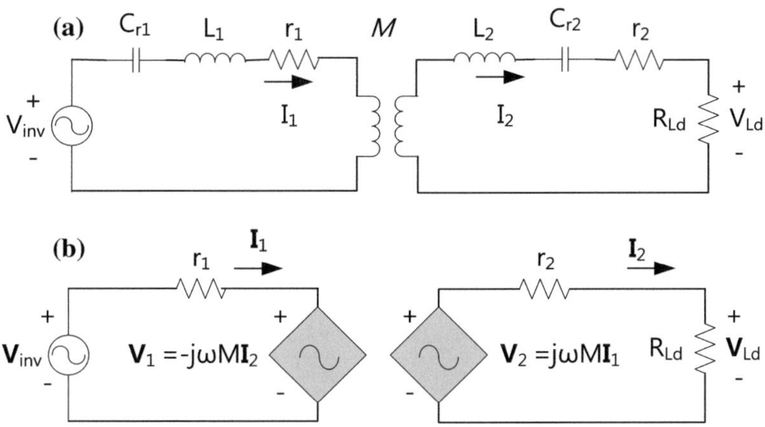

Fig. 9.3 Equivalent circuit of proposed system; **a** actual equivalent circuit and **b** equivalent circuit at perfect resonant status

becomes almost zero. In this case, the obtained equations in the time and phasor domains are as follows:

$$v_{\text{inv}} = \frac{1}{C_{r1}} \int i_1 dt + L_1 \frac{di_1}{dt} + r_1 i_1 - M \frac{di_2}{dt} \tag{9.1}$$

$$v_{\text{Ld}} = M \frac{di_1}{dt} - L_2 \frac{di_2}{dt} - \frac{1}{C_{r2}} \int i_2 dt - r_2 i_2 \tag{9.2}$$

$$V_{\text{inv}} = \frac{1}{j\omega C_{r1}} I_1 + j\omega L_1 I_1 + r_1 I_1 - j\omega M I_2 \tag{9.3}$$

$$V_{\text{Ld}} = j\omega M I_1 - j\omega L_2 I_2 - \frac{1}{j\omega C_{r2}} I_2 - r_2 I_2 \tag{9.4}$$

where M is the mutual inductance between the embedded rail and the pickup module and has a low value in an OLEV system with a high gap distance.

Thus, the equivalent circuit at the resonant frequency, which is the same as the inverter switching frequency, becomes as shown in Fig. 9.3b in the case of the phasor domain. We can see that all the inductances are cancelled by the additional resonant capacitances, and thus only, the resistance component exists in this equivalent circuit. In this case, the related equations in the phasor domain can be written as follows:

$$V_{\text{inv}} = r_1 I_1 - j\omega M I_2 \tag{9.5}$$

$$V_{\text{Ld}} = j\omega M I_1 - r_2 I_2 \tag{9.6}$$

In Fig. 9.3b, we can see that the dependent voltage sources V_1 and V_2 controlled by the secondary current I_1 and the primary current I_2 appear, respectively, in the primary and secondary circuits. Each of V_1 and V_2 is linked with I_1 and I_2 through mutual inductance M between the embedded rail and the pickup module.

Constant current control of the inverter output current I_1 becomes difficult in the case where only the resistance component exists between the inverter output voltage V_{inv} and the dependent voltage source V_1. C_{r1} is tuned so that the resonant frequency between C_{r1} and L_1 becomes slightly lower than the inverter switching frequency, and hence, the impedance made by C_{r1} and L_1 operates as equivalent inductance L_{1eq}. The resultant equation in a non-perfect resonant condition can be expressed as follows:

$$V_{\text{inv}} = j\omega L_{1eq} I_1 + R_1 I_1 - j\omega M I_2 \tag{9.7}$$

$$V_{\text{Ld}} = j\omega M I_1 - R_2 I_2 \tag{9.8}$$

In case of inverter, the inverter current I_1 is controlled to operate as a constant current source by controlling the magnitude of V_{inv}, that is, the DC-link voltage of the inverter, whose DC-link voltage is controlled by phase angle control of the phase-controlled rectifier, as already mentioned in Sect. 9.3. In the case of controlling V_{inv} under various load conditions, the phase angle between the inverter output voltage V_{inv} and the inverter current I_1 varies according to the load power, because active power increases with higher load power. For example, the inverter current I_1 has a 90° lagged phase angle compared to V_{inv} in the case of a no-load condition, but I_1 has the same phase angle with V_{inv} in the full-load case.

9.4 Design of Core Structure

9.4.1 Core Design

Wireless power transfer systems use magnetic cores to maximize the magnetic flux density. The basic core structure can be categorized into two types: EE type and UU type. The EE type is composed of a pair of E-shaped cores, and the UU type is composed of a pair of U-shaped cores as shown in Fig. 9.4 [5].

The EE type has an advantage in generating magnetic flux considering the saturation under the same pickup width constraint because the magnetic flux is divided into two separated paths. Also, the thickness of the horizontal core bar in case of the UU type must be two times greater for the same amount of magnetic flux compared to the EE type. Meanwhile, the UU type is more tolerant of lateral displacement than the EE type, because the pole distance d_p shown in Fig. 9.4b is longer.

In the EE-type structure, road embedded power cable module cables are wound around the center magnetic core pole, and two target magnetic loops and two main

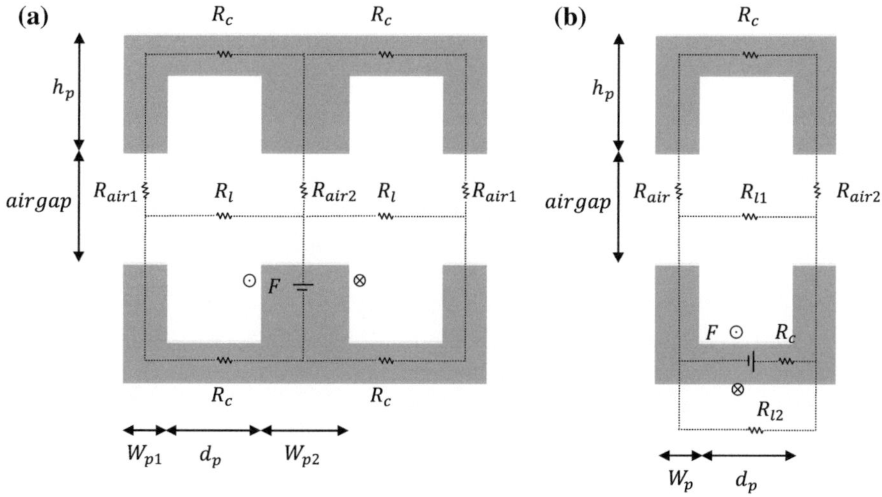

Fig. 9.4 Magnetic circuits of **a** EE-type structure and **b** UU-type structure; F Magnetomotive force in ampere-turns, R_c Magnetic reluctance in magnetic circuit, R_l Magnetic reluctance of the leakage flux, R_{air} Magnetic reluctance of air gap, h_p Height of the pole, W_p Width of the pole, and d_p Distance between pole and pole

leakage magnetic loops are formed. In the UU-type structure, road embedded power cable module cables are wound around a pole or a horizontal bar of the primary magnetic core, and a target magnetic loop and two main leakage magnetic loops are formed. In the EE type example of Fig. 9.4a, R_{target} and R_{leak}, the equivalent reluctances of the target loop and the leakage loop, respectively, can be expressed as follows:

$$R_{target} = R_{air2} + \frac{1}{2} R_{air1} + R_C \tag{9.9}$$

$$R_{leak} = R_l // R_l = \frac{1}{2} R_l \tag{9.10}$$

To transfer the maximum amount of power, the coupling coefficient k should be maximized by adjusting the reluctances, which is the ratio of effective magnetic flux through the loop to the total magnetic flux.

$$\begin{aligned} k &= \frac{\Phi_{target}}{\Phi_{target} + \Phi_{leak}} = \frac{R_{leak}}{R_{target} + R_{leak}} \\ &= \frac{R_l}{2R_{air\,2} + R_{air\,1} + 2R_C} \end{aligned} \tag{9.11}$$

Therefore, R_C should be minimized and R_l should be maximized. The reluctance can then be calculated by

$$R = \frac{l}{\mu A} = \frac{l}{\mu_0 \mu_r A} \tag{9.12}$$

The reluctance is proportional to the length l of the path and inversely proportional to the permeability μ and cross-sectional area A. Therefore, larger permeability of the core is more advantageous for reducing the core reluctance, R_C. The permeability value of normal ferrite is on the scale of hundreds and the ferrite used in the pickup, and the power supply system has a relative permeability higher than 3000.

If the permeability of the ferrite is sufficiently high, the core reluctance can be ignored, and the reluctance of the target loop is approximately R_{air2} + $1/2R_{\mathrm{air1}}$. To reduce R_{air1} and R_{air2}, it is better to increase the pole widths w_{p1} and w_{p2} to enlarge the cross-sectional area of the core poles. However, if W_{P1} and W_{P2} are lengthened, the distance d_P between the poles will be shortened by a given pickup and road embedded power cable module width constraint, and the leakage magnetic flux will be increased. This is the reason it is necessary to find the optimal pole widths W_{P1} and W_{P2} under the given pickup and road embedded power cable module width constraints.

The saturation of the core also should be considered. The saturation magnetic flux density of ferrite is about 0.4 T. As the magnetic flux density in the core becomes higher and closer to saturation, the rate of increment shrinks and the core loss rises. Heat generated due to the core loss increases the temperature of the core, which may degrade the characteristics of the core such as core loss or permeability. Accordingly, if more magnetic flux is required, the thickness of the core should be increased to reduce the magnetic flux loss, and this factor has to be considered with the pickup and road embedded power cable module width constraints.

In designing the pickup and power supply system, we suggest the use of ferrite. Due to its relatively low price and very high saturation magnetic flux density, ferrite is the most appropriate material to be used for the core in a commercialized wireless power transfer system.

9.4.2 Magnetic Core

A core material having relative permeability higher than 2000 was used as the magnetic core in the wireless power transfer system. In order to implement wireless power transfer in the OLEV system, in the early stages of the study, PL-7, which is a Korean SAMHWA electronic product of Mn–Zn ferrite series, was used because it has an excellent price–performance ratio. However, the core material was replaced with PL-13, which has a higher saturation magnetic flux density and lower core loss than PL-7, in order to achieve more efficient wireless power transfer.

Table 9.2 Properties of PL-7 and PL-13

	PL-7 Ferrite core	PL-13 Ferrite core
B_s	490 mT	520 mT
Coercivity	12 A/m	8 A/m
Loss	31 W/kg (25 kHz, 200 mT)	14.3 W/kg (25 kHz, 200 mT)
Resistivity	>500 * 10^6 μΩ·cm	>700 * 10^6 μΩ·cm
Relative permeability	2400	3200
Curie Temp.	>220 °C	>220 °C
Density	4.85 * 10^3 kg/m^3	4.90 * 10^3 kg/m^3

PL-13 has better performance than PL-7 at the operating frequency of 20 kHz. Table 9.2 illustrates differences in greater detail through a comparison of these products.

9.4.3 Core Design for Weight Reduction

In designing the core for a high-power pickup module, the key factor to consider is how much magnetic flux, which is generated from the road embedded power cable module, the core is capable of receiving. In order to increase the amount of magnetic flux the core receives, the core has to be designed with larger area. Larger cross-sectional area leads to smaller resistance and thus smaller power loss. However, a core with larger area is more expensive and heavier. To solve this problem, core parts that are not effective in receiving magnetic flux are eliminated. Elimination of interior core parts does not have a substantial effect on the magnetic flux density acquired in the pickup module.

We tested OLEV system in field environment and measured the total power transfer efficiency from the input of inverter to output of regulator, where we charged our battery with 100 kW. In this case, we obtained above 80% total power efficiency.

9.5 Conclusions

The design method for the wireless power transfer system for OLEV system is given in terms of the power circuit and core structure. The equivalent circuits and equations are obtained from the design constraints. Also, core structure is designed using core structure, the related equations, core materials, etc.

References

1. California Partners for Advanced Transit and Highways. http://www.path.berkeley.edu/. Accessed 2009
2. Huh J, Park EH, Joung GH, Rim CT (2009) High efficient inductive power supply system implemented for on line electric vehicles. Paper presented at the 2009 KPES, pp 105–110
3. Jung GH, Shin S, Lee SH, Shin JG, Kim YS, Jeon SJ (2012) High efficient inductive power supply and pickup system for on-line electric bus. Paper presented at the 2012 IEEE electric vehicle conference (IEVC), Greenville, SC: 1–5
4. Jeon SJ, Cho DH, Rim CT, Jeong GH (2011) Online electric vehicle system segment switching apparatus and a control method thereof. KS Patent WO 2011099781:A3
5. Shin JG, Shin SY, Kim YS, Ahn SY, Lee SH, Jung GH, Jeon SJ, Cho DH (2014) Design and implementation of shaped magnetic-resonance-based wireless power transfer system for roadway-powered moving electric vehicles. IEEE Trans. Industr Electron J 61(3):1179–1192

Chapter 10
Inverter and Link Road-Embedded Power with Cable Module

Gu Ho Jung

Abstract The inverter and link road-embedded power cable module for the OLEV wireless power transfer system are presented. The design specifications such as the power transfer efficiency of inverter and the rated power of inverter are given, followed by the overall block diagram of the inverter. The two-segment operation of the road-embedded power cable module is described. Finally, the control algorithm, including a block diagram, is presented with experimental results.

10.1 Introduction

As described in Chap. 4, the OLEV system is composed of an inverter, a road-embedded power cable module, a power pickup system including a pickup module and rectifier, a regulator, and a battery inside the electric vehicle [1, 2]. The inverter converts 60 Hz three-phase AC input power to 20 kHz output power. The inverter provides 200–400 A of current to the power cable modules embedded under the road, which consists of power module cables, ferrite core blocks, and mechanical components. The cables carry current and generate magnetic flux. The ferrite core blocks of the underground power supply system and the ferrite core structure of the power pickup unit maximize the active magnetic flux by shaping it upward to form a loop of the magnetic field [3, 4].

To increase the power transfer efficiency between the road-embedded power cable module and the pickup module, SMFIR (Shaped Magnetic Field In Resonance) technology was invented at KAIST to transfer a large amount of electric energy wirelessly and safely to electric vehicles in motion or stationery. SMFIR technology enables the OLEV system to transfer power through a power grid system embedded under the road. A pickup module installed under the vehicle collects the magnetic

G.H. Jung (✉)
Wireless Power Transfer Research Center KAIST, Korea Advanced Institute of Science and Technology (KAIST), 303 Truth Hall 193 Munji-Ro, Yuseong-Gu, Daejeon 305-732, South Korea
e-mail: ghjung9595@kaist.ac.kr

N.P. Suh and D.H. Cho (eds.), *The On-line Electric Vehicle*,
DOI 10.1007/978-3-319-51183-2_10

field efficiently by being in resonance with the field generated from under-road power cables and converts it into electric energy for vehicle operation [5].

This chapter introduces the power circuit and the control algorithm used for the three-phase inverter used in the OLEV system. It presents the operations of both the conventional and the segmentation type of power line and describes the design of a inverter segmentation control technology, which was verified through experimental results.

The switching frequency and the power of the inverter determine the requirements for the inverter and the road-embedded power cable module. In this chapter, the operation of the segmented OLEV system and the implementation of the OLEV road-embedded power cable module are presented, following a description of the design of the road-embedded power cable module and its segmentation. It presents the power circuit for the three-phase inverter with a description of its internal components. Then the inverter circuit for electrical segmentation and its control algorithm including a block diagram are explained.

10.2 Design Requirements

Before we start the design of inverter and link road-embedded power cable module for the OLEV system, we must first determine several design constraints and specifications, as given in Table 10.1. As shown in the first row in Table 10.1, the rated three-phase AC voltage is 380 or 480 Vac ± 10%, which can be changed depending on the main voltage of the location where the inverter is installed. For reference, we used 480 Vac ± 10% of three-phase AC voltage in the experimental laboratory but 380 Vac ± 10% in the final commercialized bus line, where 380 Vac voltage is obtained from on-ground transformer as the main power source for the inverter.

Another important requirement is the switching frequency of the inverter. The Korean government allotted two frequency bands, 20 ± 1 kHz and 60 ± 1 kHz,

Table 10.1 Design constraints for inverter and link road-embedded power cable module

Items	Values	Comments
Rated three-phase AC voltage	380 or 480 Vac ± 10%	This value can be changed depending on the main voltage of the location where the inverter is installed
Inverter switching frequency	20 kHz	20 ± 1 kHz
Rated input power of inverter	Above 125 kW	
Output AC current of inverter	200–400 Arms	This value can be changed depending on the main voltage of the location where the inverter is installed and the applied OLEV electric bus specifications
Power transfer efficiency of inverter	95%	Power efficiency measured from three-phase input of inverter to AC output of inverter

for wireless power transfer systems. As the switching frequency becomes higher, we can obtain smaller size of the embedded power module and pickup module, which reduce cost of implementation of the total system. However, 20 kHz was chosen as the switching frequency for the inverter because we must obtain higher power efficiency with lower switching loss of the available IGBT (Insulated Gate Bipolar Transistor) power device and avoid the range of audible frequencies (below 20 kHz).

On the basis of 80% total power transfer efficiency and 100 kW regulator output power, as noted in the previous chapter, the rated input power of the inverter should be above 125 kW. The 20 kHz output AC current of the inverter varies between 200 and 400 Arms depending on the main voltage of the location where the inverter is installed and the applied OLEV electric bus specifications. For example, small output current of 200 Arms can be applied in the case of a battery for a small OLEV microbus, where the battery voltage can be smaller than that of a large OLEV bus.

Finally, the target power transfer efficiency from the three-phase input of the inverter to the inverter AC output connected to the road-embedded power cable module is 95%. This value is calculated as the percentage of the active power measured at the inverter AC output divided by the active power measured at the three-phase input of the inverter.

10.3 Design of Road-Embedded Power Cable and Its Segmentation

Several road-embedded power cable module segments make up the long road-embedded power cable module of the OLEV system. The inverter only turns on the segment where a vehicle is located. By sensing the approach of a vehicle, the road-embedded power cable module segments transmit a signal to the inverter to notify where the vehicle is. The inverter turns on the segment where the vehicle is about to enter and turns off the segment where the vehicle is leaving. The segment operation method prevents the exposure of the electromagnetic field to people or electronic devices near the segments where a vehicle is absent.

10.3.1 Conventional Implementation of Segment

In the OLEV system, an inverter supports one, two or more segments. Therefore, an inverter requires extension cables to connect the inverter with distant segments. Figure 10.1 shows the conventional implementation of OLEV road-embedded power cable module segments. The solid lines are road-embedded power module cables, and the dotted lines are extension cables. The road-embedded power cable module cables generate active magnetic flux, which transfers power to the pickup modules. However, the extension cables only carry current from the inverter to the segments, and their magnetic flux should be shielded by aluminum pipes etc. If the length and number of segments increase, longer extension cables are needed.

10.3.2 Revised Implementation of Segments

This method eliminates the extension cables of conventional road-embedded power cable module segments. In this method, combinations of multiple low current cables are used for the segmentation method. Selection of the segment to be turned on is done by controlling the direction (180° phase) of AC current. Contrary to the conventional segmentation method, all the multiple cables are turned on for all segments. The directions of magnetic fields are the same and superposed on the segment that is turned on, and the directions are opposite and countervailed on the segments that are turned off.

The two segments are made up of two half current cables—A loop and B loop. For turning on the first segment, the inverter provides the same phase AC currents to the A loop and B loops. In this case, the magnetic fields from both the A and B loops are superposed on the first segment and cancel each other out on the second segment. In the case of turning on the second segment, the inverter provides opposite phase AC currents, which have 180° phase difference. The magnetic fields from both A and B loops then cancel each other out on the first segment and are superposed on the second segment. The directions of currents for segmentation are shown in Fig. 10.2. The direction of current of the A loop is forward, and only one

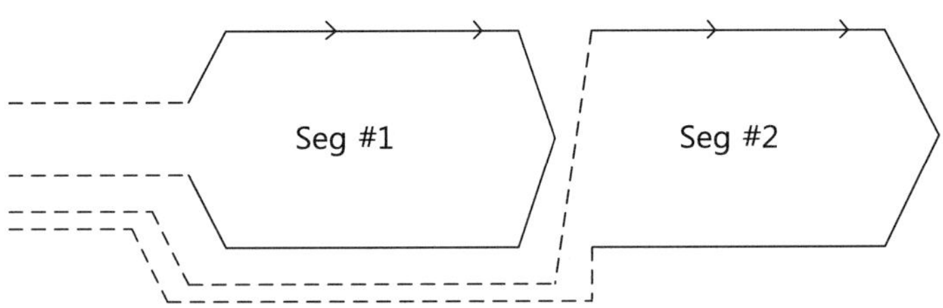

Fig. 10.1 Conventional OLEV road-embedded power cable module segments

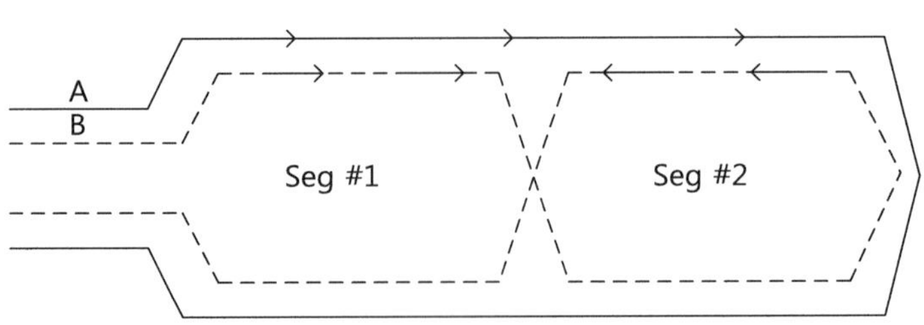

Fig. 10.2 Diagram of revised two-segment implementation

of the segments is turned on by controlling the direction of current of the B loop (Table 10.2).

Although the magnetic field in the position near to the ferrite core block is not canceled out perfectly, it is not exposed to people on the road because the ferrite core is buried 5 cm under the surface. According to the simulation results, the average magnitude of magnetic flux at 50, 100 and 150 cm from the road surface was 15.2 mG, which is below the Korean magnetic flux regulation (62.5 mG).

10.4 Design of Inverter

As shown in Fig. 10.3, the inverter for OLEV system uses a full-bridge resonance inverter to drive the primary coil. The inverter module consists of a rectifier, full-bridge inverter, blocking capacitor, output insulation transformer, and series resonance capacitor. The assembly of the inverter internal lines, primary coil, shielding, and ferrite core has an inductance value per meter. In order to maintain zero voltage switch operation of the inverter, the output block compensates the resonance point below the operating frequency. This helps the inverter work with reactance loads. The operating frequency of the inverter is 20 kHz. To keep fixed output current with various load conditions, the power supply system uses the phase controlled rectifier to control the DC-link voltage.

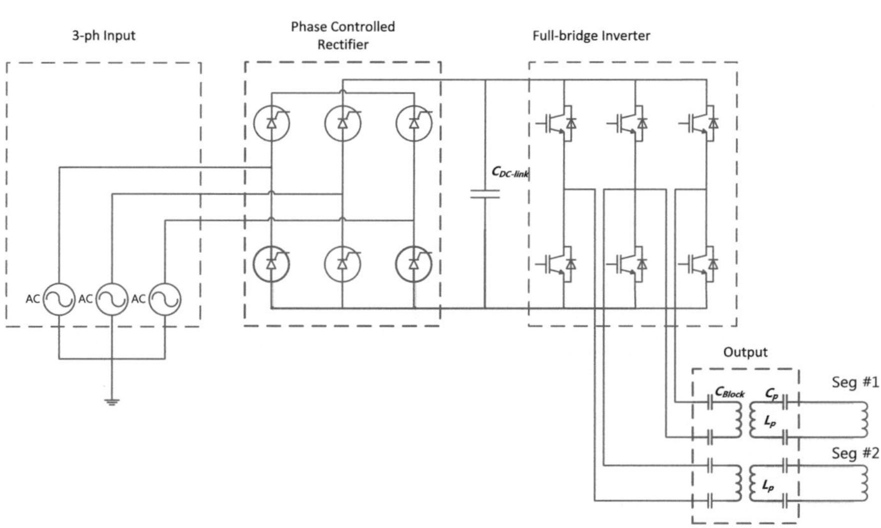

Fig. 10.3 Power circuit of OLEV infrastructure system

Table 10.2 Control of two-segment

Direction of B loop	Seg #1	Seg #2
Forward	**on**	off
Reverse	off	**on**

Control of the PCR is achieved using the PI controller. The DC-link voltage varies from 50 to 550 V, from no-load to full-load, respectively. The output module has blocking capacitors and an insulation transformer. It prevents common mode current from flowing by ground coupling between primary coil(s).

The proposed inverter system consists of a three-phase full-bridge SCR rectifier with a DC-link capacitor and a high-frequency PWM inverter using IGBTs with electrical segmentation, a high-frequency transformer, and primary line with resonance capacitors. SCR is used for rectifying three-phase AC power to DC voltage source, as commonly done. We use large capacitors, and therefore, we can regard the DC-link capacitor as a controlled DC voltage source. With this voltage source, the inverter module converts DC voltage to AC current. A transformer is used for electrical isolation between two segments via a ground. For each segment, the inverter output load consists of line inductance of the primary track segment and a resonance capacitor, which is tuned to the resonant frequency for maximum power transfer to the secondary line. We tuned the resonance frequency as 19.5 kHz. The primary track is thus seen as an inductor equivalently at the operating frequency of 20.0 kHz. In that manner, we can operate the inverter as soft switching (ZVS (Zero Voltage Switching) turn-on) mode in a steady-state.

Track segmentation is done by additional half-bridge modules. To supply current to the first segment, for example, we use the first two half-bridge networks and the others are off. In this case, the half-bridge network placed on the center is commonly used for both segmentations.

The inverter output current is changed with the load condition, for example, the state of charge (SOC) of the battery in the OLEV bus on the track. Thus, we should control the output current such that it is unchanged regardless of the amount of power that is consumed from the load, because we want uniform magnetic flux density, which is the medium of power transfer for obtaining steady induction voltage in the pickup modules. Therefore, we should have a controller for output current regulation.

To do this, we control the SCR rectifier. We use the dual PI control method shown in Fig. 10.4. The controller consists of two PI control loops. The first loop, PI_1 (z), controls the output current, and the second loop, PI_2 (z), controls the firing angle of SCR to set the DC voltage for the desired output current.

We use two different control strategies: One is a PI controller for SCR, and the other is a PI controller for IGBT. The SCR controller is used for controlling steady-state output current, and the IGBT controller is used for controlling transient response when track segmentation or rapid load change is generated.

In the proposed IGBT feedback controller, the controller receives feedback from the output rms current with a moving average through a low pass filter, and a PI controller is used to control the phase difference between two arms. If the angle is 0°, then the output current goes to zero, and if the angle is 180°, then it gives maximum output current based on the DC-link voltage.

The PWM method is used for steady-state switching of the H-bridge inverter for one segment. Two PWM pulses that have 180° phase difference force the first half-bridge arm. To protect against arm short, dead band protection is also considered. The second arm forces 180° phase shifted PWM pulses of the first arm.

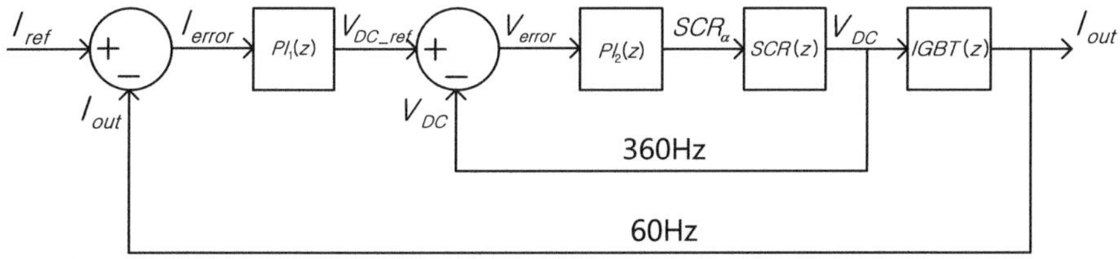

Fig. 10.4 Control algorithm block diagram

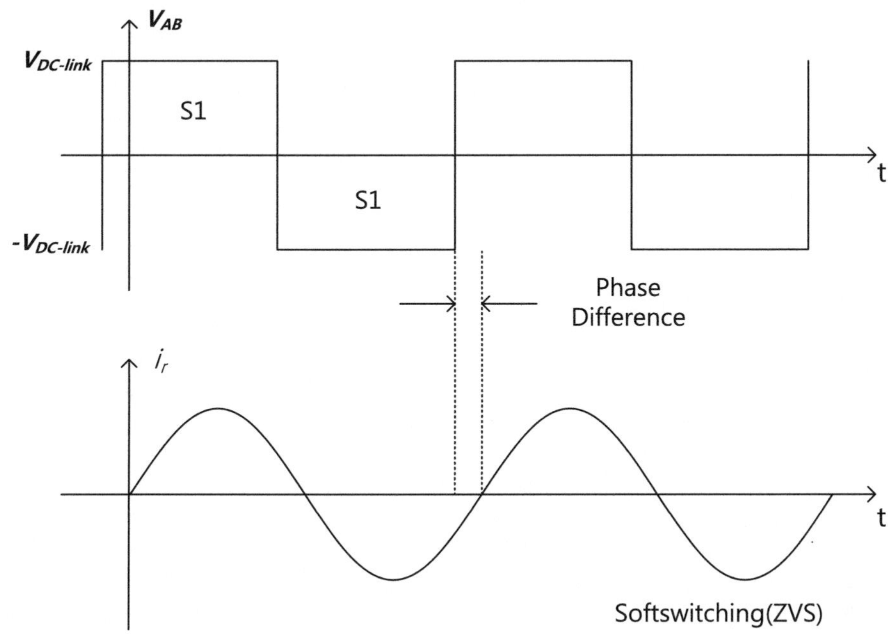

Fig. 10.5 Steady-state inverter switching

In a steady state, the inverter is operated as shown in Fig. 10.5. Because the primary road-embedded power cable module is inductively compensated, the current is slower than the voltage.

Generally, in the full-load states, the DC-link voltage is almost maximal in its safe operation region. Under this condition, if segmentation switching occurs from the full-load states to no-load states, extremely large current runs into the primary road-embedded power cable module if immediate output current control is not operated. This can cause severe damage not only to the track, but also to the IGBT itself. However, a slow response of the SCR controller cannot regulate the output current immediately when segmentation or a sudden load change happens. Because the DC-link capacitor has very large capacitance generally, it requires a lot of time to decrease the DC-link voltage below a safe level. However, we cannot stop switching until the DC voltage is decreased to satisfy the segmentation time limit.

The DSP controller in inverter controls the PCR and inverter part, which is composed of four parts: a sensing part, communication part, cooling part, and power control part.

There are several sensing parts such as sensors for current, voltage, and temperature. The current sensor is used to obtain output current level information of the inverter, and the voltage sensor is used for sensing the DC-link voltage. Based on both types of information, the system controller determines the action for the full-bridge inverter. The inverter has a CAN communication network, which is used to share status information between the inverter and vehicle recognition sensors. The cooling sensor operates when heat in the heat sink increases beyond the restricted value. It continuously monitors temperature inside the inverter and ensures that the inverter operates in the proper temperature range. The OLEV inverter with a two-segment controller and primary road-embedded power cable modules is designed.

10.5 Experimental Results

We implemented the inverter for the OLEV system as shown in Fig. 10.6. We used parallel IGBTs to ensure the current limit margin. With AC capacitors, we com pensate the resonance frequency as 19.5 kHz to each segment so that the road-embedded power cable module is seen to be inductance at 20.0 kHz.

We tested OLEV system in field environment and measured the total power transfer efficiency from the input of inverter to output of regulator, where we charged our battery with 30 kW and used an additional 70 kW static resistive load bank to obtain 100 kW output power. In this case, we obtained above 80% total power efficiency and 95% power efficiency for the inverter.

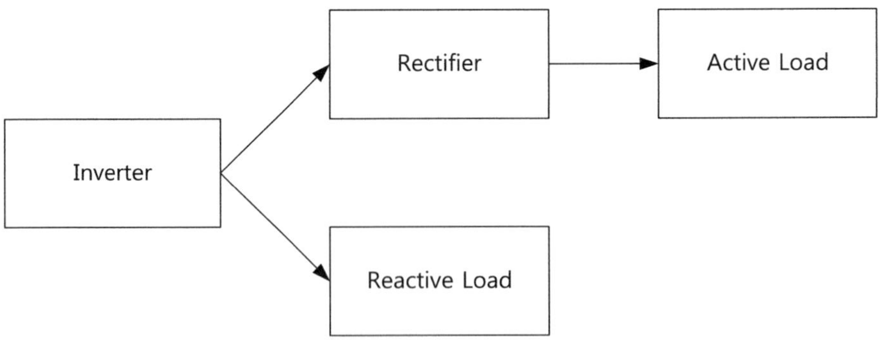

Fig. 10.6 Experimental setup block diagram for test of inverter

10.6 Conclusions

The power circuit and the control algorithm for the three-phase inverter for OLEV system were presented based on the given design constraints. Also, we explained the operation of the road-embedded power cable module with the conventional and revised segmentation type. Finally, a inverter segmentation control method was proposed and verified through experimental results.

References

1. California Partners for Advanced Transit and Highways (2009) http://www.path.berkeley.edu/. Accessed 31 Mar 2009
2. Huh J, Park EH, Joung GH, Rim CT (2009) High efficient inductive power supply system implemented for on line electric vehicles. Paper presented at the 2009 KPES, 105–110
3. Jung GH, Shin BS, Lee SH, Shin JG, Kim YS, Jeon SJ (2012) High efficient inductive power supply and pickup system for on-line electric bus. Paper presented at the 2012 IEEE Electric Vehicle Conference (IEVC), Greenville, SC, 1–5
4. Jeon SJ, Cho DH, Rim CT, Jeong GH (2011) Online electric vehicle system segment switching apparatus and a control method thereof, KS Patent WO 2011099781 A3
5. Shin JG, Shin SY, Kim YS, Ahn SY, Lee SH, Jung GH, Jeon SJ, Cho DH (2014) Design and implementation of shaped magnetic-resonance-based wireless power transfer system for roadway-powered moving electric vehicles. IEEE Trans Ind Electron J 61(3):1179–1192

Chapter 11
Installation of Road-Embedded Power Cable

Gu Ho Jung

Abstract The installation of road embedded power cable for the OLEV wireless power transfer system is presented. The installation requirements such as the width of road embedded power cable are given, followed by the description of the internal structure of the road embedded power supply system with high-frequency Litz wire and a ferrite core structure. The installation steps of the underground power supply system are then presented. The safety issues related to the tests carried out at the test bed of Seoul National Park are introduced with test standards and test results.

11.1 Introduction

This chapter covers four issues related to the installation of underground power supply system [1, 2]: first, the details of the internal structure of the road embedded power supply system consisting of high-frequency Litz wire and a ferrite core structure; second, the installation steps of the underground power supply system; third, the initial tests carried out at the Seoul National Park, comparing the test results with the test standards; fourth, the safety issues [3, 4].

11.2 Main Components for Installation

The main role of the underground power supply system is the creation of a magnetic field that generates power, which is transmitted to the vehicle. A power supply system consists of primary coils, ferrite cores, signal cables, steel reinforcements, EMF shielding, and common coils [5]. The primary coil is one-turn winding Litz

G.H. Jung (✉)
Wireless Power Transfer Research Center KAIST,
Korea Advanced Institute of Science and Technology, (KAIST),
303 Truth Hall 193 Munji-Ro, Yuseong-Gu, Daejeon 305-732, South Korea
e-mail: ghjung9595@kaist.ac.kr

N.P. Suh and D.H. Cho (eds.), *The On-line Electric Vehicle*,
DOI 10.1007/978-3-319-51183-2_11

wire with 5 mm^2. The coil has a shield that is made of FRP (Fiber-Reinforced Plastic). The detailed specification of the Litz wire is given in Table 11.1.

E-type ferrite core structure installed for the underground power supply system is shown in Fig. 11.1. The ferrite core has an E-type structure. The center core has

Table 11.1 Specifications of Litz wire coil for high-frequency application

Item			Specified values (Nom. 95 mm^2)
1	Insulation core	Conductor	– Nom. diameter: 0.4 mm
		Coating	– Urethane material coating – Thickness: Nom. 0.02 mm – Nom. diameter: 0.44 mm
2	Assembly	Stranding #1	7 strands of Insulation core – Binder taping (optional) – Nom. diameter: 1.3 mm
		Stranding #2	6 strands of insulation core – Binder taping (optional) – Nom. diameter: 4.0 mm
		Stranding #3	18 strands of insulation core – Binder taping (optional) – Nom. diameter: 21.0 mm
3	Insulation		– Heat-resistant cross-linked polyolefin (150 °C) – Thickness: 1.0 mm – Binder taping (Optional) – Cable diameter: Nom. 23 mm
4	Jacket		– Tracking resistant cross-linked polyolefin (150 °C) – Thickness: 1.5 mm – Cable diameter: Nom. 26 mm (±1.5 mm)

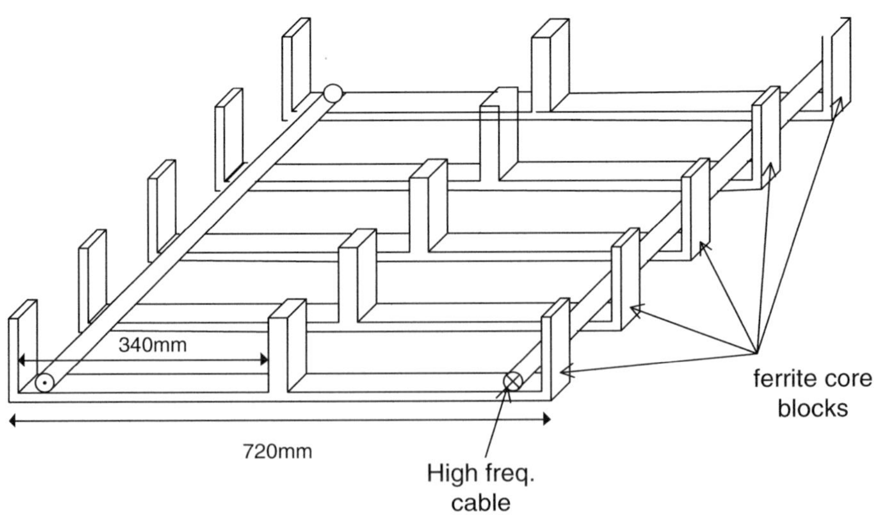

Fig. 11.1 Ferrite core structure for high-frequency application

twofold greater thickness than that of the other cores. For the ferrite core, PL-13 (Mn–Zn ferrite) is used. Copper mesh with 30-mm mesh is used as the EMF shielding material.

The power supply system requires reinforced mechanical structures to keep its original shape to produce consistent AC current from the road embedded power cables. Steel reinforcement of rebar can result in larger tensile stress of the power supply system. When installing the steel reinforcement, the position of the reinforcement is very important. This is because the shape of the magnetic field should not be affected by the steel reinforcement. The position of the steel reinforcement was selected as shown in Fig. 11.7.

11.3 Installation Requirements

As shown in Fig. 11.2, the road embedded power cable module consists of the primary coil cable and FRP cable tube, ferrite core, common coil and aluminum cable tube, signal cable, FRP cable tube, etc. The primary coil cables are located on the primary ferrite, and 20 kHz AC current flows through this cable. The cross-sectional area of this cable is above 95 mm^2, where an area of 1 mm^2 can tolerate electric current of 3A.

Fiber-reinforced plastic (FRP) tubes are used to protect the road embedded power cables. FRP is durable enough to tolerate the weight of vehicles and does not affect electrical or magnetical fields. The FRP tubes physically protect the road embedded power cable module cables or primary cables from underground pressure, heat, and water.

For safety, the embedded power cable module is segmented. For this purpose, the magnetic sensors are located at the entrance of the segment to detect the arrival

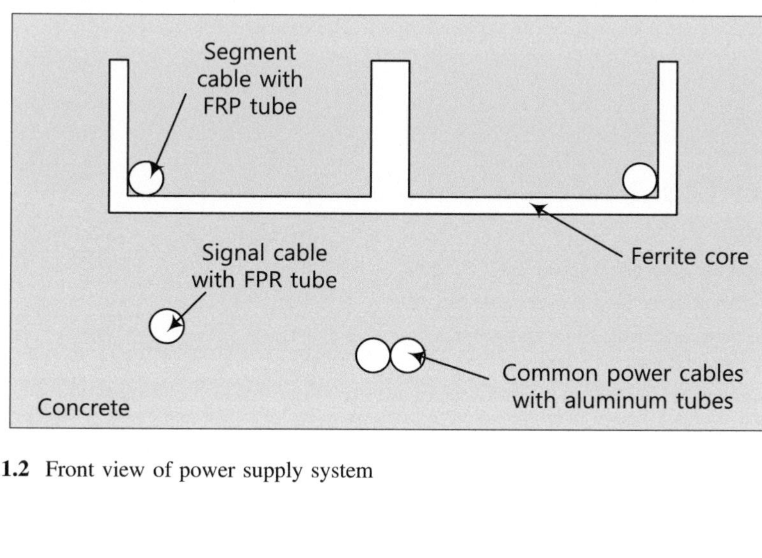

Fig. 11.2 Front view of power supply system

of electric vehicles. When the sensor of the first segment detects the coming of an electric vehicle, the first segment of the embedded power cable module segment is activated. When the second sensor senses the electric vehicle, the first segment turns off and the second segment turns on. The signal cable sends a signal from the sensor of the embedded power cable module to the inverter controller. FRP cable tube also protects this signal cable.

The common power cables or coils are used to connect each segment to the inverter. They must be shielded so as not to affect the segment power cables magnetically. Thus, aluminum tubes are used for the common power cables. Aluminum has low permeability and high conductivity, which are good shielding characteristics against magnetic flux.

Before we start designing the road embedded power supply system for the OLEV electric bus system, we have to determine several target installation requirements and specifications, as shown in Table 11.2. The maximum width and height of the road embedded power supply system should be below 800 and 350 mm; these values are determined by considering the internal and external structures and materials of the road.

Also, the height f_h and width f_w of the ferrite cores are 110 and 720 mm. Note that we can obtain higher output voltage of the pickup module as the height and width of the ferrite core become greater.

Figure 11.3 and Table 11.2 are the geometric parameters and values between each component for the actually installed road embedded power system.

11.4 Installation of Road Embedded Power Cable

The first step for the installation of the road embedded power supply system was to inspect the status of the existing road to identify and avoid potential risks and problems. The location for installation of the road embedded power supply system was selected in a region without underground objects such as water pipes or gas pipes.

Table 11.2 Installation requirements and parameters of power supply system

Parameter	Symbol	Value (mm)
Width of road embedded power cable module	Wp	800
Height of road embedded power cable module	Hp	350
Distance between ferrite cores and ground	T_0	60
Height of ferrite cores	f_h	110
Width of ferrite cores	f_w	720
Thickness of center ferrite cores	C_c	20
Thickness of side ferrite cores	C_s	10

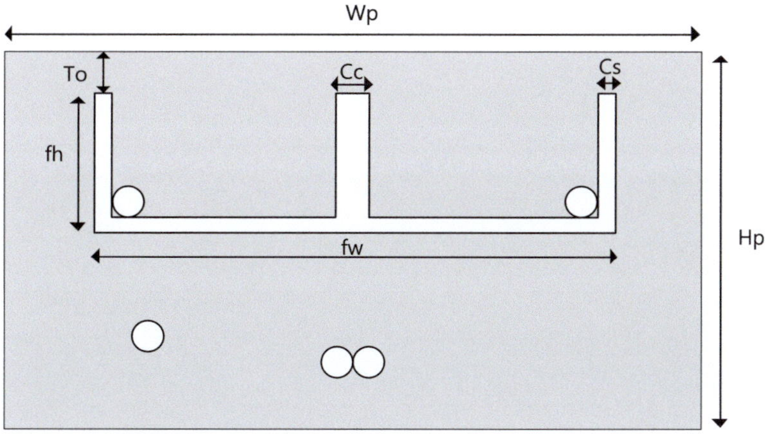

Fig. 11.3 Geometric parameters of power supply system

Fig. 11.4 Cutting work of selected road

After the location was selected, work to cut and excavate the existing road proceeded. As shown in Figs. 11.4, 11.5, 11.6, 11.7, and 11.8, it entailed excavating the existing road.

Fig. 11.5 Process for excavating existing road

After these steps were carried out, the coil and core were installed in the structure prepared to install the power supply system. That is, components for the underground road such as the supporting structure, ferrite core, and sensors were installed under the road. In this case, distances among each component and structures were maintained by considering mechanical and environmental solidities, EMF problems, and so on.

We used a shielding coil and net to reduce the EMF (electromagnetic field) level to below the Korean government guideline. EMF shielding work was carried out separately from the core and coil work because there was an unspecified region on-site in excess of the EMF safety level, unlike in the laboratory. As shown in the Fig. 11.8, EMF shielding work was carried out in these regions of a specially designed manholes and leading wire.

After the installation was completed, the final step of installation of the road embedded power supply system was finished through the process of pouring concrete. In this case, a vinyl sheet was situated between the existing road and the power rail to maintain moisture to securing mechanical strength of the concrete structure.

Fig. 11.6 Structure prepared to install power supply system

Fig. 11.7 Installation of the road embedded power supply system

Fig. 11.8 EMF shielding work at regions in excess of EMF safety level

11.5 Safety Issues

Safety tests for the power cable, the induced voltage between earth and metal, the mechanical strength of the embedded road module, the electromagnetic field (EMF), and the power quality were carried out at the test bed located in Seoul National Park. Below are the test items, test standards, and test results for the applied OLEV infrastructure and vehicle.

In the case of the road embedded power cable module, it should be manufactured to satisfy several related Korean standards including copper area and resistance, hot set, allowable current, and non-combustibility. From the test results, it was confirmed that this manufactured cable satisfies all of these standards (Table 11.3).

The induced voltage between earth and metal should be below 57.5 V, 100 mA at 575 Ω of human resistance. The test result was a maximum of 2.55 V at about 4.4 mA, which satisfies the relevant standard level (Table 11.4).

The measured EMF level should be below the Korean government regulation level, that is, 6.25 uT @ 20 kHz, and the test measurement method is based on the IEC 62110 method, where the EMF is measured at heights of 0.5, 1, and 1.5 m. As seen in the following table, the test result is below 6.1 uT, which satisfies the relevant EMF level (Table 11.5).

In the meantime, power quality at the AC main of inverter should satisfy V_{THD} and I_{TDD} levels based on Korean and IEEE standards, where V_{THD}(voltage harmonic distortion) and I_{TDD} (current total demand distortion) are defined as follows:

$$V_{\mathrm{THD}} = \frac{\left(V_2^2 + V_3^2 + \ldots + V_N^2\right)^{1/2}}{V_1} \times 100\% \qquad (11.1)$$

Table 11.3 Test items, test standards, and test results of power cable

Test item	Test standard		Test results	Comment
Copper area andresistance	KS C IEC 60228	Below 0.204 Ω/km (90 mm^2 cable)	– Copper area: 90 mm^2 – Copper resistance: 0.194 Ω/km,	
Hot set	KS C IEC 60502-1	– Below 175% at load – Below 15% at no load	• Insulation Material – 90% at load 2.5% at no load • Sheath – 79% at load, 12% at no load	
Allowable current	JASO D609 IEC	Above 200 A @ 40 °C	362.0 A (JASO), 366.8 A (IEC)	
Non-combustibilty	KS C IEC 60502-1 (IEC 60332-1)	Below 540 mm	118 mm	

Table 11.4 Test items, test standards, and test results of induced voltage between earth and metal

Test item	Test standard		Test results	Comment
Induced voltage between earth and metal	KS C IEC 60479 @ 20 kHz	Below 57.5 V, 100 mA @ 575 Ω of human resistance	Max. 2.55 V, about 4.4 mA	

Table 11.5 Test items, test standards, and test results of electro magnetic field (EMF)

Test item	Test standard		Test results
EMF	IEC62110*	$\leq$6.25 uT @ 20 kHz	Below 6.1 uT

**IEC62110* Magnetic field levels generated by AC power systems
Measurement procedures with regard to public exposure
Mean of EMF measured at heights of 0.5, 1, and 1.5 m

Table 11.6 Test items, test standards, and test results of power quality

	Test item	Test standard		Test results	Comment
1	V_{THD} (%)	Korean Standard IEEE Std. 519	Below 3% (Korean Standard), Below 5% (IEEE)	Before active filter installation: – 1.52% (R), 1.55% (S), 1.61% (T) After active filter installation: – 0.73% (R), 0.77% (S), 0.74% (T)	
2	I_{TDD} (%)	IEEE Std. 519	$\leq$8%	Before active filter installation: – 17.41% (R), 17.90% (S), 18.27% (T) After active filter installation: – 3.01% (R), 3.15% (S), 3.21% (T)	

where V_2, V_3, …, V_N are voltage harmonics, and V_1 is the fundamental voltage.

$$I_{\text{TDD}} = \frac{\left(I_2^2 + I_3^2 + \ldots + I_N^2\right)^{1/2}}{I_{\text{L}}} \times 100\% \tag{11.2}$$

where I_2, I_3, …, I_N are the current harmonics, and I_L is the max. load current.

Standards for V_{THD} are below 3% 5% in the case of the Korean and IEEE standards, respectively, where the test result was below 1% after applying an active power filter to AC mains, which satisfies the related standards. Also, the standard for I_{TDD} is below 8% in the case of IEEE standard 519, where the test result was about 3% after applying active power filter to the AC mains, which satisfies the related standards (Table 11.6).

11.6 Conclusion

We tested OLEV system in field environment as shown in Sect. 11.4 and confirmed that measured EMF is below ICINRP guideline. Also, the $V_{THD}$and I_{TDD} measured after active filter installation were below 3 and 8%, respectively.

Road embedded power supply system using a power cable with high-frequency Litz wire and a ferrite core was designed, installed, and tested with good results. The installation steps were designed and implemented. Safety tests were conducted, which satisfied international and Korean standards.

References

1. California Partners for Advanced Transit and Highways (2009) http://www.path.berkeley.edu/. Accessed 31 Mar 2009
2. Huh J, Park EH, Joung GH. Rim CT (2009) High efficient inductive power supply system implemented for on line electric vehicles. Paper presented at the 2009 KPES, 105–110
3. Jung GH, Shin BS, Lee SH, Shin JG, Kim YS, Jeon SJ (2012) High efficient inductive power supply and pickup system for on-line electric bus. Paper presented at the 2012 IEEE Electric Vehicle Conference (IEVC), Greenville, SC, 1–5
4. Jeon SJ, Cho DH, Rim CT, Jeong GH (2011) Online electric vehicle system segment switching apparatus and a control method thereof, KS Patent WO 2011099781 A3, 10
5. Shin JG, Shin S, Kim YS, Ahn SY, Lee SH, Jung GH, Jeon SJ, Cho DH (2014) Design and implementation of shaped magnetic-resonance-based wireless power transfer system for roadway-powered moving electric vehicles. IEEE Trans Ind Electron J 61(3):1179–1192

Chapter 12
Pickup and Rectifier

Uooyeol Yoon

Abstract The consequence of misalignment of the wireless power transfer system is described based on the simulation of the effect of the misalignment of the pickup module for various pickup coil models. The effects of misalignment for the OLEV system and Qualcomm Halo are compared. The conversion of AC voltage to DC voltage by the rectifier is explained using the waveforms. The design of the capacitor and diode is mathematically explained in terms of the capacitance, rated voltage, and ripple current capacity. In designing the diode, important variables are the rated voltage, rated current, and recovery time. Also the method of pickup design is described, including the core material selection, multi-winding, system packaging, and pickup shielding.

12.1 Introduction

This chapter presents the effect of misalignment on the power transfer efficiency of the wireless power transfer system. An electromagnetic simulation is used to explain the effect of the misalignment of the pickup module of the WPT system. The simulation model is shown, and the magnetic field distribution is identified under various cases of the pickup coil model. In addition, the DDQ model developed by Qualcomm Halo and OLEV system is introduced to solve the misalignment problem.

The design of the pickup unit of the OLEV system is also described. The induced voltage of the pickup unit is mathematically defined according to the current flowing in the power supply system. The topology of the wireless power transfer system, especially the series–series resonant circuit, is presented, and the equations for power transfer from the primary side and the quality factor are derived. The resonance frequency is an important factor in achieving maximum power transfer efficiency.

U. Yoon (✉)
Wireless Power Transfer Research Center KAIST, Korea Advanced Institute of Science and Technology (KAIST), 303 Truth Hall 193 Munji-Ro, Yuseong-Gu Daejeon 305-732, South Korea
e-mail: uyoon@kaist.ac.kr

N.P. Suh and D.H. Cho (eds.), *The On-line Electric Vehicle*,
DOI 10.1007/978-3-319-51183-2_12

The effect of capacitance at resonance is defined. The resonance frequency and power transfer efficiency as functions of the resonance capacitance are shown. Finally, core material selection, multi-winding, system packaging, and pickup shielding are briefly described.

This chapter also presents the conversion of AC voltage to DC voltage by the rectifier, using the waveforms of each part to aid in understanding. The design methods of the capacitor and diode are mathematically explained in terms of the capacitance, rated voltage, and ripple current capacity. In designing the diode, important variables are the rated voltage, rated current, and recovery time.

12.2 Tolerance for Left and Right Shift (Misalignment)

It is well known that the efficiency of power transfer of the OLEV system is a function of alignment during driving and stopping of the vehicle. In this section, the effect of the misalignment of the pickup module with respect to the underground power transfer module of the wireless power transfer system is presented. The solution presented is based on electromagnetic simulation.

The specifications used in the simulation of the pickup efficiency are given in Table 12.1.

Figure 12.1 shows the simulation model of the wireless power transfer system. The wireless power transfer system consists of the road-embedded power cable module, the ferrite core of the road-embedded power cable module, the pickup module, and the ferrite core of the pickup module. The other structures include stainless steel protector component and the chassis of the electric bus. The road-embedded power cable module part includes a coil and an E-shaped ferrite core. The coils of the pickup part can be divided into three parts, the left part, right part, and center part. The ferrite core in the pickup module has a meander-line shape across the loop coil. The ferrite material used here is PL-13 ferrite with a relative permeability of 3200. The utilized road-embedded power cable module and the pickup module model were simulated by Ansys Maxwell, which is based on the finite element method (FEM). The physical parameter for misalignment in this

Table 12.1 Assumed electric specifications of a pickup module for misalignment simulation

Specifications	Value
Power capacity of a pickup module	20 kW
Number of channel	2
Air gap (Core-to-core)	170 mm (230 mm)
EMF	Under 62.5 mG
Operating frequency	20 kHz
AC output voltage	400–450 V
AC output current	19.5–22 A
DC output voltage	400–450 V
DC output current	17.4–20 A

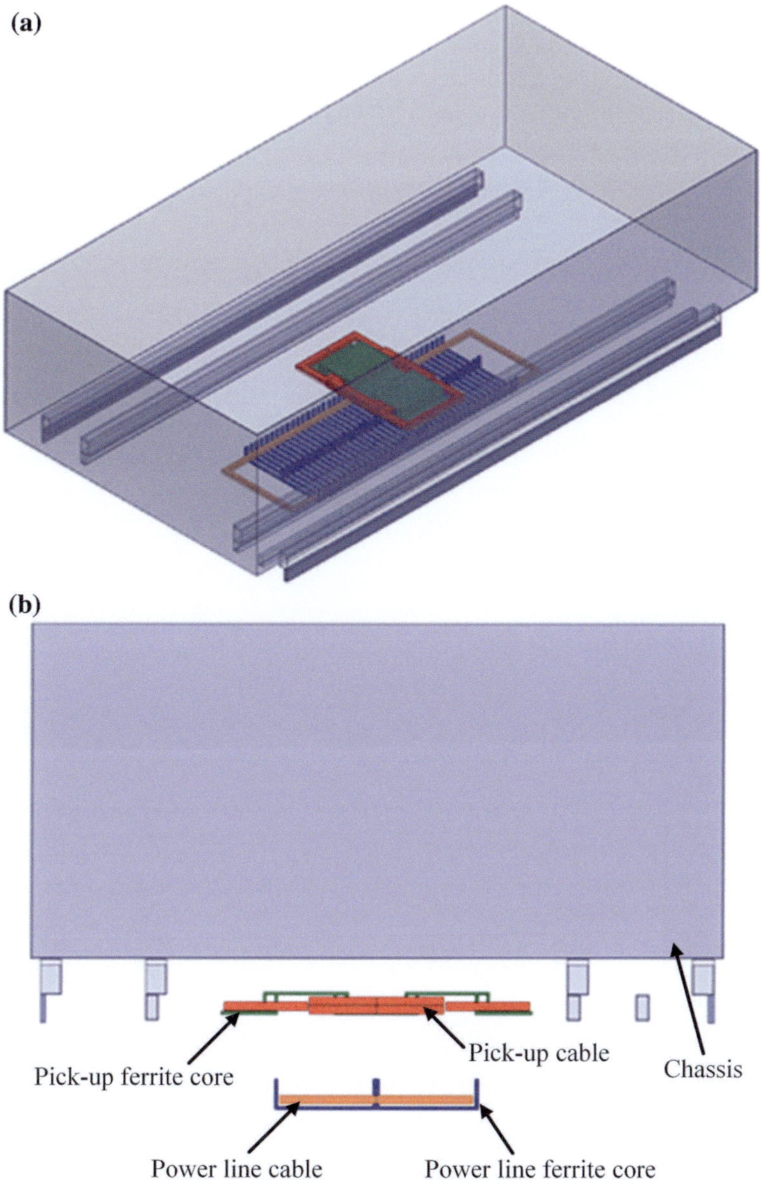

Fig. 12.1 Simulation model of OLEV system. **a** Perspective view, **b** Side view

chapter is the induced voltage of the pickup module, which depends on the magnetic field generated by excited AC current of the road-embedded power cable module. This voltage is an important parameter in evaluating the performance of wireless power transfer system. The electromagnetic field (EMF) value is then considered.

Based on the solution of the effect of misalignment in the WPT system, the number of turns for the three coils in the pickup module is optimized. The simulation results for these three coils are as follows: 28-64-28, 46-46-46, and 64-28-64.

Figure 12.2a, b shows the simulation results for the 28-64-28 case. Figure 12.3a, b is the simulation result for the 46-46-46 case. Figure 12.4a, b is the simulation results for the 64-28-64 case.

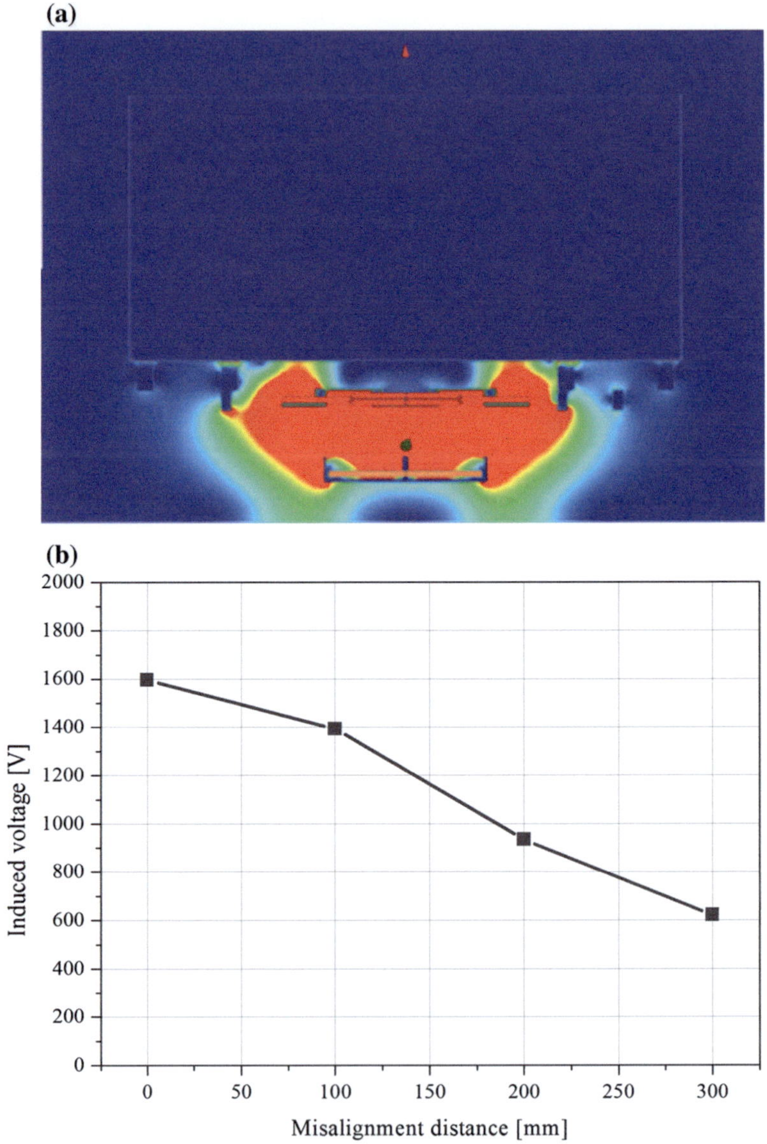

Fig. 12.2 Simulated results of **a** magnetic field distribution and **b** induced voltage in terms of misalignment for 28-64-28 case

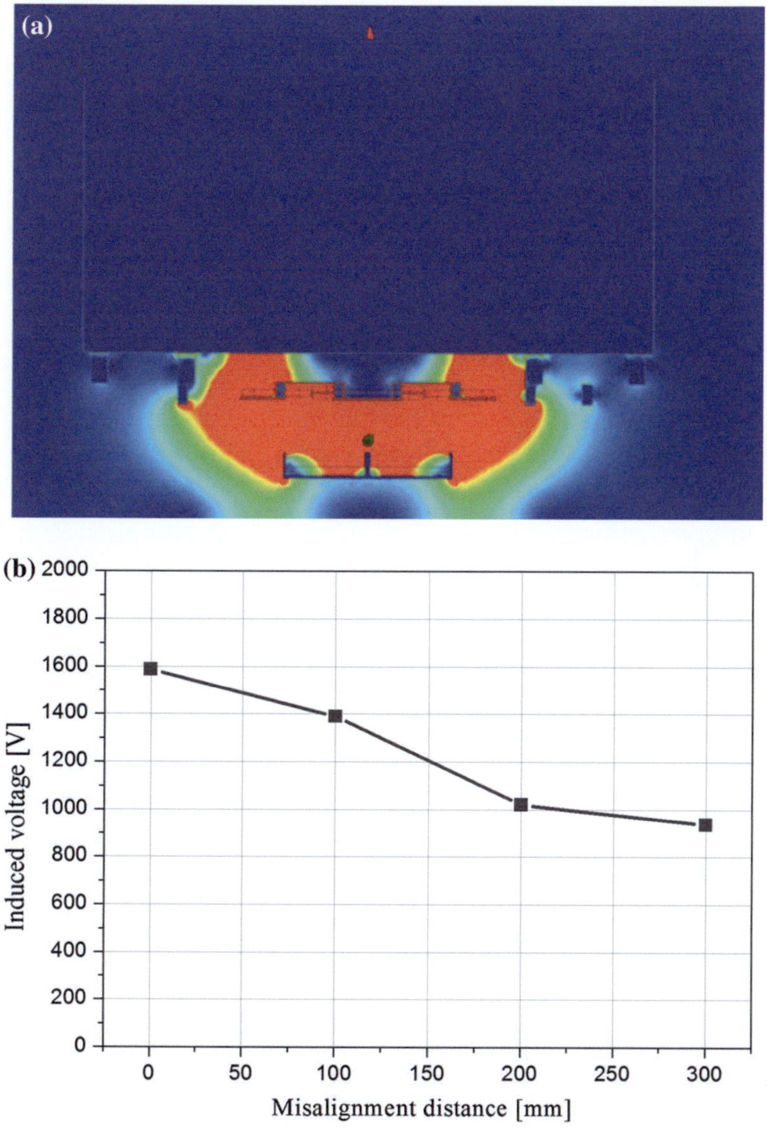

Fig. 12.3 Simulated results of **a** magnetic field distribution and **b** induced voltage in terms of misalignment for 46-46-46 case

Table 12.2 shows the simulation results of EMF for the three aforementioned cases. In summary, the case of 46-46-46 is the optimized structure in terms of performance, misalignment, and EMF. In this structure, there are two paths for the magnetic field in the wireless power transfer system. If the system makes the misalignment, there will be an unbalance between two magnetic field paths.

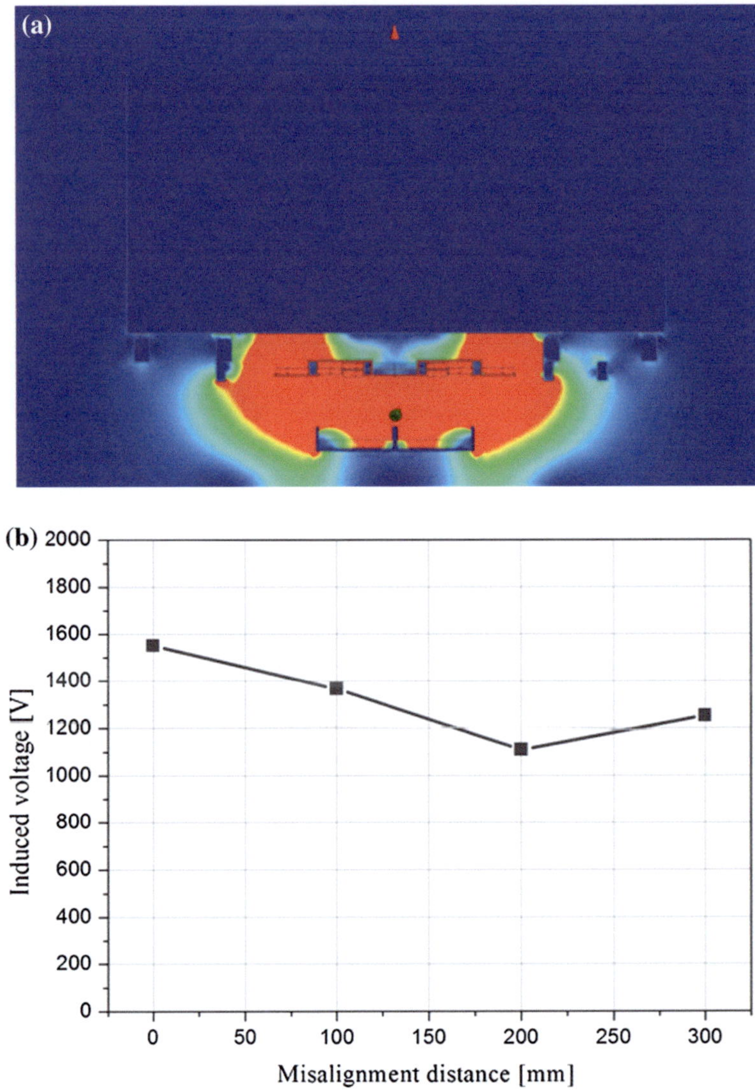

Fig. 12.4 Simulated results of **a** magnetic field distribution and **b** induced voltage in terms of misalignment for 64-28-64 case

Table 12.2 Simulation results of EMF for three cases

Arrangement of the number of turns	EMF (μT)
28-64-28	7.8
46-46-46	9.7
64-28-64	11.3

From the results above, adjusting the arrangement of the number of turns in pickup module will be a good solution for the misalignment problem that causes the unbalance of magnetic field paths in the pickup system.

12.2.1 Issues of Misalignment Tolerance in Qualcomm Halo

In case of Qualcomm Halo, Qualcomm Halo introduced a wireless power transform pad called 'DD' quadrature design [1]. The 'DD' pad is able to produce a single-side flux with a path height twice as large as the conventional circular pad. Therefore, the 'DD' pad has a higher power transfer efficiency and more robust performance under misalignment than that possible with circular pads for a similar material cost. A 0.31-m^2 'DD' enables 2 kW power transfer over an oval area measuring 540 mm × 800 mm with a 200-mm air gap.

12.3 Design of Pickup System

12.3.1 Current in the Power Supply System and Induced Voltage in the Pickup System

The magnetomotive force (MMF) at the pickup module voltage is defined as

$$V_{\mathrm{mmf}} = NI = \int \underline{H} \cdot dl \tag{12.1}$$

where N is the number of turns and I is the current in the road-embedded power cable module side. $\underline{H}$ is magnetic field vector. Equation (12.1) states that V_{mmf} can be represented as a curvilinear integral of the magnetic field. The reluctance, which is the resistance of the magnetic circuit, can be defined as

$$R_m = \frac{V_{\mathrm{mmf}}}{\Phi} = \frac{NI}{\Phi} \tag{12.2}$$

where Φ is the flux. It is assumed that the reluctance is unchanged. Then, (12.2) delineates that if the road-embedded power cable module current is decreased, the flux is decreased, and V_{mmf} is also decreased.

The road-embedded power cable module current, the number of turns, and the size of the pickup side.

(1) Topology of Contactless Power Transfer System

Series–series, series–parallel, parallel–series, and parallel–parallel resonance circuits are available. As shown in Fig. 12.5, the series–series resonance circuit has been applied to the OLEV system [2].

The lumped impedance Z_s is described as

$$Z_S = j\omega L_S + \frac{1}{j\omega C_S} + R_S \tag{12.3}$$

where ω is the resonant frequency, L_S is the inductance of the secondary side, C_S is the capacitance of the secondary side tuned for resonance, and R_S is the load.

Reflected impedance is described as

$$Z_R = \frac{\omega^2 M^2}{Z_S}; \quad M : \text{Mutual inductance} \tag{12.4}$$

where M is the mutual inductance.

The real part and imaginary part of (12.4) can be obtained as (12.5) and (12.6) by substituting (12.3) into (12.4).

$$\mathrm{Re}Z_R = \frac{\omega^2 C_S^2 M^2 R_S}{(\omega^2 C_S L_S - 1)^2 + \omega^2 C_S^2 R_S^2} \tag{12.5}$$

$$\mathrm{Im}Z_R = \frac{-\omega^3 C_S^2 M^2 (\omega^2 C_S L_S - 1)}{(\omega^2 C_S L_S - 1)^2 + \omega^2 C_S^2 R_S^2} \tag{12.6}$$

Then, power transferred from primary to secondary is

$$P = (\mathrm{Re}Z_R) I_P^2 \tag{12.7}$$

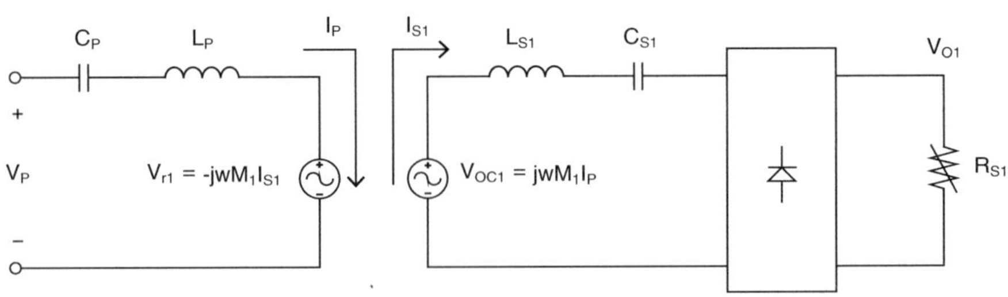

Fig. 12.5 Series–series resonance circuit

Also, power transferred from primary to secondary at the resonance frequency can be described as

$$P = (\mathrm{Re}Z_{R,\omega=\omega})I_P^2 \quad (12.8)$$

$$\mathrm{Re}Z_{R,\omega=\omega} = \frac{\omega_o^2 M^2}{R_S} \quad (12.9)$$

Here, the quality factors are

$$Q_P = \frac{L_P R_S}{\omega_o M^2}, \; Q_S = \frac{\omega_o L_S}{R_S} \quad (12.10)$$

where L_p is the inductance of the primary side and ω_o is the resonant frequency.

(2) Resonant frequency depending on the Resonant Capacitance (C_s)

The resonant frequency can be described as

$$f_{\mathrm{res}} = \frac{1}{2\pi\sqrt{LC}} \quad (12.11)$$

where L is the pickup inductance and C is the capacitance installed to resonate. A graph of the resonant frequency alteration depending on the capacitance is shown in Fig. 12.6a.

(3) Output power depending on the Resonant frequency

The output power is changed according to the resonant frequency, as shown in Fig. 12.6b. It is verified that the output power is maximal at 20 kHz.

As a result, it is important to tune the resonant frequency of the road-embedded power cable module side and the pickup side so as to obtain the highest power transfer efficiency.

(4) Consideration of core material selection

* To maximize the flux density (ferromagnetic)
* To minimize the core loss (hysteresis loss and eddy current loss)
* To minimize the cost

(5) Multi-Winding

The number of turns should be increased to maximize the flux density between the power supply system and pickup system. However, the increased number of turns causes a withstand voltage problem between pickup coils due to the large inductance.

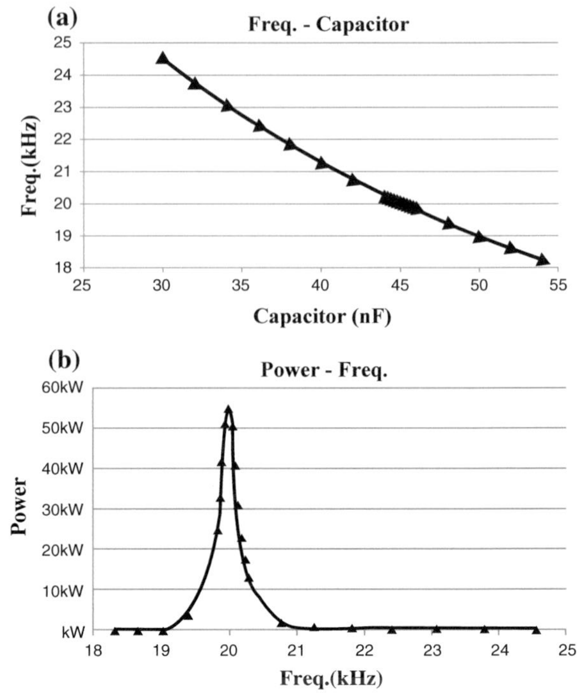

Fig. 12.6 **a** Resonant frequency versus capacitance. **b** Output power versus resonant frequency

To solve the problem, the multi-winding technique is applied to the pickup design. As shown in Fig. 12.7a, the multi-winding technique can reduce the inductance and the withstand voltage. Therefore, the high power transfer efficiency can be realized by removing the withstand voltage problem.

(6) System Packaging

The OLEV system is installed in the vehicle and operates regardless of the climatic conditions. Therefore, waterproof packaging is an essential requirement. As shown in Fig. 12.7b, the material should have robust strength mechanically against external impact and it should prevent deformation. Suitable dielectric and thermal properties are also required.

(7) Pickup shielding

In order to prevent outflow of the high-frequency EMF, which causes conductor heating and affects the human body, the pickup device has to be shielded. Figure 12.7c shows the pickup shielding.

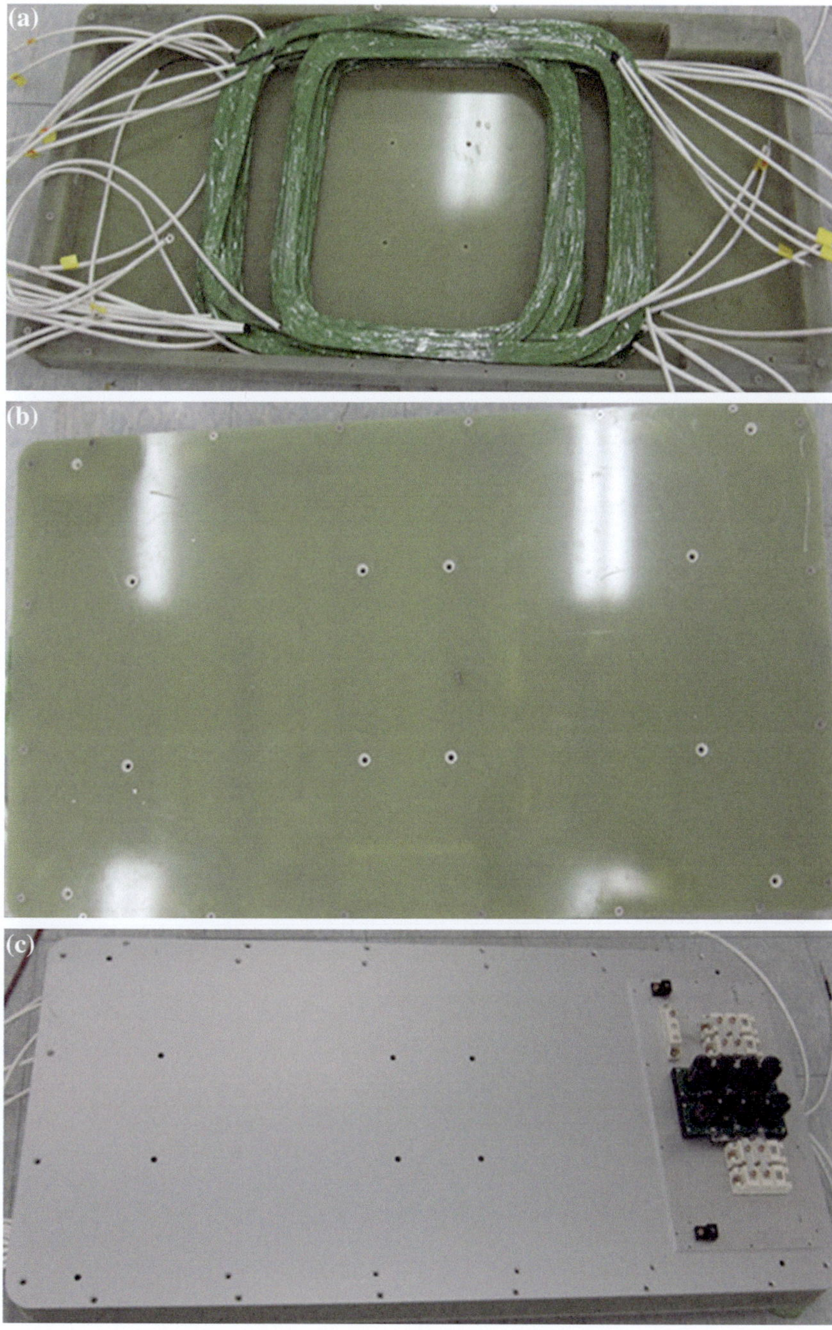

Fig. 12.7 **a** Multi-winding structure. **b** Waterproof packaging. **c** Pickup shielding

12.4 Design of Rectifier

12.4.1 Principle of the Single-Phase Diode Rectifier

The single-phase diode rectifier for the high-frequency application consists of four diodes, a dc-link capacitor filter, resistors, and the protection diode.

The four diodes of the rectifier carry out the transformation function from AC input voltage to DC output voltage. The dc-link capacitor filter can reduce the high-frequency ripple at the DC output voltage. Finally, the protection diode (called blocking diode) is required to prevent reverse current.

The operation principle can be divided into two steps depending on the sign of the AC input voltage, Vac.

In case of the positive sign, as shown in Fig. 12.8a, diodes D1 and D2 are conducted. In the case of the negative sign, D3 and D4 are conducted, as shown in Fig. 12.8b.

The input voltage of the rectifier is AC voltage. The AC voltage is transformed into full wave rectified voltage. The high-frequency component of the rectified voltage is reduced by the capacitor filter. As a result, the output voltage of the rectifier has only a DC component.

12.4.2 Design of the Diode Rectifier

(1) Capacitor design [3]

The capacitance should be designed with consideration of the rectified voltage ripple factor because the ripple factor depends on the capacitance. To simply analyze the rectified voltage ripple factor, an approximated rectified voltage can be utilized. The approximated rectified voltage waveform of the capacitor is presented in Fig. 12.9.

The output power, $P\Delta t$, can be expressed as

$$\frac{1}{2}C(V_{\text{peak}}^2 - V_{\min}^2) = P\Delta t \tag{12.12}$$

where V_{peak} is the maximum voltage, $V_{\min}$ is the minimum voltage, and C is the capacitance.

From (12.12), we can obtain following equation:

$$V_{\text{peak}}^2 - V_{\min}^2 = \frac{2P\Delta t}{C} \tag{12.13}$$

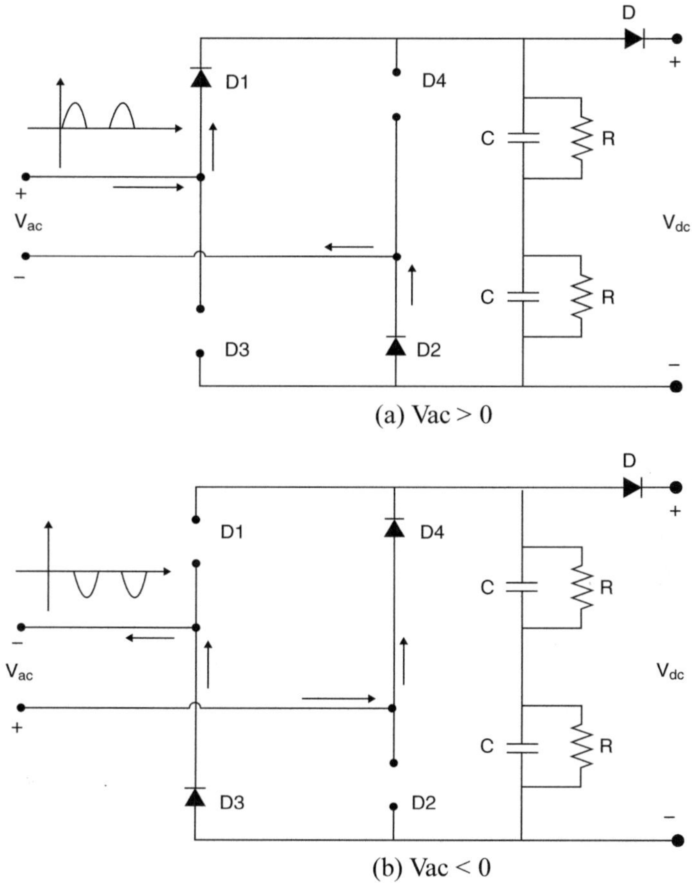

Fig. 12.8 Operation principle of diode rectifier

Fig. 12.9 Approximated rectified voltage

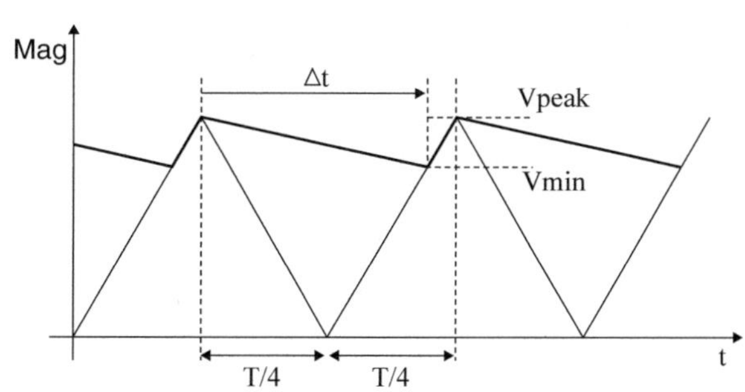

Thus,

$$V_{\text{peak}} - V_{\min} = \frac{2P\Delta t}{C(V_{\text{peak}} + V_{\min})} \tag{12.14}$$

Then, Δt can be expressed as

$$\Delta t = \frac{T}{4V_{\text{peak}}}(V_{\text{peak}} + V_{\min}) \tag{12.15}$$

Thus, by combining (12.14) with (12.15)

$$V_{\text{peak}} - V_{\min} = \frac{2P\frac{T}{4V_{\text{peak}}}(V_{\text{peak}} + V_{\min})}{C(V_{\text{peak}} + V_{\min})} = \frac{PT}{2CV_{\text{peak}}} \tag{12.16}$$

Here, (12.16) is the ripple component in the DC voltage. Therefore, the capacitance for satisfying the proposed ripple factor can be calculated as

$$C = \frac{P}{2fCV_{\text{peak}}(V_{\text{peak}} - V_{\min})} \tag{12.17}$$

The rated voltage of the capacitor should be designed larger than the DC voltage.

$$V_{c(\text{rated})} \geq V_{dc} \tag{12.18}$$

Finally, the ripple current capacity of the capacitor can be considered. The rated current of the capacitor should be larger than the ripple current.

$$i_{c(\text{rated})} \geq \Delta i_{dc} \tag{12.19}$$

(2) Diode design

First, the diode rated voltage should be larger than the input AC voltage magnitude.

$$V_{D(\text{rated})} > |V_{ac}| \tag{12.20}$$

Second, the diode rated current should be larger than the input AC current magnitude, which can be obtained by dividing the rated power into the input AC voltage.

$$I_{D(\text{rated})} > \frac{P_{\text{rated}}}{|V_{ac}|} = |I_{ac}| \tag{12.21}$$

Finally, the recovery time of the diode can be considered [4]. A diode having faster recovery characteristics than the AC input frequency is recommended.

12.5 Conclusions

The misalignment issue of the wireless power transfer system was presented. As the method to solve the misalignment problem, the DDQ model developed by Qualcomm Halo was introduced as well as the pickup module of OLEV system was presented and explained by an electromagnetic simulation under various cases of the pickup coil model. Also, the pickup design method was described and the effects of resonance frequency were also defined. To implement in OLEV system, core material selection, multi-winding, system packaging, and pickup shielding were briefly described. Moreover, the design method of high-frequency rectifier was mathematically presented. The consideration of each part selection was also described.

References

1. Budhia M, Boys JT, Covic GA, Huang C-Y (2013) Development of a single-sided flux magnetic coupler for electric vehicle IPT charging systems. IEEE Trans Ind Electron 60(1):318–328
2. Cho SY, Lee I.-O, Moon SH, Kim B-C, Kim KY (2013) Series-series compensated wireless power transfer at two different resonant frequencies. ECCE ASIA, pp. 1052–1058
3. Kwasinski A (2012) EE642L, Power electronics, Capacitor filtered diode bridge rectifier. Technical Document
4. Solid State Electronics Co. P. LTD(2014) Importance of incorporating fast recovery diodes in rectifier bridge. Technical Document

Chapter 13
Regulator

Gu Ho Jung

Abstract This chapter presents the regulator for the OLEV wireless power transfer system. The design requirements such as the rated output voltage and power of the regulator are given, followed by description of the operation principle, power circuit, and dual-loop control block for the regulator. Finally, a feed-forward control method to balance the input voltage of each boost converter module is described, followed by explanation of the voltage and current of the regulator connected to the battery using the related waveforms.

13.1 Introduction

This chapter introduces the operating principle, power circuits, and dual-loop control block for the regulator with the governing equations and waveforms. It also describes the role of the regulator in providing stable battery charging voltage and its dual-loop control structure. The operational principle and the related equations for the boost converter are also given. The rated output voltage and power for the regulator are then described, the values of which depend on the rated specifications of the motor driving inverter in the electric vehicle. This chapter also provides the power circuits and the dual-loop control loop for the regulator and pulse width modulation (PWM) signals. Pickup coils and rectifiers in a single pickup module are presented in a simplified model, and the mutual interference between pickup coils is discussed. It also presents a feed-forward control method to balance the input voltage of each boost converter module. Finally, the output voltages and current of the regulator connected to the battery are described based on the related waveforms [1, 2].

G.H. Jung (✉)
Wireless Power Transfer Research Center KAIST, Korea Advanced Institute of Science and Technology (KAIST), 303 Truth Hall 193 Munji-Ro, Yuseong-Gu Daejeon 305-732, South Korea
e-mail: ghjung9595@kaist.ac.kr

N.P. Suh and D.H. Cho (eds.), *The On-line Electric Vehicle*,
DOI 10.1007/978-3-319-51183-2_13

13.2 Overall Wireless Power Transfer System

As stated in Chap. 1, vehicles with internal combustion (IC) engines have been a major source of air pollution. In order to alleviate this issue associated with CO_2, battery-powered electric vehicles have been introduced. However, battery-powered electric vehicles (BPEVs) have their own shortcomings such as large volume occupied by the battery, long charging time, lack of charging infrastructure, and the cost of batteries.

The OLEV system described in this book provides an effective solution to many of the problems associated with vehicles with IC engines and BPEVs. As described in preceding chapters, OLEV receives electric power while in motion or stationary via Shaped Magnetic Field in Resonance (SMFIR) [3, 4]. The SMFIR system is a wireless power transfer system to electric vehicles that utilizes a magnetic field generated by power cable modules buried underground. The OLEV system is composed of an power inverter, a road embedded power cable module or track, a pickup system including a pickup module and a rectifier, a regulator, and a battery inside the electric vehicle. The inverter supplies high-frequency current to the rail, which is composed of high-frequency cable and a magnetic core. The inverter converts 60 Hz three-phase AC input power to 20 kHz output power. The inverter provides 200–400 A consistent current to road embedded power cable modules.

When the inverter supplies high-frequency current to the rail, the rail generates a high-frequency magnetic field through the road embedded power cable module with ferrite core, and then, the pickup module under the vehicle with the rectifier converts this magnetic field to low DC voltage. The regulator then outputs high DC voltage appropriate for a battery using this low DC voltage.

Stable output voltage cannot be acquired when a vehicle moves on a rough road. A regulator is thus necessary for the OLEV system to make the output voltage stable. We use a dual-loop proportional integral (PI) controller for this regulator as a DC–DC converter, which reduces the output voltage ripple. By adding a feed-forward control loop that controls the input current in the boost converter modules, the regulator maintains a balance between the rectified current of the first pickup coil and that of the second pickup coil. Also, the rectified voltages between the first pickup coil and the second pickup coil are balanced.

From the experimental results, we obtained 100 kW of output power for the regulator and it was possible to charge the battery on a segment in both transient and steady-state conditions. Finally, the developed regulator was converted to a commercialized version and it is operating inside the commercialized OLEV bus.

13.3 Design Requirements

The OLEV inverter converts 60 Hz three-phase current to a 20 kHz single-phase current. This current flows in the road embedded power cable module and transfers the power to the five pickup modules with 20 kHz magnetic flux frequency. The five rectifiers convert the AC current of the pickup modules to DC current. The regulator controls the current to boost the input voltage up to the reference voltage. In this case, the distance between the ferrite cores in the pickups and those in the road embedded power cable module is 260 mm. The reliable operation current range of each pickup module is below 25 A.

A boost converter is presented in Fig. 13.1. Equations (13.1), (13.2), (13.3), and (13.4) below delineate operation of the boost converter presented in Fig. 13.1 [5, 6].

$$v_{L(\text{average})} = v_i \bullet D + (v_i - v_o) \bullet (1 - D) \quad (0 \leq \mathrm{t} \leq \mathrm{T}) \tag{13.1}$$

$$v_{i(\text{average})} = \frac{1}{T}\int_0^T v_s \mathrm{d}t = \frac{1}{T}\int_{DT}^T v_o \mathrm{d}t = (1 - D)v_o \tag{13.2}$$

$$v_o = \frac{1}{(1 - D)} v_i \tag{13.3}$$

$$v_L = L\frac{\mathrm{d}i_L}{\mathrm{d}t} \tag{13.4}$$

D represents the duty ratio of switching device in these equations. Equation (13.1) gives the average voltage between the inductor, where the end-to-end voltage of the inductor becomes the input voltage *Vi* during turn-on time of duty *D* and the voltage difference *Vi* − *V*out during turn-off time of duty 1 − D.

In steady state, the average voltage of the inductor is zero per one cycle. Equation (13.2) is related to the average input voltage. The average input voltage is equal to the average voltage in the semiconductor device of the insulated-gate bipolar transistor (IGBT). The output voltage is expressed by Eq. (13.3). From these equations, we can see that the output voltage of the boost converter is controlled by duty *D*. The change in the inductor current is expressed by Eq. (13.4).

The rated output voltage and power of the regulator are 620 V and 100 kW, respectively, and they depend on the rated specifications of the motor driving inverter in the electric vehicle.

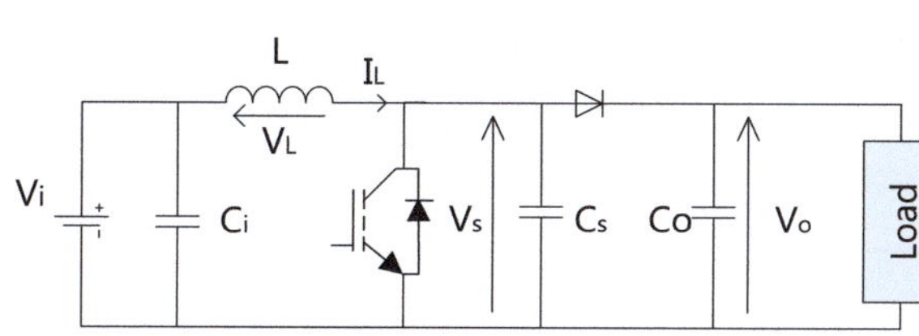

Fig. 13.1 Boost converter

13.4 Design of Regulator

The pickup modules and regulator in the OLEV system are shown in Fig. 13.2. They are composed of 5 pickup modules with 10 pickup coils, 10 rectifiers, and 10 boost-type regulators. Two pickup coils in each pickup module are series-connected with compensating capacitors, two rectifiers, and two boost converter modules. The rectifiers convert the AC current to DC current, and the boost converter module boosts the output voltage of each rectifier up to the rated voltage appropriate for the battery. The regulator has ten boost converter modules with IGBT switching devices. The gates of the IGBTs are switched by PWM signals, and a DSP board controls the duty ratios of the PWM signals. From this process, we can obtain the rated output voltage for the battery.

Figure 13.3 is the control block diagram with the dual control loop for the OLEV regulator, where the duty D is controlled by the PI controller of the inner second current loop and the outer first loop controls the output voltage of the regulator as the reference voltage. The switching frequency of IGBT devices is

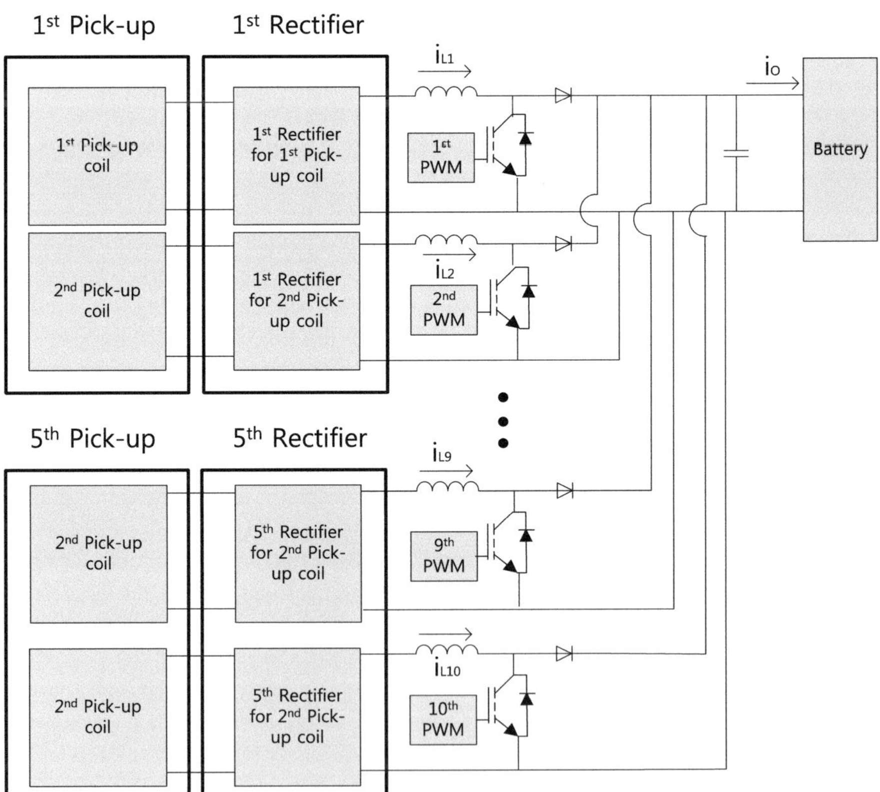

Fig. 13.2 Pickup modules and regulator in OLEV system

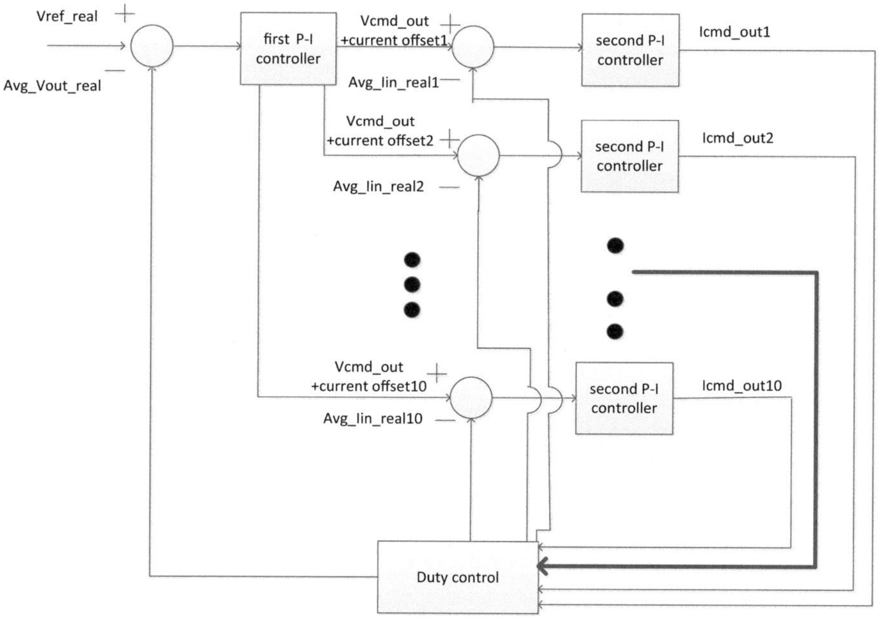

Fig. 13.3 Control block diagram of OLEV regulator

Fig. 13.4 PWM signals in each boot converter module corresponding to interrupt counter in DSP controller

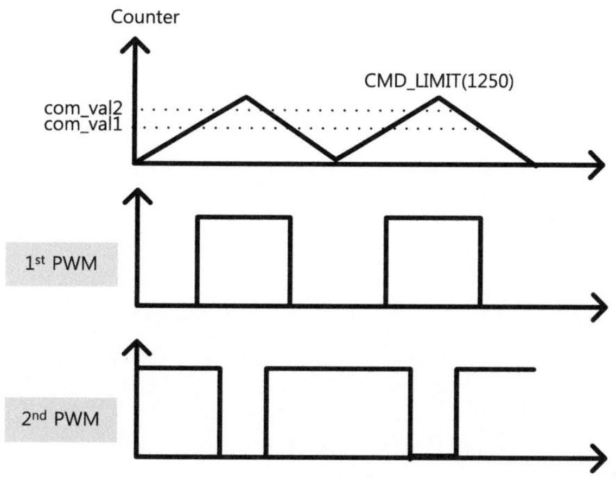

15 kHz. The sampling frequency of the inner current control loop and the outer voltage loop is 7.5 kHz and 750 Hz, respectively. The PWM duty of each boost converter module is determined and calculated by the current command Icmd out, as shown in Fig. 13.3.

Figure 13.4 is the PWM signal in the boost converter module corresponding to the interrupt counter in the DSP controller. The counter clock is operated at

37.5 MHz with an up–down counter according to the DSP register configuration. This clock counts up and down between 0 and 1250, and the period of the PWM signal becomes 15 kHz. The first PWM signal is 0 V while the counter is lower than com_val1 and becomes 3.3 V if the counter is greater than com_val1. Meanwhile, the second PWM signal is 3.3 V while the counter is lower than com_val2 and becomes 0 V if the counter is greater than com_val1.

These values of com_val1 and com_val2 are calculated by Eqs. (13.9) and (13.10). Also, the corresponding duty is calculated by Eqs. (13.11) and (13.12). In this case, the phase difference of both PWM signals becomes 180°.

$$\text{current\_offset1} = (\text{avg\_Vin1} - \text{avg\_Vin2}) * \text{scaling factor} \tag{13.5}$$

$$\text{current\_offset2} = (\text{avg\_Vin2} - \text{avg\_Vin1}) * \text{scaling factor} \tag{13.6}$$

$$\text{reference current1} = \text{Vcmd\_out} + \text{current\_offset1} \tag{13.7}$$

$$\text{reference current2} = \text{Vcmd\_out} + \text{current\_offset2} \tag{13.8}$$

$$\text{com\_val1} = \text{CMD\_LIMIT} - \text{Icmd\_out1} \tag{13.9}$$

$$\text{com\_val2} = \text{Icmd\_out2} \tag{13.10}$$

$$\text{duty\_percent1} = (\text{Icmd\_out1}/\text{CMD\_LIMIT}) * 100 \tag{13.11}$$

$$\text{duty\_percent2} = (\text{Icmd\_out2}/\text{CMD\_LIMIT}) * 100 \tag{13.12}$$

$$w_{rp} = \frac{w_r}{\sqrt{1+k}} \tag{13.13}$$

$$V_{1-\text{out}} = V_1 - i_{L_1}(R_{C_1} + R_{S_1}) - L_1\frac{\mathrm{d}i_{L_1}}{\mathrm{d}t} - M_{12}\frac{\mathrm{d}i_{L_2}}{\mathrm{d}t} - \frac{1}{C_1}\int i_{L_1}\mathrm{d}t - e_1 \tag{13.14}$$

$$V_{2-\text{out}} = V_2 - i_{L2}(R_{C_2} + R_{S_2}) - L_2\frac{\mathrm{d}i_{L_2}}{\mathrm{d}t} - M_{12}\frac{\mathrm{d}i_{L_1}}{\mathrm{d}t} - \frac{1}{C_2}\int i_{L_2}\mathrm{d}t - e_2 \tag{13.15}$$

13.5 Pickup Interface

According to Eq. (13.13), if each pickup coil's resonant frequency is adjusted between 25 and 26 kHz, one pickup's resonant frequency will be about 20 kHz, where w_{rp} is the resonant angular frequency of the pickup, w_{r} is the resonant frequency of each pickup coil, and k is a coupling coefficient. We cannot adjust the resonance of the five pickups correctly at 20 kHz because the inductances and capacitances in the pickup coils vary slightly depending on the characteristics of the cores and each pickup coil is affected by the operation of the others. Thus, the output voltages and the output currents in the ten pickup coils become different from each other.

If the inner second PI controllers make the input currents of the ten boost converter modules be the same as each other, the difference in rectified voltages between two pickup coils increases because the resonant frequencies of the two pickup coils are somewhat different.

As the output power increases, the input voltage of the boost converter modules decreases because of core loss, coil loss, and magnetic interference caused by the other pickup module. Also, as the duty ratio increases, the efficiency of the boost converter module declines.

We can solve the problem of the other pickups' interference to some degree through a resonant frequency fine-tuning technique. A single pickup module is constructed by two pickup coils, where both pickup coils are located in close proximity. The impedance caused by mutual inductance in these two pickup coils is affected more by operation of the other coil. Therefore, we must maintain a balance between the first pickup coil's rectified current and the second pickup coil's rectified current to obtain the balanced rectified voltage between the first pickup coil and the second pickup coil. We thereby increased the maximum output power and the efficiency of the OLEV regulator.

Figure 13.5 shows a simplified model of two pickup coils and rectifiers in a single pickup module, where R_{C_1}, R_{C_2} and R_{S_1}, R_{S_2} are resistances representing core losses and winding losses, respectively. Also, L_1 and L_2 are self-inductances and M_{12} is the mutual inductance between both coils. V_1, V_2, V_1_out, and V_2_out are the voltages induced in one pickup created from the road embedded power cable module and the input voltages of the boost converter modules, respectively. C_1 is a compensating capacitance for L_1 and M_{12}, and C_2 is a compensating capacitance for L_2 and M_{12}. R_{L_1} and R_{L_2} are load resistances.

Equations (13.14) and (13.15) are expressed in the time domain for Fig. 13.5, where e_1 and e_2 are interferences of the other pickups. We assume that e_1 is nearly equal to e_2 because the pickup coils in Fig. 13.7 are in one pickup. According to Eqs. (13.14) and (13.15), when V_1 and i_{L_1} are equal to V_2 and i_{L_2}, respectively, if the impedance of R_{C_1}, R_{S_1}, L_1, and M_{12} is lower than the impedance of R_{C_2}, R_{S_2}, L_2,

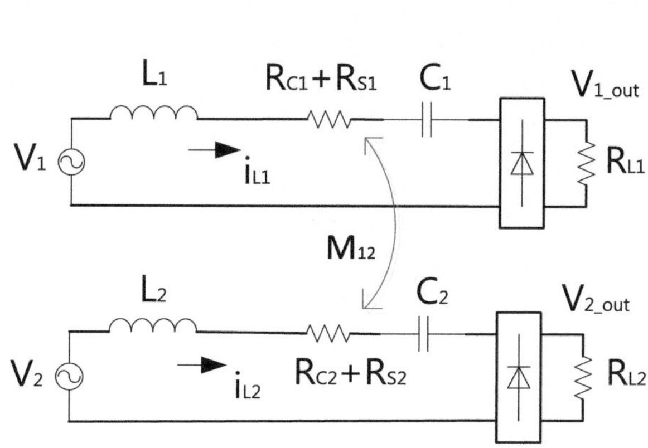

Fig. 13.5 Simplified model of two pickup coils and rectifiers in a single pickup module

and M_{12}, V_1_out is larger than V_2_out. In this situation, as i_{L_1} and i_{L_2} increase, if we control i_{L_1} and i_{L_2} to be equal to each other, the difference between V_1_out and V_2_out is increased. Therefore, if the second PI controllers make the input currents of the ten boost converter modules the same as each other, the difference between the two pickup coils' rectified voltages increases. In the previous situation, if we try to increase i_{L_1} and decrease i_{L_2}, because of decreased i_{L_2}, in L_1, the effect of M_{12} is decreased. However, according to increased i_{L_1}, the voltage drop of V_1 increases. In L_2, the effect of M_{12} is increased. However, according to decreased i_{L_2}, the voltage drop of V_2 decreases. Thus, V_1_out decreases and V_2_out increases. However, if we use Eqs. (13.5) and (13.6), as the difference between the two pickup coils' resonant frequencies increases, the difference between V_1_out and V_2_out rises slightly. Also, the difference between i_{L_1} and i_{L_2} slightly increases. As a result, we can balance the voltages of the two boost converter modules.

In short, our control method is explained as follows. If the input voltage of the first boost converter module is larger than the input voltage of the second boost converter module, the first pickup coil can absorb more power than the second pickup coil can. For this reason, we can try to make the current offset of the first boost converter module be a positive value and the current offset of the second boost converter module be a negative value. Thus, the reference current of the first boost converter module is larger than the reference current of the second boost converter module. The voltage and current balance of the input stage voltages can be maintained through this operation.

13.6 Battery Interface

The duty ratios of the PWM signals are controlled through the PI controllers. Using interleaving of the PWM signals, the ripples of the output current and the output voltage are reduced.

Figure 13.6 shows the input currents of two boost converter modules connected to one rectifier when the duty ratio is 0.5. The phase difference of the modules is 180°. Figure 13.7 shows the input currents of ten boost converter modules when the duty ratio is 0.5. Between the first boost converter module and the third boost converter module, the phase difference is 72°. Also, the phase difference between the ninth boost converter module and the first boost converter module is 72°. Therefore, the phase difference of each boost converter module is 36°. As a result, the DC output current and the DC output voltage ripples are zero theoretically.

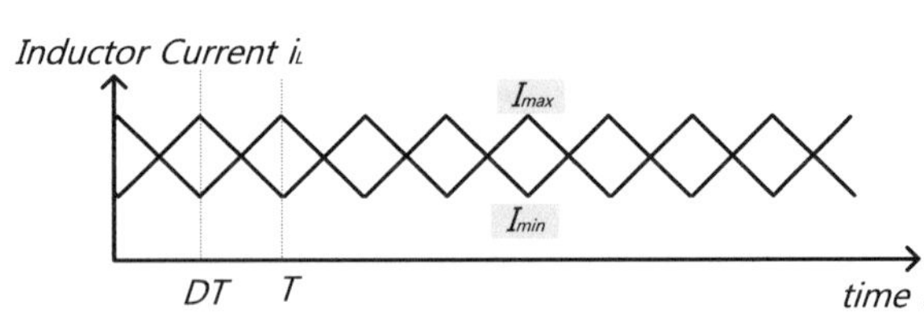

Fig. 13.6 Input currents of two boost converter modules connected to rectifier

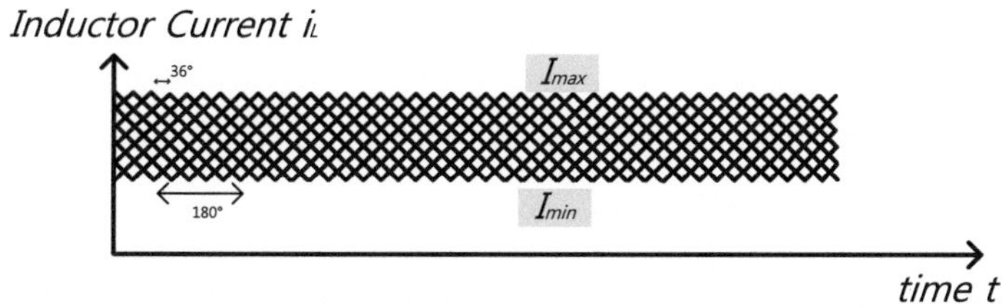

Fig. 13.7 Input currents of 10 boost converter modules

Fig. 13.8 Output voltage ripple of 10 boost converter modules with respect to duty ratio according to phases

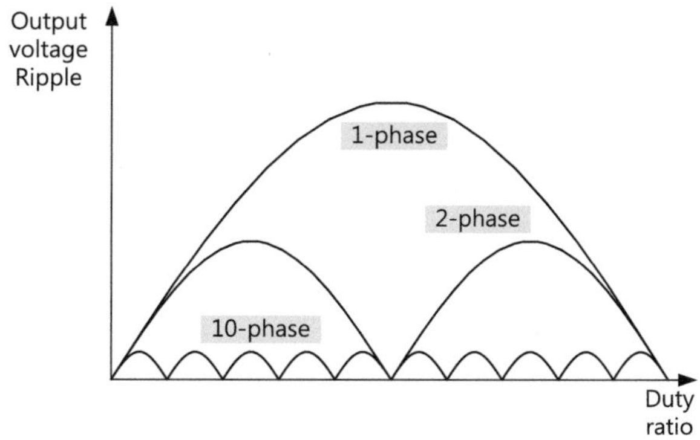

Figure 13.8 shows the output voltage ripple of ten boost converter modules as s function of the duty ratios according to phases. The output current ripple is proportional to the output voltage ripple.

The battery is connected to the output line of the regulator, and the battery can receive power with 600–650 V stable voltage from the regulator.

13.7 Conclusions

This chapter introduced the operation principle, power circuits, and dual control loop for regulator with the related equations and waveforms. Also, a feed-forward control method to balance the input voltage of each boost converter was discussed. Finally, the output voltages and current of the regulator connected to the battery were explained through the related waveforms. We tested OLEV system in field environment and measured the total power transfer efficiency from the input of the inverter to output of the regulator, where we charged our battery with 30 kW and used an additional 70 kW static resistive load bank to obtain 100 kW output power. In this case, we obtained above 80% total power efficiency and above 96% power efficiency for regulator.

References

1. Kim YS, Son YD, Shin SY, Shin JG, Song BY, Lee SH, Jung GH, Jeon SJ (2012) Design of a regulator for multi-pick-up systems through using current offsets. In: 2012 IEEE electric vehicle conference (IEVC), Greenville, SC, pp. 1–6
2. Jung GH, Shin BS, Lee SH, Shin JG, Kim YS, Jeon SJ (2012) High efficient inductive power supply and pickup system for on-line electric bus. In: IEEE Electric Vehicle Conference (IEVC), Greenville, SC, pp. 1–5
3. Shin JG, Song BY, Lee SH, Shin SY, Lee SH, Kim YS, Jeon SJ (2012) Contactless power transfer systems for on-line electric vehicle (OLEV). In: IEEE electric vehicle conference (IEVC), Greenville, SC, pp. 1–4
4. Shin JG, Shin SY, Kim YS, Ahn SY, Lee SH, Jung GH, Jeon S-J, Cho DH (2014) Design and implementation of shaped magnetic-resonance-based wireless power transfer system for roadway-powered moving electric vehicles. IEEE Trans. Ind Electron J 61(3):1179–1192
5. Mohan N, Undeland TM, Robbins WP (2003) Power electronics. Wiley
6. Madawala UK, Thrimawithana DJ (2011) A bidirectional inductive power interface for electric vehicles in V2G systems. IEEE Trans Ind Electron J 58(10):4789–4796

Chapter 14
Shielding of Magnetic Field

Seungyoung Ahn

Abstract As the number and magnitude of radiated electromagnetic fields from electronic devices increase, the radiation of electromagnetic field must be carefully considered because it can cause physiological harm on human body. It may also lead to malfunctioning of other electronic systems. Since OLEV involves power of over kilowatts, the transmitted and received power and the electromagnetic compatibility problem require careful considerations. Various shielding methods are presented, including passive and active shielding. A reactive shield is also introduced and explained as a possible solution for reduction in magnetic field in OLEV applications.

14.1 Introduction

This chapter discusses the radiation of the electromagnetic field of the wireless power transfer system and the means of shielding it in order to minimize any adverse effects of the electromagnetic field on human physiology and on other electronic devices and systems. All electric or electronic systems must be tested for electromagnetic compatibility (EMC) for commercialization. Electromagnetic interference (EMI) to other electronic systems and the effect of electromagnetic field (EMF), i.e. radiation, on human must be less than those established international standards. In a wireless power transfer system, these issues must be addressed since the system generates a strong electromagnetic field.

Traditionally, there are many ways to reduce unwanted electromagnetic fields, such as shielding, balancing, grounding, filtering. In many electronic systems, the integrated circuits, packages, and printed circuit boards are designed to have minimal electromagnetic radiation. In the case of wireless power transfer systems

S. Ahn (✉)
The Cho Chun Shik Graduate School for Green Transportation KAIST,
Korea Advanced Institute of Science and Technology (KAIST),
291 Daehak-ro, Yuseong-gu Daejeon 3414, South Korea
e-mail: sahn@kaist.ac.kr

N.P. Suh and D.H. Cho (eds.), *The On-line Electric Vehicle*,
DOI 10.1007/978-3-319-51183-2_14

used in SMFIR for OLEV, it is difficult to eliminate the leakage electromagnetic field completely because the system generates a large field to transfer sufficient power wirelessly to vehicles.

In this chapter, the need to reduce unintended electromagnetic radiation is explained and various shielding methods to reduce the leakage electromagnetic field are presented. In contrast to a general electronic system, the wireless power transfer system for OLEV uses a low-frequency (i.e. tens of kHz) magnetic field. Hence, shielding of low-frequency magnetic fields is presented in this chapter. Both passive and active shielding methods for EMF reduction in the wireless power transfer system are presented. Then, the specific shielding methods applied to the OLEV system are introduced and explained.

14.2 Need for Shielding of Electromagnetic Field

The generation of a strong magnetic field from transmitting and receiving coils causes two potential problems from the regulatory perspective: first, the effect of generated magnetic field on human body and second damage to electronic circuits or logic systems. If the frequency of the magnetic field is very low and the wavelength is much larger than the distance from the magnetic field source, the wave impedance will be very low, as shown in Fig. 14.1. In Fig. 14.1, wave impedances with different frequencies are shown as a function of the distance from the source. Up to 10 MHz, the electromagnetic field is considered as a near field, and the magnetic field decreases rapidly [1, 2]. Therefore, the strength of the magnetic field itself must be measured in the vicinity of the magnetic field source [3]. Each country regulates the allowable strength of the EMF. In the following sections, the methodologies of shielding the electromagnetic field are discussed.

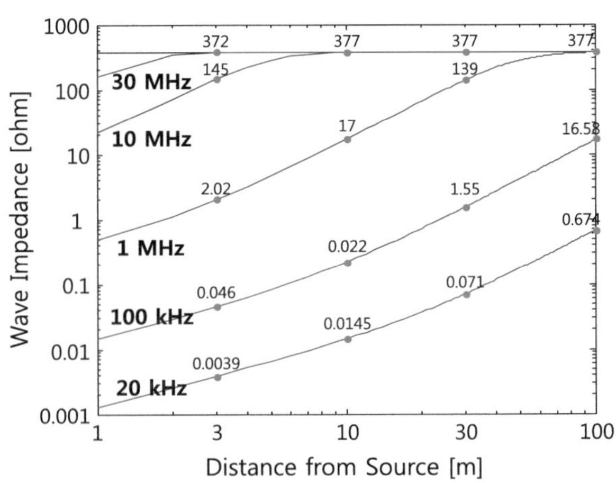

Fig. 14.1 Calculated wave impedance of loop antenna with respect to distance from source for different frequencies [2]

The electromagnetic interference at a long distance must be determined to prevent any disturbances to the communication system. As shown in Fig. 14.1, the wave impedance, which is the ratio of the electric field over the magnetic field, increases when the distance increases. When the distance increases and the impedance reaches close to 377 Ω, the field is considered a far field. When the frequency exceeds 30 MHz, the far-field condition is assumed in the current standard measurement of EMI. For the OLEV, the EMI involves a far-field situation, and therefore, the validity of this assumption in the current EMI test should be discussed.

To reduce the leakage magnetic field from the wireless power transfer coils, one or a combination of various shielding methods can be applied according to the frequency, coil type and shape, distance, and regulation to be satisfied. The conductive shield in Fig. 14.2a is the most common method for low-frequency reduction [4]. However, the shielding effectiveness is limited because the eddy current is not controlled to be optimal. Magnetic shielding is also a common method to reduce radiation, as well as to enhance the magnetic field, because the magnetic material can change the path of the leakage magnetic field, as shown in Fig. 14.2b. The main disadvantages of the magnetic shield are heavy weight and limited shielding effectiveness. Furthermore, it is difficult to apply to vehicular wireless power transfer systems as the passive shield cannot be attached under the

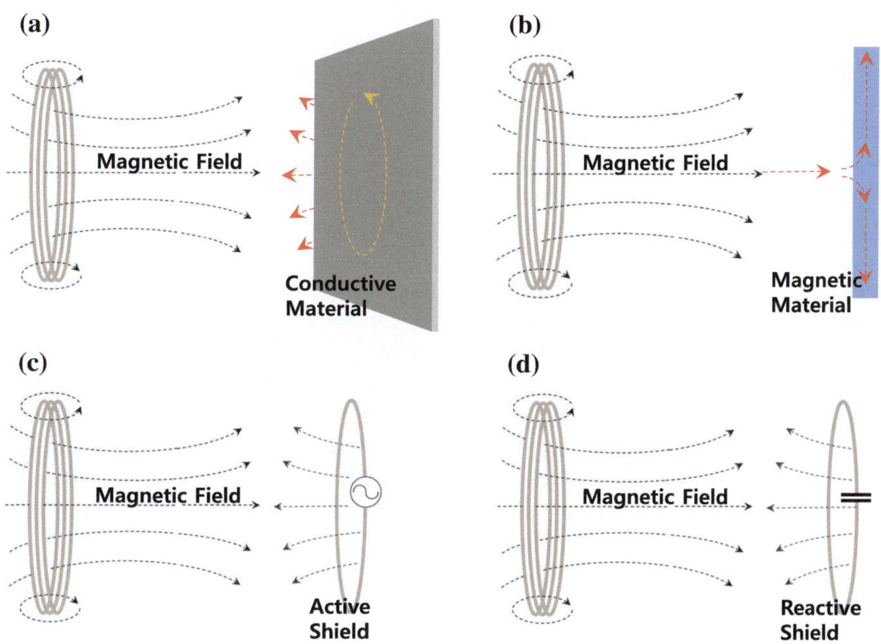

Fig. 14.2 Various shielding methods, using **a** conductive shield, **b** magnetic shield, **c** active shield, and **d** reactive shield

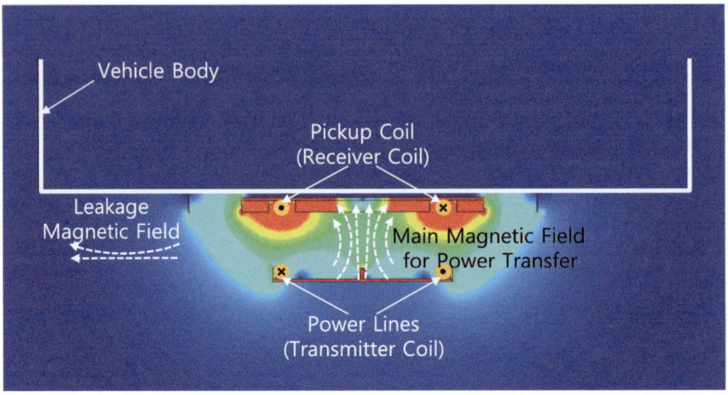

Fig. 14.3 Distribution of magnetic field for OLEV

car while it is moving. The active shield in Fig. 14.3c mitigates the limitation of effectiveness of the conductive shield by controlling the magnitude and phase of the cancelling magnetic field. As a cancelling magnetic field can be generated from a coil positioned not on the line of sight between the wireless power transfer system coil and measurement position, we can have more freedom to place the active shield and can increase the applicability to vehicle systems. However, the active shield requires an additional power source to generate the cancelling magnetic field. Also, precise real-time control of the cancelling magnetic field in terms of magnitude and phase is very important. Error in the phase or magnitude of the cancelling magnetic field can result in an even larger EMF than the original magnetic field. The reactive shield in Fig. 14.3d can solve the problem of the active shield because it uses the leakage magnetic field to reduce the incident magnetic field. As the reactive magnetic shield generates a cancelling magnetic field by adding a capacitor to the shielding coil, it requires no additional power source.

Figure 14.3 shows the magnetic flux density distribution of the OLEV. In the case of the vertical magnetic flux type, there is one magnetic flux path between the road embedded power cable modules and pickup coils where the power is transferred. The return flux comes back to the road embedded power cable modules via the sides of the main flux path. The horizontal magnetic flux type has two magnetic flux paths. The side road embedded power cable modules of this type have return flux paths on the side of the main flux path. The return flux path creates fringing magnetic flux, and this flux is measured as the EMF level of the OLEV. In this work, the target EMF level of the OLEV is 6.25 μT according to the regulation of the Korea Communications Commission, which follows the ICNIRP guideline given in Table 14.1 [5].

As the power supply system of the OLEV generates large amounts of magnetic field to transfer 100 kW of power, the amount necessary for the vehicle, magnetic flux is on the order of thousands of μT between the road embedded power cable modules and pickup coils beneath the vehicle while power is transferred. Therefore,

Table 14.1 Limit of magnetic flux density based on ICNIRP guideline

Frequency range	B-field (μT)
Up to 1 Hz	4×10^4
1–8 Hz	$4 \times 10^4/f^2$
8–8000 Hz	5/f
0.8–150 kHz	6.25
0.15–10 MHz	0.92/f
10–400 MHz	0.092
400–2000 MHz	$0.0046f^{1/2}$
2–300 GHz	0.20

if even 0.1% of the magnetic field leaks from the OLEV system, the EMF level could exceed the regulation of 6.25 μT.

14.3 Passive Shielding

Basically, passive shielding using metal plates is applied to the first test version of the OLEV for reduction in the electromagnetic field. For protection of passengers from the magnetic field, a metal plate is applied to the bottom of the vehicle. As the road embedded power cable modules are the source of the magnetic field, vertical plate shields are applied, as shown in Fig. 14.4a [6]. To improve the shielding effectiveness of the passive shield, soft contacts are additionally applied between the bottom plate and the vertical ground plate by metal brushes, as shown in Fig. 14.4b. The metal brush is a bundle of thin metal wires attached beneath the bottom plate and connects the current path between the vehicle body and the ground plate underneath the road surface. A photograph of an implemented metal brush is shown in Fig. 14.4c. The number of connections using metal brushes is a significant factor to improve the shielding effectiveness of the passive shielding. The EMF level is decreased from 14.4 to 3.5 μT when the number of connections using metal brushes is increased from 2 to 8.

14.4 Active Shielding

The passive shield can significantly reduce the electromagnetic field; however, this can be utilized only in limited applications because the metal brush will be worn due to the friction and the metal ground shield at the road surface can be covered with non-metal material, disturbing the connection between the vehicle body and ground plate. The cost and labour for maintenance would therefore become significant disadvantages.

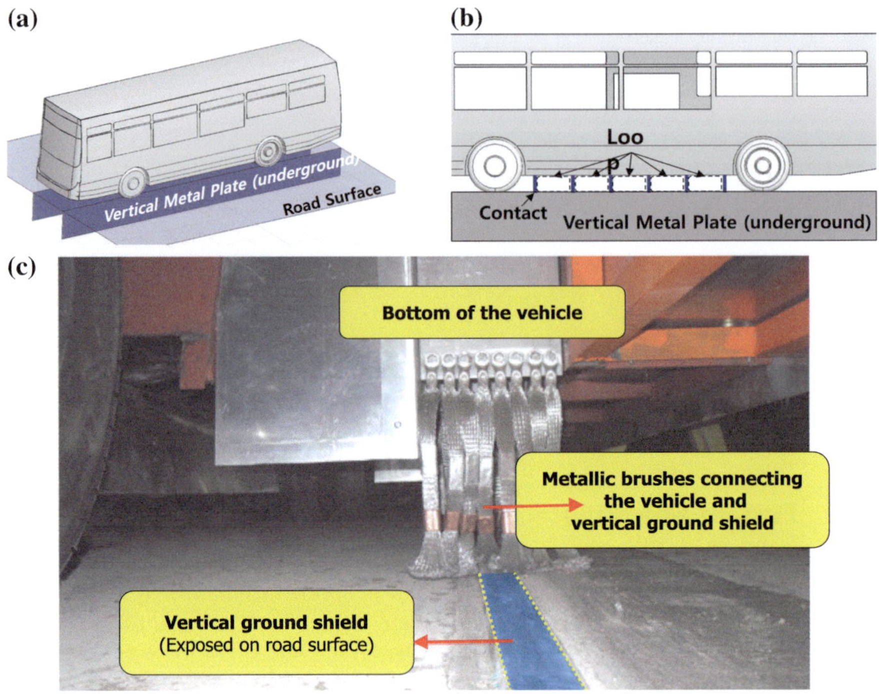

Fig. 14.4 Passive shielding for OLEV **a** vertical ground plate buried underground for reduction of magnetic field. **b** Connections between vehicle body and underground vertical metal plate for passive shield. **c** Implemented metal brush in passive shield at the bottom of OLEV [6]

Fig. 14.5 Concept of active shield for OLEV

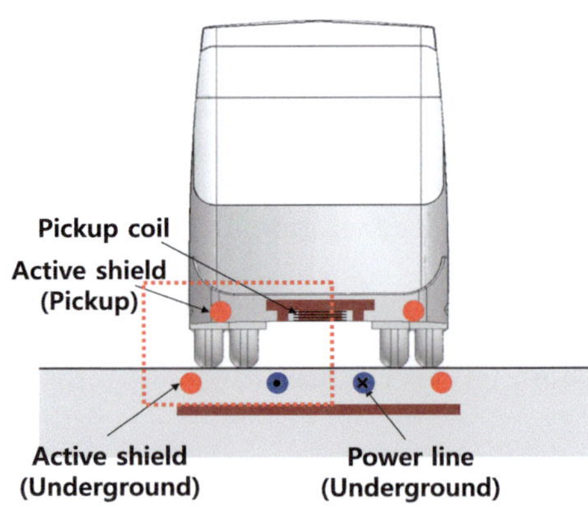

The EMF can be minimized by active shielding with or without passive shields independently. The basic concept of the active shield is shown in Fig. 14.5. Similar to road embedded power cable modules, the active shield is also a metal wire that carries the same frequency with current, but the phase is the opposite of the current in the pickup coil [7].

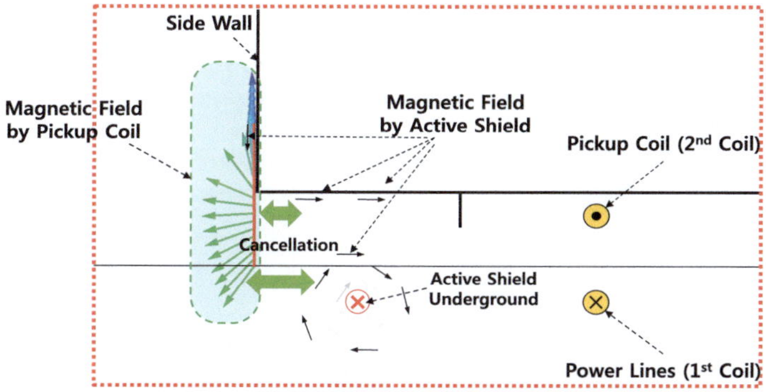

Fig. 14.6 Direction of magnetic field from pickup coils and active shield

In the design of the active shield, the directions of magnetic fields by the source and active shield should be carefully considered. In Fig. 14.6, which is the magnified view of the dotted rectangle in Fig. 14.5, the direction of the magnetic field is shown. To reduce the EMF level below the regulation at all positions, the magnetic field from the active shield should be almost the same as that from the pickup coils at all positions. At a position above 20 cm from road surface, the magnetic field vector is parallel to the metal plate because of the metallic shield at the bottom of the vehicle. It is hence more effective to place the active shield close to the pickup coil. However, if the active shield is moved closer to the pickup coil, the current of the active shield should be larger. For this reason, the placement of the active shield is compromised considering the shielding effectiveness and current magnitude. At the optimal value of current, the magnetic flux density is reduced to 1/10 of the density without the active shield.

Figure 14.7 shows the measured EMF after the active shield is applied to the OLEV. In the measurement, the IEC 62110 standard, which defines the magnetic field measurement of electric systems, has been selected as a standard measurement method [8]. According to IEC 62110, the average value of measurement at three points is reported as the final measurement. The three points are 0.5, 1.0, and 1.5 m above the ground or floor when the height of the test vehicle is greater than 1.5 m. By applying the active shield to the OLEV system, the leakage magnetic field has been reduced from 10.3 to 5.3 μT.

14.5 Reactive Shielding

The main problem of the active shield is the necessity of an additional power source to generate the cancelling magnetic field. The reactive shield in Fig. 14.2d solves this problem by using the leakage magnetic field to reduce the incident magnetic field, thus requiring no additional power source. Instead of using an additional

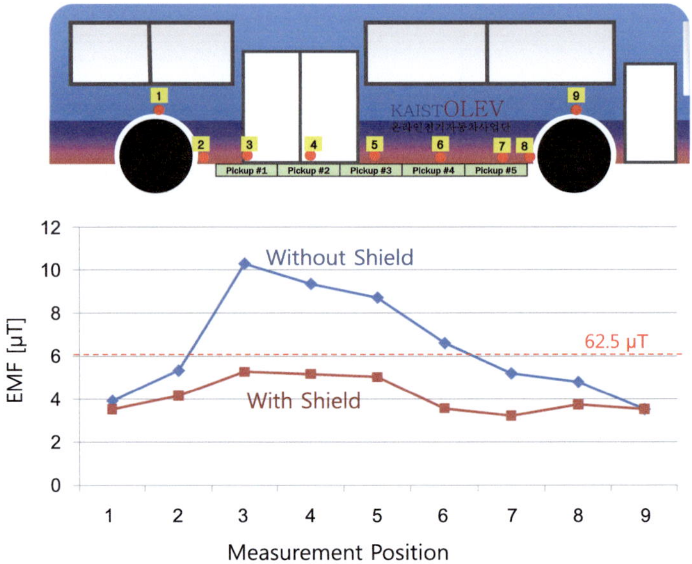

Fig. 14.7 Measured EMF level applied to OLEV

power, the reactive magnetic shield generates the cancelling magnetic field by adding a capacitor in the shield coil [9].

When an incident magnetic field from the wireless power transfer system coils passes through a shield coil, a cancelling magnetic field is induced and the induced magnetic field cancels the incident magnetic field, as shown in Fig. 14.8. The induced voltage at the shield coil is expressed as follows:

$$V_{\mathrm{ind}} = -\frac{\mathrm{d}\varphi}{\mathrm{d}t} = -\frac{\mathrm{d}B \cdot S}{\mathrm{d}t} = -j\omega B_0 e^{\mathrm{j}\omega\mathrm{t}} \cdot S \tag{14.1}$$

where S is the area of the shield coil.

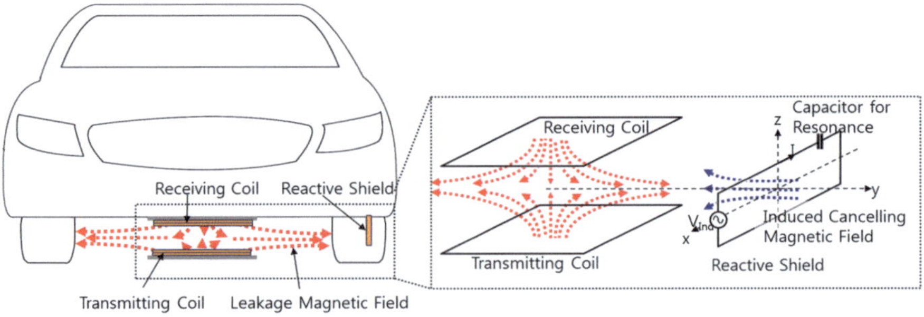

Fig. 14.8 General structure of wireless power transfer system for electric vehicle and reactive shield to reduce leakage magnetic field [9]

The equivalent circuit of the reactive shield coil consists of the shield inductance, L_{sh}, shield capacitance, C_{sh}, and parasitic shield resistance, R_{sh}. The shield inductance and shield capacitance determine the series resonance frequency. When the resonance frequency becomes close to operating frequency, the impedance of the loop becomes at the operating frequency. The shield capacitance using a lumped capacitor component is adjusted to control the resonance frequency. When the resonance frequency is close to the wireless power transfer frequency, even small induced voltage V_{ind} can generate large shield current, I_{sh}, and hence a magnetic field that is large enough to cancel the incident magnetic field, because the shield current, I_{sh}, is determined by the impedance of the shield coil.

The generated cancelling magnetic field can be determined by the current flowing in the shield coil, which is calculated as follows:

$$I_{sh} = \frac{V_{ind}}{Z_{sh}} = \frac{V_{ind}}{\left(j\omega L_{sh} + \frac{1}{j\omega C_{sh}}\right) + R_{sh}} \tag{14.2}$$

Finally, the cancelling magnetic field from the reactive shield coil is calculated as follows

$$B_{sh} = \frac{\mu_0}{2r} n_{sh} I_{sh} = -\frac{\mu_0}{2r} \frac{n_{sh} B_0 S}{L_{eq}} e^{j\omega t} \tag{14.3}$$

where n_{sh} is the number of turns in the shield coil.

The generated cancelling magnetic field is proportional to the equivalent inductance L_{eq}, and this value can be controlled by determining the shield capacitance C_{sh}.

A 3D magnetic field solver, ANSYS Maxwell, is used to simulate the effects of the magnetic field. As seen in the cross-sectional view presented in Fig. 14.9, the reactive shield is applied only at the right side of the wireless power transfer system. The reactive shield coil significantly reduces the magnetic field of the right part compared with the left part where no shield is applied.

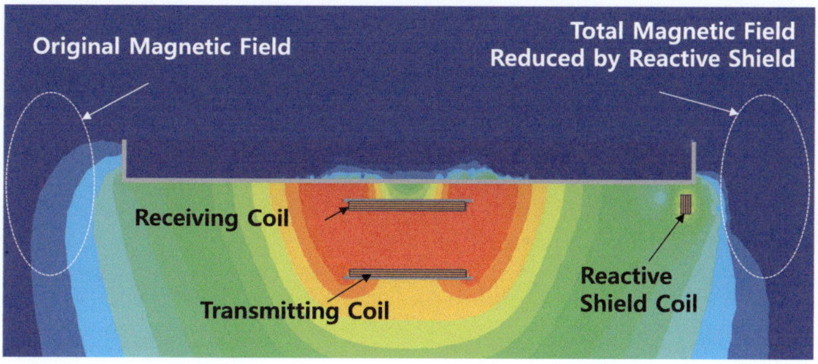

Fig. 14.9 Simulated magnetic flux density of wireless power transfer system with and without reactive shield [9]

14.6 Conclusions

Although the wireless power transfer system is very convenient and safe compared with the system using wire and connectors, there is some possibility of harmfulness of the unwanted electromagnetic field to human body. For OLEV system, the electromagnetic field issue could be more serious than small electronic products because the power is much larger. However, the electromagnetic compatibility has been a research topic with a long history since tens of years ago, and there have been lots of research activities to find the solution of the similar problems.

The passive, active, and reactive shielding methods introduced in this chapter are the good solutions to reduce the leakage electromagnetic field from the wireless power transfer system. Besides, there are more solutions to reduce the electromagnetic interference, such as low-noise inverter design, noise filtering, magnetic field balancing, grounding. Each solution and the combination of these solutions are expected to reduce the possible electromagnetic danger and enhance the value of OLEV system and the other wireless power transfer systems.

References

1. Ott HW (2008) Noise reduction techniques in electronic systems, 2nd edn. Wiley-Interscience
2. Chun Y, Park S, Kim J, Kim J, Kim H, Kim J, Kim N, Ahn S (2014) Electromagnetic compatibility of resonance coupling wireless power transfer in on-line electric vehicle system. IEICE Trans Commun E97-B(2): 416–423
3. Kong S, Kim M, Koo K, Ahn S, Bae B, Kim J (2011) Analytical expressions for maximum transferred power in wireless power transfer systems. Paper presented at IEEE electromagnetic compatibility symposium, Long Beach, California
4. Kim M, Kim S, Chun Y, Park S, Ahn S (2014) Low frequency electromagnetic compatibility of wirelessly powered electric vehicle. Paper presented at EMC Tokyo, Tokyo, Japan
5. International Commission on Non-Ionizing Radiation Protection (1998) Guidelines for limiting exposure to time-varying electric, magnetic, and electromagnetic fields (UP TO 300 GHz). Health Phys 74(4)
6. Ahn S, Pak J, Song T, Lee H, Byun J, Kang D, Choi C-S, Chun Y, Rim C, Yim J, Cho D, Kim J (2010) Low frequency electromagnetic field reduction techniques for the on-line electric vehicle (OLEV). Paper presented at IEEE electromagnetic compatibility symposium, Port Lauderdale, Florida
7. Buccella C, Feliziani M, Fuina V (2002) ELF magnetic field mitigation by active shielding. Paper presented at IEEE international symposium on industrial electronics, L'Aquila, Italy
8. International Electrotechnical Commission (2009) Magnetic field levels generated by a.c. power systems–measurement procedures with regard to public exposure. IEC 62110
9. Kim S, Park HH, Kim J, Kim J, Ahn S (2014) Design and analysis of a resonant reactive shield for a wireless power electric vehicle. IEEE Trans Microwave Theory Tech 62(4):1057–1066

Chapter 15
High Power and Energy Management System in OLEV

In-Soo Suh

Abstract The fundamental architecture of the wireless charging system for OLEV is presented. The electric power is supplied either from the on-board energy storage system and/or the wirelessly delivered electric power. Super-capacitor source can also be used. This chapter presents the control of the power flow for the dynamic wireless charging system used for OLEV and other vehicles.

15.1 Introduction

This chapter covers the energy management in OLEV. OLEV is designed to receive electric power from batteries when the vehicle is on roads without the underground power supply system. When it is on roads with underground power supply, the vehicle would normally use the electric power delivered to it wirelessly from the underground power supply system. When OLEV is climbing a hill or need additional power for faster acceleration, it would draw its electric power from both the battery and the underground power supply system. Conversely, when excess energy is delivered to OLEV wirelessly while in motion, the excess energy would be diverted to the battery for storage. A similar energy management strategy is used in hybrid electric vehicles (HEVs), which also have two power sources, i.e., batteries and an internal combustion engine [1]. The main objective of the energy management strategy for these vehicles is to maximize the energy efficiency while delivering the required performance under various driving conditions. The general concept of energy management system presented in this chapter can be extended when there are more than two energy sources.

I.-S. Suh (✉)
Korea Advanced Institute of Science and Technology (KAIST),
5338 Capri Dr., Troy, MI 48098, USA
e-mail: insoo.suh@kaist.ac.kr

Present Address:
I.-S. Suh
General Motors, 30500 Mound Rd., Warren, MI 48092, USA

N.P. Suh and D.H. Cho (eds.), *The On-line Electric Vehicle*,
DOI 10.1007/978-3-319-51183-2_15

Fig. 15.1 OLEV vehicles introduced to the public in spring, 2010

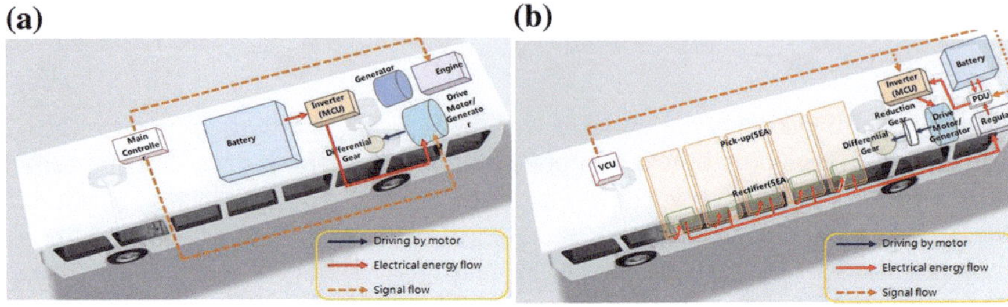

Fig. 15.2 Schematic comparison of example powertrain layout of a bus application **a** series hybrid system, **b** OLEV system

Specific power flow management of OLEV can be different depending on applications. In this chapter, a typical power flow management architecture is discussed for the OLEV vehicles built and demonstrated at KAIST, which are shown in Fig. 15.1. Required battery performance and safety issues are also discussed.

Typical powertrain layouts of a series hybrid and an OLEV system for a commercial bus application are shown in Fig. 15.2. Major powertrain components, main power flow, and control signal flow are shown in the figure.

15.2 Overall Power Control Architecture of OLEV System

The OLEV system can be divided into two sub-systems: the power supply system (embedded into the road) and the electric vehicle (equipped with power pickup modules). The power supply system embedded in the road can be considered as the primary circuit that generates magnetic field above the ground. The power pickup system mounted on the vehicle may be considered as the secondary electromagnetic circuit. These two systems operate at a common resonant frequency [2–4].

The schematic architecture of the OLEV system on the overall power control is shown in Fig. 15.3.

In this figure, the power supply system is described as the track inductance of a certain powered track distance, where the electricity is supplied through the track

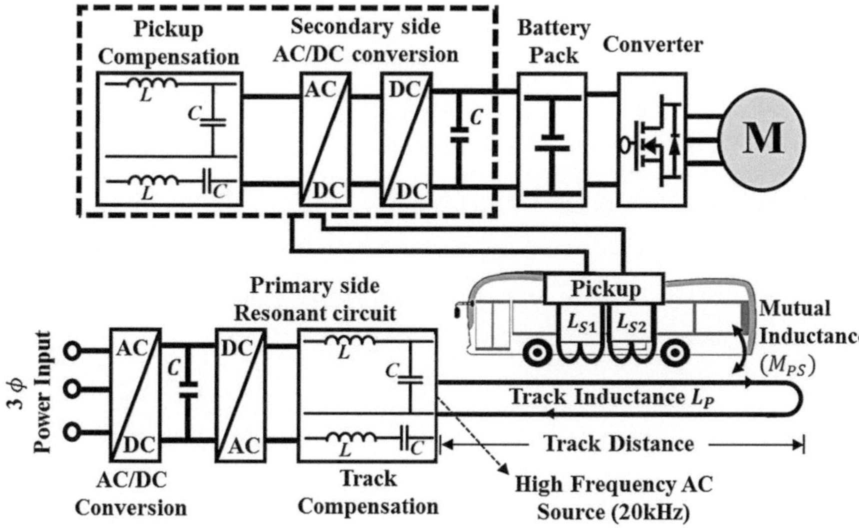

Fig. 15.3 Overall OLEV system architecture: vehicle and power supply system

compensation circuit via AC–DC and DC–AC converters for the required operational frequency for the resonant circuits from the typical 3-phase AC grid power input. The compensation circuit is to tune the primary circuit into the required resonance of the various track distance of the power supply side of primary coil. The vehicle side of power pickup system consists of the pickup device, which is attached underneath of the vehicle. It connected to the pickup coil compensation unit and the secondary side of AC–DC and DC–DC converter in order to supply the required power to the battery and/or the motor driver unit. The voltage regulating control of this DC–DC converter (or called a regulator) is used to regulate the voltage of the power received from the road.

For the bus shown in Fig. 15.1, the 600-V AC induction motor was the main propulsion system. The bus had a long 168-cell (621V) Li-polymer battery system, which had 14 cells in series in each pack and 12 packs in parallel connection. The maximum energy storage capacity was 25 kWh, which was a minimally required for a pure electric bus. The DC–DC converter in the pickup system side provided the proper voltage input to the battery. It communicated with the BMS (battery management system) to control the charging and discharging operation of the battery.

The wirelessly delivered power can be supplied either to the battery or directly fed into the motor drive unit, depending on the motor power requirement, through a PDU (power distribution unit) which is a set of circuit breaker to switch the power flow [5, 6].

An example of power flow control is shown in Fig. 15.4. When the wireless power is supplied, battery charge and discharge as well as the brake regeneration must be managed as shown in a flow chart of Fig. 15.4.

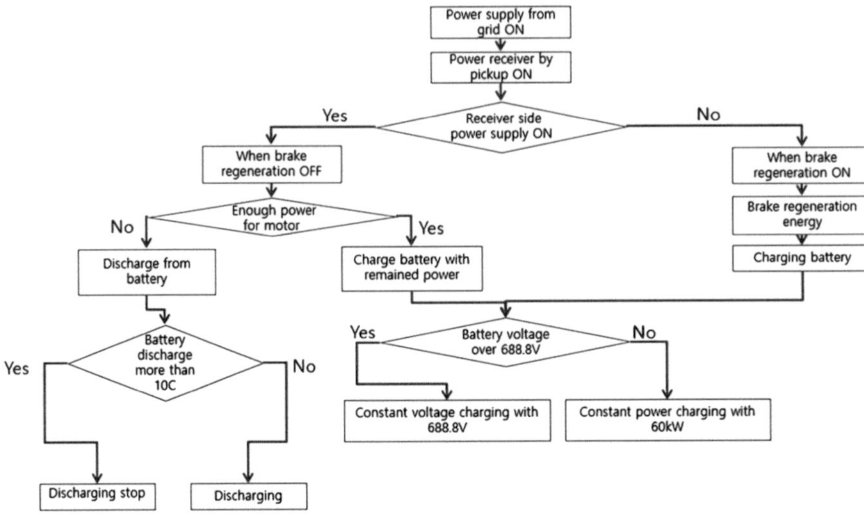

Fig. 15.4 Battery charging and discharging control algorithm. The wireless power is delivered to the brake regeneration system, based on an assessment of the required motor power for given driving conditions

15.3 On-Board Architecture of OLEV Bus System

The architecture of OLEV is similar to that of a general plug-in electric vehicle, but there is an additional power source that transfers energy from the ground while the vehicle is in motion. The additional functional blocks of this system are the power pickup module attached to the bottom of the vehicle, a bank of capacitors, and a regulator, as shown in Fig, 15.3. The drive motor can be driven either by battery power or by power from the pickup system. The pickup power capacity is increased by adding additional pickup modules. OLEV buses had five modules, each with 20 kW capacity, providing up to 100 kW of power.

Generally, the system provides the required power to the propulsion motor directly from the received power, whenever the underground power supply is available. By placing the underground power supply systems at right intervals, the charge in the battery oscillates about a set point. This increases the battery life by eliminating the need for complete charge and discharge cycles, as described in Fig. 15.4. Typically the state of the charge in these batteries is maintained, on average, at about 50% level, about which the charge fluctuates cyclically. The capacitor bank is necessary to tune the secondary circuit into the proper resonating frequency as a secondary side of a compensation circuit or a pickup-side compensation circuit.

The electrical system architecture of OLEV is presented schematically in Fig. 15.5. It shows the major function controllers of OLEV. The functional blocks in the diagram in Fig. 15.5 represent the motor control unit (MCU), the battery

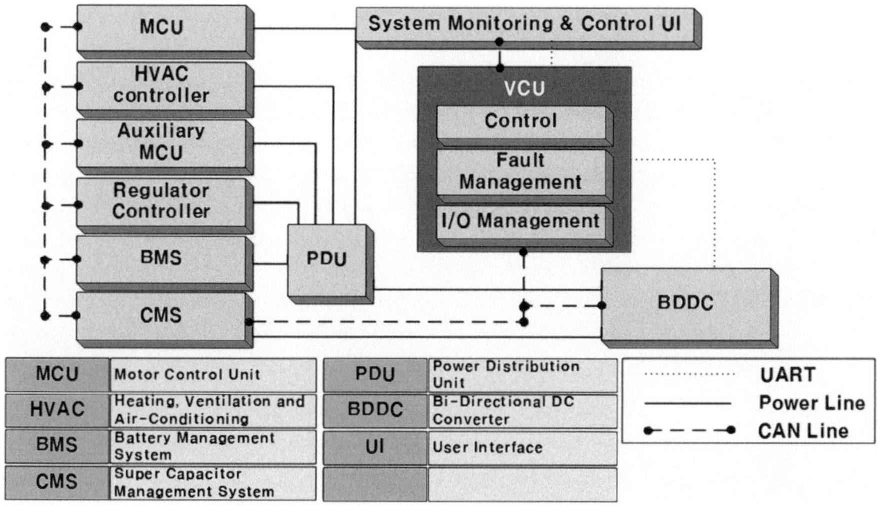

Fig. 15.5 OLEV vehicle system architecture in control network

management system (BMS), the controller for the super-capacitor (denoted as CMS), the bidirectional DC–DC converter (BDDC), the main vehicle controller unit (VCU), and the power distribution unit (PDU). The PDU controls the distribution and conversion of power between the sources (i.e., the battery, the capacitor bank and the pickup regulator, the traction motors, and the accessory loads such as the HVAC, and the battery charging circuitry). Among the controllers, the CAN protocol has been applied, except the BDDC is communicated through a serial connection as denoted UART (universal asynchronous receiver/transmitter) in Fig. 15.5.

Additional components necessary for communication and control integration within a vehicle are also added. Overall, when converting from a typical diesel engine operated bus (12,000 kg curb weight), the weight can be reduced by about 1000 kg when compared with a pure electrical bus. As a rule of thumb, the weight reduction can be estimated as being about 25% of the weight of the typically required battery.

15.4 Road-Embedded Power Supply Architecture of OLEV System

The OLEV infrastructure system (i.e., powered road) is composed of a set of powered tracks along the length of the road traveled. A powered track is composed of a set of segments, which can have different lengths; these segments in a powered track are fed by a set of inverters controlled through a switch box. Sufficient distance between the segments to accommodate bridges, pedestrian crossings, etc., can be provided.

The power cables are inserted inside the Fiber Reinforced Plastic (FRP) tubes that are buried in concrete and laid on the rib structure of the assembled ferrite core frames. The FRP tubes are used to protect the cable from possible damages during pouring concrete while providing extra strength to the concrete. If the target life expectancy of the road construction is set as 20 years of continuous use, the typical annual maintenance cost is about 8% of the initial expense with proper procedures for maintenance. The power supply cable adopts a Litz cable for high-frequency application and is wrapped around the ferrite core with a single turn. The Litz wire minimizes power loss caused by skin and proximity effects. The Litz wires for transferring power from the inverter are typically installed on a rail.

15.5 Optimizing Magnetic Flux Field

The design variables include the geometric layout of coils and the ferrite shapes in the transverse and longitudinal directions, as seen from a vehicle-based coordinate system. The tuning parameters are inductance, capacitance of both circuits including the air gap (or ground height of the vehicle) between the road surface and the bottom of the pickup modules. For maximum power transfer efficiency and minimal leakage field flux of the magnetic field, the magnetic field shapes must be optimized in three dimensions. Shaped magnetic field in resonance (SMFIR) is used. In an SMFIR system, the magnetic field is controlled through frequency tuning circuitry to remain in resonance. Ferrite cores are used to shape the magnetic field, directing it toward the pickups and away from pedestrians, vehicles, and other objects on the road. With SMFIR, it is possible to transfer power efficiently across a 20-cm air gap. The shapes of the magnetic field are shown in Fig. 15.6.

Magnetic saturation should also be considered. The closer the magnetic density gets to saturation, the more core losses increase. In addition, the heat caused by core loss will lead to changes in the core characteristics. For these reasons, the core diameter should be enlarged when a higher magnetic flux is required. When adjusting the core diameter, the limitations for the core width for a given vehicle should be considered.

A material that has high initial permeability typically has lower magnetic resistance. For the case with 20 kHz resonant frequency, it is recommended that the initial permeability should be higher than 3200. Also the core loss, density, and resistance should be as low as possible. The materials used in the KAIST OLEV demonstration have saturation magnetic flux density over 490 mT at 25 °C, and the density should be retained at over 390 mT at 100 °C. The Curie temperature of the ferrite core is approximately 220 °C, and the relative permeability is 2400.

The 2-D schematic model in Fig. 15.7 shows the primary and secondary magnetic flux density in an OLEV system. The road-embedded power cable module is installed along the vehicle longitudinal driving direction (under the road), and the pickup

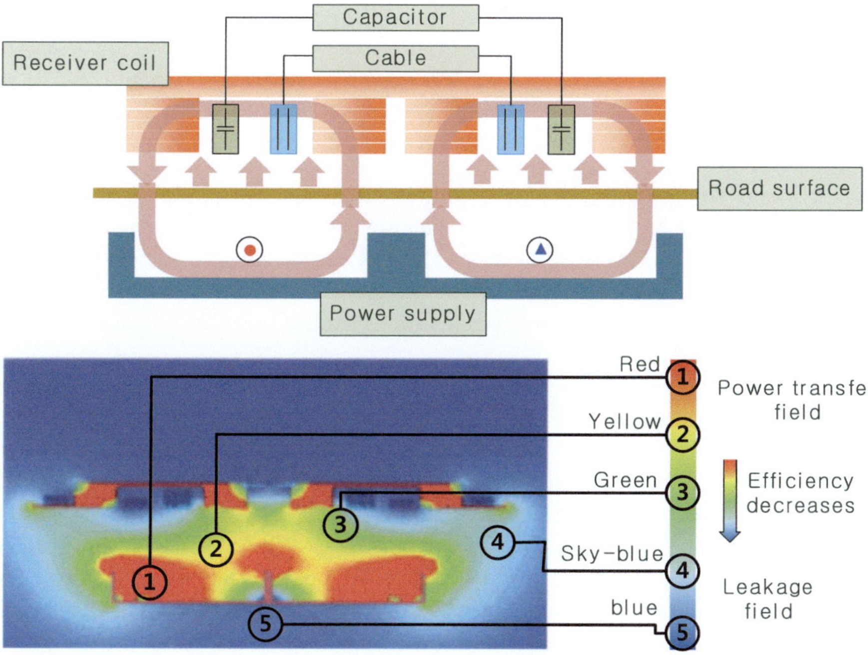

Fig. 15.6 Schematic diagram of cross-sectional view of magnetic flux density for SMFIR

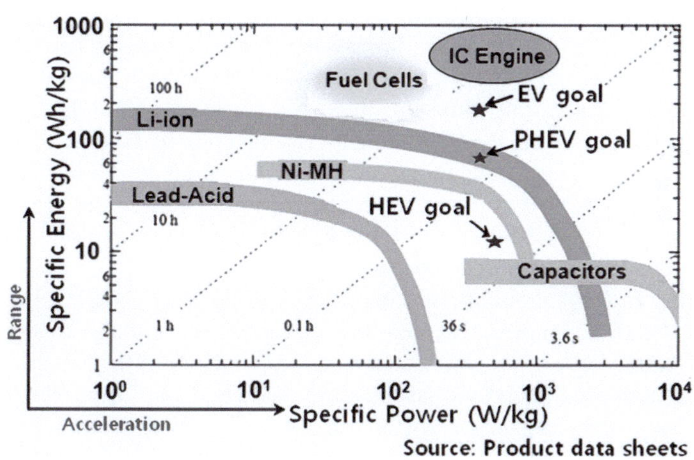

Fig. 15.7 Ragone plot of various electrochemical energy storage and conversion devices

modules are attached to the vehicle body (under the vehicle). Both modules, the supply and the pickup, are equipped with specific arrangements of ferrite cores for optimal shaping of the magnetic field. The magnetic field in Fig. 15.6 shows a dual elliptical shape, which is known as a dual-mode, wireless power transfer mechanism. Several such shapes exist and are described in detail in Chap. 5 of this book.

15.6 General Requirements of an Energy Storage System (ESS)

With wireless charging during vehicle operation, the battery energy storage capacity can be smaller (depending on the vehicle travel pattern, route elevation and distance, and power supply infrastructure installed rate (PIR) along the traveled route). In general, a battery that is about a quarter of the size of a typical EV battery is required, while the PIR is about 20%. The PIR can be considered zero for a pure battery-operated EV (BEV). Thus, when designing an OLEV system, the battery capacity and PIR will be design variables for optimization of installation cost and vehicle performance.

Most electric vehicles (EVs) obtain the electric energy needed for operation from on-board energy storage (ESS) devices (i.e., batteries). However, current battery technology provides limited driving range per charge, relatively long charge duration, and high cost, mainly driven by the lower specific energy and energy density of current battery technology. While intensive research into improving battery technology is ongoing, these issues must be addressed before future battery technologies become available in order to increase public adoption and acceptance of EVs in both public and personal transportation. OLEV's major advantage is that it eliminates the need for a large number of expensive batteries.

The Ragone plot in Fig. 15.7 shows specific power and energy of various battery technologies and energy storage and generation device [7]. For pure electric vehicle or hybrid electric vehicle applications, Li-ion batteries are promising, even with current technology levels. However, the specific power and energy capacity of these batteries still significantly fall short of IC engines' capacity because of the limited driving range and charging time. In 2003, the United States Advanced Battery Consortium (USABC) set the requirements for batteries of EVs to achieve broad market penetration [7]. These requirements, summarized in Table 15.1, cover a wide range of issues. For example, the required specific power and specific energy are 400 W/kg and 200Wh/kg, respectively. With current technology, Li-ion batteries cannot meet the required specifications.

According to the National Renewable Energy Laboratory (NREL), the expected battery life varies significantly with the swing of the battery state of charge (SOC) or battery charging and discharging cycle depth. The batteries in BEV applications are constrained by design to discharge depths from 80 to 20% of capacity. The battery capacity and PIR can be flexible design variables, and thus, the charge-discharge duty cycles can be varied over the vehicle's typical driving pattern and distance in order to optimize cost and performance. The battery duty cycle is typically set for shallow cycles with an SOC swing of 20–40% capacity, depending on the combined design of battery capacity and PIR. For example, in the case of the Seoul Grand Park OLEV train, the SOC swing is limited to about 10% of capacity because of the short distance required to complete a single round trip. The battery cycle life in this case will reach several tens of thousands of

Table 15.1 USABC battery capacity requirements for widespread EV introduction [7]

Parameter (units) of fully burdened system	Minimum goals for long-term commercialization	Long-term goal
Power density (W/L)	460	600
Specific power—discharge, 80% DOD/30 s (W/kg)	300	400
Specific power—regen, 20% DOD/10 s (W/kg)	150	200
Energy density—C/3 discharge rate (Wh/L)	230	300
Specific energy—C/3 discharge rate (Wh/kg)	150	200
Specific power/specific energy ratio	2:1	2:1
Total pack size (kWh)	40	40
Life (years)	10	10
Cycle life—80% DOD (cycles)	1000	1000
Power and capacity degradation (% of rated spec)	20	20
Selling price—25,000 units @ 40 kWh ($/kWh)	<150	100
Operating environment (°C)	−40 to +50 20% performance loss (10% desired)	−40 to +85
Normal recharge time	6 h (4 h desired)	3–6 h
High rate charge	20–70% SOC in <30 min @ 150 W/kg (<20 min @ 270 W/kg desired)	40–80% SOC in 15 min
Continuous discharge in 1 h—no failure (% of rated energy capacity)	75	75

charge–discharge cycles or practically unlimited lifetime, thus providing a viable solution to concerns of battery durability.

15.7 Considerations of Electrical Safety in OLEVs

15.7.1 Electrical Safety of Power Drive System in a Bus

An example of an OLEV powertrain wiring network is shown in Fig. 15.8. An OLEV powertrain is controlled by many electrical devices and circuit elements of both high and low voltage electrical system. For a commercial bus application of OLEV, a 620 V high voltage is applied since the required voltage for the AC induction motor has been applied to the bus or people mover. Figure 15.8 shows the high-voltage road-embedded power cable modules and the low voltage

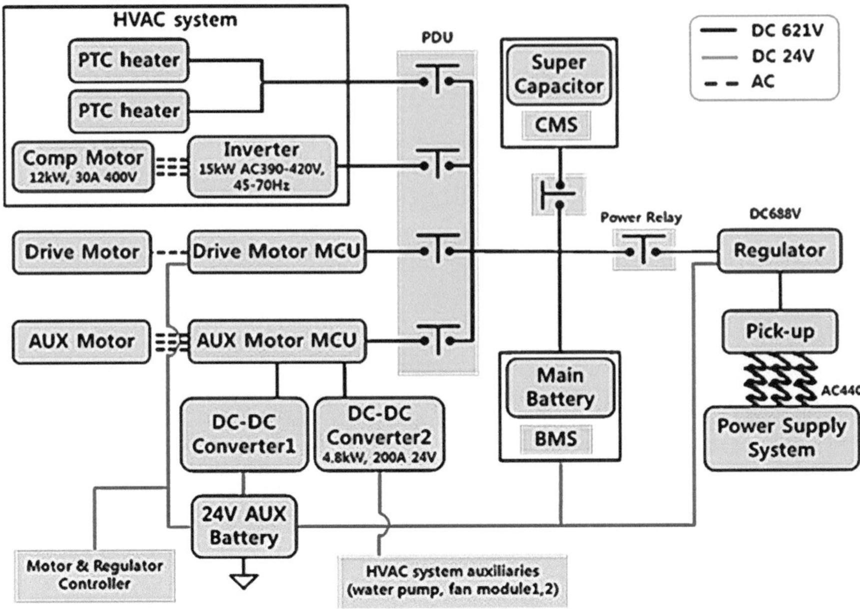

Fig. 15.8 Example of concise wiring diagram of OLEV powertrain

road-embedded power cable modules, which are used mainly for the traction power of the motor drive and the auxiliary power source in the vehicle, respectively. As discussed earlier in this chapter, the main functions of powertrain control are the high power flow control, motor direction control, and the motor speed control, while properly distributing the energy source from the wirelessly delivered power and the on-board ESS. The power relay and PDU (power distribution unit) perform a very important function of the electrical safety of the whole vehicle, in conjunction with the charging control, battery management control, and required driving power control for the motor.

The implementation of specific electrical systems in order to enhance the stability and safety of the high power bus in an electric vehicle has four basic goals:

- To protect persons, either inside or outside the vehicle, from electric shock;
- To prevent possible damage to on-board systems from the accidental release of electric energy;
- To ensure reliability and prolonged function of the vehicle's electrical and electronic components;
- To limit damage to the vehicle's systems in the event of a fault.

This protection must be guaranteed even in the event of a failure(s) of the basic protection systems, such as a cable's primary insulation, or in the event of a vehicular collision. Postcrash safety has also been drawing increased attention in recent years, as evidenced by the ISO TC22/SC21 in-progress proposal for post-crash safety regulations, set to become the fourth section of ISO 6469.

15.7.2 Electrical Grounding of OLEV System

(1) Metallic Chassis

Most EVs, like internal combustion engine vehicles (ICEVs), have chassis frames primarily made of conductive materials. In the case of low voltage circuits, like those used in ICEVs, it is acceptable to use this frame as a part of the electric current return path. This is because the voltage levels in the circuit are not high enough to present a serious risk to persons who might be exposed to them.

On the other hand, the electrical circuits in EVs may carry voltage levels that can be harmful to persons, defined by both SAE J2344 and ISO 6469-3 as voltages above 30 V AC or 60 V DC. Circuits with these or higher voltages may not use the vehicle chassis as a current return path. Figure 15.9 shows three different schemes of grounding commonly applied to EVs.

This grounding provides an extra layer of protection from electric shock for the occupants of the vehicle in the case of a fault. It also protects emergency responders who might come in contact with the car, in a postcrash condition, where a nonevident first ground fault might have occurred and someone could inadvertently close the circuit by touching a part of the car. Therefore, EVs typically use fully isolated (or floating ground) systems for their electrical traction power per ISO 6469-3.

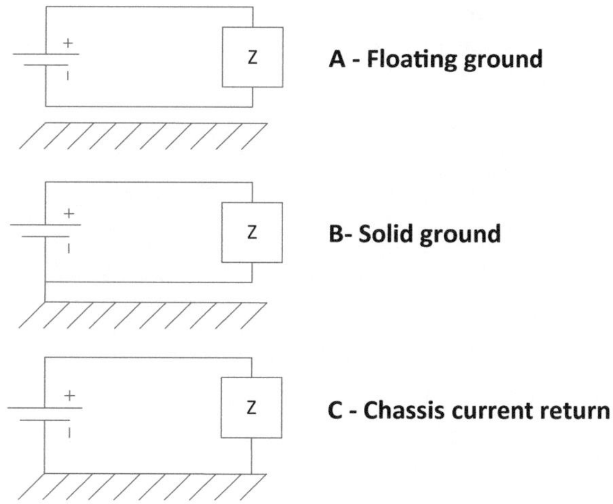

Fig. 15.9 Three different grounding schemes used in electric vehicles

(2) Floating Ground Systems

In a floating ground system, the electrical supply terminals are not conductively connected to the earth ground or chassis ground at any point. In EVs, this means that neither polarity of the battery may come into conductive contact with the vehicle frame.

Since fuses and circuit breakers may not be sensitive enough to interrupt a double ground fault, especially one that flows through a high impedance (i.e., a person), the insulation resistance must be continuously checked by a monitoring system to ensure that either terminal of the battery has not become connected to the vehicle's conductive frame.

(3) High Power and Energy, On-board Rechargeable Energy Storage System (RESS)

Depending on the vehicle, the RESS inside an EV stores several kWh of energy and delivers short-circuit current of several tens of kA. Additionally, having an on-board power supply means that there is no means of protection or disconnection outside the vehicle. If the protection systems fail, there may not be another way to stop the fault until one of the elements involved is completely destroyed.

Furthermore, many recent electric vehicles utilize lithium-ion batteries, which can be highly unstable if not properly protected. As these batteries are inside the vehicle, they present further risks of fire, explosion, and electrolyte or gas expulsion in the event of a short circuit.

(4) Mobility

A vehicle that undergoes an electrical fault while moving is also under risk of loss of control. Because of this, it is not always wise to disconnect the electrical system during a non-critical fault. For example, SAE J2344 recommends against disconnecting the system in the event of a first ground fault if the vehicle is in "motoring mode." Additionally, any components necessary to assure the basic function of the vehicle must be carried on-board at all times. Electric systems in vehicles must also account for the possibility of a collision, with the vehicle's occupants still on board. Safety after a crash including towing and storage/disposal of vehicle has to be considered as well.

15.7.3 Electrical Issues and Possible Solutions

(1) Short Circuit and Overload

While the size and type of batteries found in EVs can vary greatly, short-circuit currents can be expected to range from a few to several tens of kilo amperes. A short circuit of this magnitude, even for a few tenths of a second, can cause important injuries and destroy the entire electrical system inside a vehicle. Additionally, batteries output energy in DC, which is harder to interrupt than AC, as

DC currents do not cross through zero twice every cycle. An overload is a different type of overcurrent condition and can damage controllers or motors if present for an extended period of time. Overload currents range from values above acceptable continuous full load current (FLA), up to about 8 times FLA. Modern solid state motor drivers typically include overload protection functions.

Several devices for the interruption of DC fault currents exist:

- Fuses: Fuses are low initial cost, one-time-use protection devices. Fuses intended to protect against short circuit, overload or both are available. They are reliable and simple, but they must be replaced every time a fault occurs. Fuses can operate under every orientation and do not suffer from false trips due to vibration. They are relatively compact and have the lowest initial cost. Because of their fast response, semiconductor protecting fuses can limit the total energy that flows during a short circuit, minimizing possible damage in the event of a fault. However, in the case of a vehicle, a blown fuse might mean getting stranded on a road if spare fuses are not available on-board. The act of changing a fuse can also possibly bring an untrained operator into physical proximity with an electrical system that recently underwent a fault, whose cause may or may not yet be known. Even if safety is assured through the usage of interlocks that disconnect the supply when a fuse panel is opened, the operator may be unwilling or unable to change them.
- Circuit breakers: These are reusable short-circuit and overload protections, with two triggering elements. A thermal relay, which works with accumulated heat to protect against overloads and magnetic, will trigger instantaneously if a specific, much higher current value is exceeded, thereby protecting against short circuits. Models with a single element, a shunt trip coil or under-voltage release are also available.
- Positive temperature coefficient devices (PPTC): These are devices whose resistance increases significantly with heat. During normal functioning, they have very low resistance and are virtually invisible to the circuit; however, when excessive current flows during a short circuit, they increase their resistance as they heat up, limiting the flow of fault current. When the fault is cleared, they cool down and restore the circuit to normal function. Adequate heat dissipation, proximity to other heat sources, and operating temperature range must be carefully considered when these devices are used.
- Overload relays and advanced overload protection devices: These relays are used to protect against overload, over temperature, and some otherwise hard to detect faults, such as the loss of a single phase. They are often built in into motor drivers or battery management systems (BMS). Since the repeated interrupt capacities of relays are much lower than the system's full short-circuit current, they are usually inadequate as the sole protective element in a circuit.

(2) Protection Coordination and Selectivity

Possible damage in the event of a fault can be limited by using several types and layers of protection. In order for such a system to be effective, the designer must assure that for any given fault, the ideal protection will be the first to be triggered and will be capable of clearing the fault properly before any other protective elements start to act. The general principle is that smaller branch protections must clear faults before trunk protections start to act. Additionally, it is more convenient and economical to interrupt slower and lower current faults with reclosable protective elements whenever possible, while leaving fuses for higher current level faults. In addition, because fuses have a minimum breaking current rating, a fuse that has been exposed to a current larger than its pre-arcing value but below its minimum clearing Joule integral value (current squared multiplied by time, or I^2t, which represents the heating of the fuse element) may fail to clear the fault and continue to arc indefinitely.

The design of protection systems with multiple series elements that meet these criteria is known as protection coordination. A system where elements in branches clear faults before systems in larger upstream trunks start to open is said to be selective. In selective systems, the fault clearing Joule integral (I^2t) values on downstream fuses must be lower than the upstream pre-arcing I^2t values.

(3) Other Factors in the Selection of Short-Circuit Protective Devices for EVs

The interrupt rating (IR) on a short-circuit protective device is the maximum fault current value for which a given fuse or circuit breaker is effective. The IR of the short circuit protective element must be greater than the prospective current (potential maximum short-circuit current of the system) at the point where the element will be installed, including all current contributions from the battery, motors, and capacitors. Short-circuit protection devices must be chosen considering the peak voltage rather than the supply nominal voltage. The DC voltage and IR rating of a given protective device will be significantly lower than its AC rating. Only DC-rated elements should be used on a DC bus. Furthermore, the DC nominal operating voltage for protective devices given by manufacturers typically corresponds to their function in electrical networks with very low inductance, i.e., a 2 ms or lower inductive time constant, and must be down-rated as the relative inductance of the network increases. Typical time constants for motor circuits are in the 20–50 ms range.

Fuses must also be sized below the maximum ampacity of the cables they protect, in accordance with IEC 60364-4-43. This also means that the speed of protection is a trade-off. On the other hand, faster protections minimize possible damage; however, fast protections must be sized with larger margins in order to prevent them from opening during normal motor start-ups. This means that cables and switches must also be sized to be able to carry these higher currents without damage.

(4) Shock Protection

Typical EV energy storage systems have voltages and/or current levels that can be dangerous to persons. Under normal operating conditions, electrical systems are isolated by such factors as cable insulators; however, additional protection systems should also be installed to protect people from shock when basic insulation fails. These shock protection systems typically include the following:

- Insulation resistance monitors: These are devices that continuously measure the resistance between the vehicle chassis and both terminals of the RESS. If the resistance falls below a certain value, the RESS is disconnected. An insulation resistance of 500 Ω/V for AC circuits of 30 V or above, and 100 Ω/V for DC circuits with voltages above 60 V are required by ISO 6469-3. SAE J2578 describes additional barriers that might be needed for mixed AC/DC circuits. These levels are determined by the maximum current levels considered as harmless for humans, according to IEC 60479-1.
- High-voltage interlock loops (HVIL): These are systems that send small current pulses through loops containing potentially harmful voltages. If the system detects a lack of continuity, it assumes the loop is open and disconnects the electrical supply.
- Enclosures: The risk of direct contact of persons with live parts of the electrical system can also be reduced through proper enclosure design. Broadly speaking, ISO 6469-3 recommends enclosures that meet an ingress protection rating equivalent or higher than IPXXD (as described by IEC 60529) when closed, and continue to meet IPXXB even when open. The specifics depend on the contents, location, and locking mechanisms, etc., of the enclosure, and detailed information can be found in the aforementioned standards.
- Hazardous Voltage Bus Discharge relays: Capacitors retain their electrical potential even after the supply is turned off. This may present a risk of shock, especially to service personnel. A discharge relay provides a discharge path by temporarily connecting a shunt discharge resistor.
- Crash sensors: EVs often use crash sensors, like those used for airbags, to disconnect the supply from the electrical system in the event of a crash.

(5) Protection of Electronic Components from Transients

EVs, like internal combustion engine vehicles, are expected to run with little need of maintenance for long periods of time. Switching transients are some of the principle culprits of reducing the lifetimes of electronic components. Vehicles must also be capable of passing electromagnetic compatibility tests for both susceptibility and emissions, such as those described in ISO 11451 and CISPR 12.

The current flowing across a capacitance depends on the rate of change of the voltage across its terminals multiplied by the value of the capacitance. This means that a theoretically infinite level of current can flow through a discharged capacitance for the instantly short period after it is connected to a voltage supply. This is known as an inrush current. Similarly the voltage across an inductor depends on the negative of the rate of change of current across its terminals, multiplied by the inductance value. The negative sign means that when a switch opens and attempts to interrupt the flow of current, the theoretical voltage across an inductance tends toward infinity. Inductances are the electrical equivalent of mass and kinetic energy. It is not possible to completely stop the flow of current through an inductance until the energy stored in it has been dissipated or stored elsewhere. This means that during a switching event, the voltage will increase until it ionizes the air between the contacts and produces an electric arc. The energy will then be dissipated in the form of very high voltages and low current arc crossing though an opening contact. Real-life systems do not have infinite currents or voltages; however, actual capacitance-charging inrush currents and voltage spikes produced by switching transients are often greater than those of the RESS, and capacitors, inductors or other components on the bus are rated to withstand. Equipment failures due to transients are particularly difficult to track, as they will rarely destroy a component in a single switching cycle. Transients can also induce or be induced as electrical noise, which can come from outside sources or from a different circuit within the same vehicle. Noise induced from electromagnetic interference is usually not a problem in high-voltage buses; however, transients from the switching components in the high power bus may induce noise in control circuits. Since the energy transferred through inductive and capacitive coupling depends on the derivatives of current and voltage, respectively, reducing the rate of change of these variables during transient events in the power bus, i.e., smoothing them out, results into reduced noise emission and thus induces low voltage power and control circuits.

Typical systems used to provide transient and noise protection include:

- Precharge relay or battery disconnect unit (BDU): A precharge relay charges the capacitances on the bus through a current limited circuit when the battery is initially connected to the bus. This is done for about 5 capacitive (R/C) time constants, considered as the time necessary for inrush current to fully subside. The current limited circuit is then bypassed. Alternatively, the precharge relay can be controlled with current measurements, if they are available. A BDU has the added benefit of disconnecting both terminals of the battery, as suggested by SAE J2344.
- Transient voltage suppression diodes (TVS): An avalanche breakdown device, which exhibits very high resistance under normal working conditions and rapidly reduces the resistance of the power relay switch once a voltage threshold is exceeded. They are very fast and capable of dissipating large amounts of power, but only for a very short period of time. The purpose of a TVS diode is to prevent delicate components on the power bus from being exposed to high-voltage peaks during a transient event caused by switching an inductance, lightning, etc., by clamping the bus to ground. Other similar devices such as metal-oxide varistors, spark gaps, and gas tubes have similar functioning and may be used to achieve this. TVSs are often built into individual components (motor controllers, etc.); however, it may be necessary to install them if they are absent or if excessive peak voltages are observed during switching.
- X/Y capacitors: These are nonpolarized capacitors, designed to be able to withstand high-voltage transients. They absorb noise on the bus and flatten peaks on transients. X capacitors are connected between phases and are effective against symmetrical noise or interference, whereas Y capacitors, connected between a phase and ground act against asymmetrical noise. The ratings X and

Y are related to the safety level of the capacitor according to EN60384-14. Since Y capacitors are connected between the line and chassis ground, they present a risk of shock if they fail to be closed. X capacitors on the other hand are placed between live phases and would therefore cause a fuse or similar protection system to open when they fail.

15.8 Conclusions

The energy management and power flow management strategy has been discussed in this chapter for wireless dynamically charging electric vehicles. A fundamental architecture of power flow management has been described along with their individual functions. Similar to the hybrid electric vehicle case, which splits the power sources during the various driving condition, the OLEV system have a hybrid electrical power sources from the on-board battery and wirelessly delivered power from the road-embedded power supply system. Considering the motor drive power requirement under various driving conditions, the energy management and power flow management control should be performed to protect the system and optimized operational efficiency of the whole system operations. The on-road power supply system and on-board power pickup system are described in view of their major power components and power flow management. Considering the specified battery charging and discharging characteristics, the power flow control over the battery power, wirelessly delivered power, and brake regeneration power are managed though the given control algorithm responding to the demand of motor driving power depending on the driving conditions. The accurate power distribution control is implemented through the vehicle level integrated control architecture, as discussed in this chapter.

In addition to the magnetic field optimization approaches, related to the high power consideration, the battery performance requirements for OLEV system are discussed, followed by the discussions on possible issues and remedies of on-board electrical safety for OLEV application.

References

1. McKinsey & Company (2009) Electrifying cars: how these industries will evolve McKinsey quarterly, Nov 3, 2009
2. Suh NP, Cho DH, Rim CT (2010) Design of on-line electric vehicle (OLEV) plenary presentation. CIRP design conference Nantes 19–21 Apr 2010
3. Suh IS (2010) KAIST on-line electric vehicle as a future transportation technology with low carbon and green growth international forum on public transportation policy hosted by KOTI, Korea, Mar 2010
4. Suh IS (2010) On-line electric vehicle development invited speaker at the 2010 international forum on electric vehicle, Korea, 17–19 June 2010

5. Suh IS (2011) On-road electrification for optimized power supply in OLEV application. Trans Soc Des Process Sci 15(3):13–27
6. Suh IS et al (2011) Control algorithm of HVAC system power management in OLEV application. Trans Soc Des Process Sci 15(3):29–42
7. http://www.uscar.org/guest/article_view.php?articles_id=85, 2003

Chapter 16
System Structure and the Allocation of Wireless Charging Power Supply Systems for OLEV System

Young Jae Jang

Abstract The system structure of OLEV is analyzed in terms of the charging logistics of electric vehicles, followed by a mathematical optimization model for an economical design. Since the battery cost constitutes a large fraction of the total vehicle cost, the capacity of the battery is the key factor in the design of an OLEV system. Another critical important issue is the allocation of the charging infrastructure, including a set of wireless power supply systems installed throughout the route. It is based on a mathematical model and optimization method developed earlier. A simple numerical case with hypothetical data is presented to verify the model.

16.1 Introduction

16.1.1 Overview

The goal of this chapter is to present the system structure of the OLEV system from a logistic perspective and a logical design framework for the commercialization and deployment of that system. A mathematical optimization model for an economical design is presented. Given that the battery cost accounts for a significant portion of the total vehicle cost, the capacity of the battery is the key factor in the initial design of an OLEV system. Another critical design parameter in an OLEV system is the allocation of the charging infrastructure, which consists of a set of wireless power supply systems installed underground throughout the route. This chapter builds upon earlier work in [1–3] by the author and his colleagues on the mathematical model and optimization methods that simultaneously determine the optimal allocation of the charging infrastructure and the optimal size of the battery for an OLEV system. A simple numerical case with hypothetical data is presented to verify the

Y.J. Jang (✉)
Industrial and Systems Engineering, Korea Advanced Institute of Science and Technology (KAIST), 291 Daehak-ro, Yuseong-gu, Daejeon 34141, South Korea
e-mail: yjang@kaist.ac.kr

N.P. Suh and D.H. Cho (eds.), *The On-line Electric Vehicle*,
DOI 10.1007/978-3-319-51183-2_16

model. The optimization models presented in this chapter can be used to design an OLEV-based mass transportation system. In addition, the models provide insight into the design of an OLEV-based advanced transportation system.

16.1.2 Electric Transit Bus System and Current Issues

The application of the conventional system structure of electric vehicles such as battery-powered electric buses—which use an electric motor as a power source with an in-vehicle battery pack—to electric transit bus systems has a number of serious technical and economic drawbacks. High battery costs, a long recharging time, and the limited availability of charging stations are among the well-known problems of current battery electric vehicles. In addition to the existing problems of conventional electric vehicles, further limitations and issues arise when these vehicles are used for transit bus systems. Considering that a mass transit bus needs a much larger battery pack than a small passenger vehicle, the initial investment needed to deploy electric buses for a mass transit system is significant. A larger battery also requires more recharging downtime, which refers to the nonoperational period of the electric vehicle during battery charging [3]. As an electric transit bus is likely to require a larger battery pack, the recharging downtime would be several hours with current charging solutions. If the bus system requires around-the-clock service, more buses may be needed to compensate for the lengthy downtime. The last and most serious problem in the application of current electric solutions to mass transportation is the limited driving range between of the charge. Current battery limitations prevent all-electric transit buses from operating all day after an overnight charge [4].

The online electric vehicle (OLEV) system, in which a wireless power transfer infrastructure installed under the road, charges a fleet of electric buses that operate on that road. The technology is innovative, in that the battery in the bus is charged while the bus moves or stops over the charging infrastructure. The basic concept of the OLEV system was discussed in Part I. As shown in Fig. 16.1, the vehicle collects electric energy from the underground power supply system using SMFIR. Because charging takes place while the vehicle is in motion or stationary, the system eliminates the major problem of conventional electric vehicles, the need to discontinue bus operation to charge the battery.

16.2 System Design Structure

The OLEV system is interpreted from the perspective of energy logistics combined with vehicle utilization. An structural analysis of the OLEV system was first proposed in [2]. Public transportation is defined in this chapter as a fleet of vehicles that circulate on fixed routes and carry passengers from one place to another. The

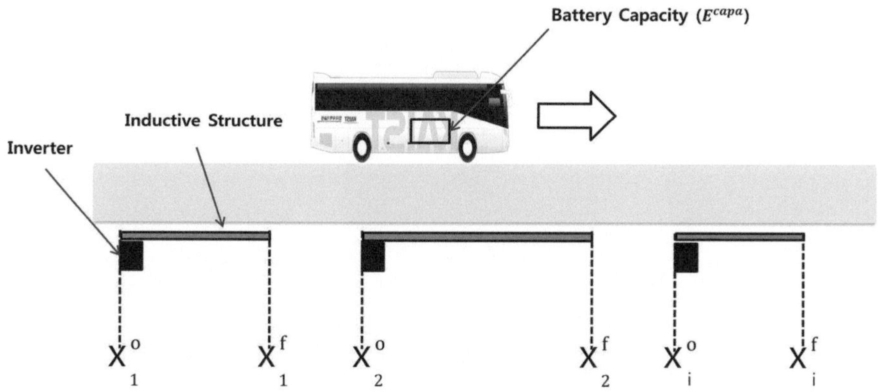

Fig. 16.1 OLEV system configurations

demands of the passengers are known, and it is assumed that sufficient buses operate to meet these demands. There is a predetermined daily operating time, during which the buses are in continuous service.

The parameters for a general electric transit bus, which is a conventional battery electric bus, are defined first. The benefits of the OLEV system are then explained by a comparison of its energy logistics with those of other types of electric buses. For our system structure analysis, the following terms are defined.

- **Mean charging time** (T_c): the average charging time of a vehicle.
- **Mean downtime for energy refueling** (T_d): the average time a vehicle is not in operation during battery recharging or energy refueling.
- **Mean per-charge operation time** (T_u): the operational uptime of the vehicle.
- **Vehicle utilization** (U_v): the fraction of time the electric vehicle is operational over all available time.

Figure 16.2 illustrates the structures of various types of electric transportation systems. For simplicity, it is assumed that a transportation system is in service around the clock. The horizontal axis in each sub-figure represents the time frame. The white and red bars represent the operating and charging times, respectively. Figure 16.2a shows the charging logistics of a conventional electric vehicle such as a battery plug-in-type vehicle, in which the battery is recharged via a cable connection between the vehicle and the charging station. As a cable connection is required in this case, the vehicle cannot operate while the battery is being charged. As a result, the mean downtime for energy refueling for the plug-in-type structure, $\overline{T}_d^{\text{plugin}}$, is equal to the mean charging time $\overline{T}_c$, and the vehicle utilization is expressed as follows:

$$\overline{T}_d^{\text{plugin}} = \overline{T}_c, \tag{16.1}$$

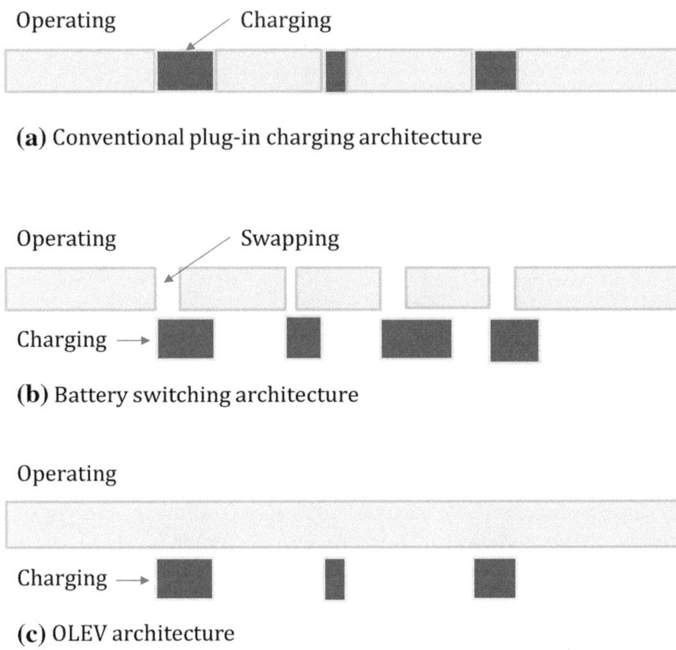

Fig. 16.2 System structures of electric buses

$$U_{\mathrm{v}}^{\mathrm{plugin}} = \frac{\overline{T}_{\mathrm{u}}}{\overline{T}_{\mathrm{u}} + \overline{T}_{\mathrm{d}}^{\mathrm{plugin}}} = \frac{\overline{T}_{\mathrm{u}}}{\overline{T}_{\mathrm{u}} + \overline{T}_{\mathrm{c}}}. \tag{16.2}$$

Figure 16.2b is the case of an electric vehicle with a battery-swapping solution. Firms such as *Better Places*™ provide battery-switching stations that allow drivers to pull in and swap their batteries as easily as filling up with gas. With this structure, the battery-switching time is equivalent to the downtime for energy refueling, $\overline{T}_{\mathrm{d}}^{\mathrm{switch}}$. This solution is effective only if the energy refueling time is less than the battery charging time, i.e.,

$$\overline{T}_{\mathrm{d}}^{\mathrm{switch}} < \overline{T}_{\mathrm{c}}. \tag{16.3}$$

As a consequence, the vehicle utilization for this case is expressed as

$$U_{\mathrm{v}}^{\mathrm{switch}} = \frac{\overline{T}_{\mathrm{u}}}{\overline{T}_{\mathrm{u}} + \overline{T}_{\mathrm{d}}^{\mathrm{switch}}}. \tag{16.4}$$

If the condition in Eq. (16.3) is met, the following inequality can be stated:

$$U_{\mathrm{v}}^{\mathrm{switch}} > U_{\mathrm{v}}^{\mathrm{plugin}}. \tag{16.5}$$

The final system structure shown in Fig. 16.2c depicts the wireless power transfer solution applied in the OLEV system. The electric energy is transferred wirelessly with no physical constraints, so the operation of the vehicle is not interrupted. As long as the power supply systems are appropriately installed along the vehicles' route, constant operation is possible. Therefore, in theory, without consideration of the cost issue, the mean downtime for energy recharging is zero and vehicle utilization is equal to 1.

$$\overline{T}_{\mathrm{d}}^{\mathrm{olev}} = 0, \tag{16.6}$$

$$U_{\mathrm{v}}^{\mathrm{olev}} = \frac{\overline{T}_{\mathrm{u}}}{\overline{T}_{\mathrm{u}} + \overline{T}_{\mathrm{d}}^{\mathrm{olev}}} = 1. \tag{16.7}$$

The OLEV system effectively eliminates the problem of recharging time by decoupling the physical charging event and vehicle operation. However, there is one crucial cost issue that must be considered in this case. That installation of the power supply systems in the road requires an initial investment cost, an issue that is dealt with in the next section.

16.3 OLEV System Modeling

16.3.1 Optimization Issue

As discussed in the previous section, the operating performance is unaffected in the wireless charging structure as long as there is sufficient charging infrastructure in the vehicles' path. The question then becomes how much charging infrastructure is sufficient. As the cost of the battery and the charging infrastructure accounts for a significant proportion of the total investment cost, economical evaluation of these parameters is critical to the success of the commercialization of an OLEV system. The general optimization problem formulation for economical determination of the battery size and allocation of charging infrastructure is first introduced. An optimization model is required to determine the most economical parameter design [1–3]. The design parameters are the size of the battery and the allocation of the charging infrastructure along the route. The basic system requirement is the provision of sufficient power to complete the service. To better illustrate the optimization problem, the following two extreme cases are considered.

16.3.2 Operational Rules and Assumptions

Before a mathematical model is constructed, the notations and operational rules for the OLEV system are first defined. In the model, a public transportation system with buses circulating on a single loop is considered. The model can be extended to multiple loops, which is discussed later. The buses operate with the following rules.

(1) Multiple buses operate on the route.
(2) Each bus is equipped with a battery of identical size.
(3) There are multiple stations and each bus stops at every station.
(4) There is one base station at which the buses are idle when they are not in service. Each service begins and ends at the base station.
(5) Once a bus completes a loop (or completes a service), it remains idle at the base station for a certain amount of time, referred to as the resting time.
(6) During the bus resting time at the base station, the battery is charged to its maximum level before service resumption.

A charger is installed at the base station. To fully charge the batteries at the base station, the charger requires a sufficient power supply. The charging infrastructure system in our model refers specifically to the power supply systems, each of which comprises one inverter and one power cable. The length of the power cable may vary. Figure 16.1 illustrates the design variables for system optimization. We need to determine the optimal allocation of the power supply systems, by identifying the number of power supply systems required and the length of each power supply system.

We define the base station as the reference point. Any point on the route is described by the travel distance x from the reference point. As the route is circular, the starting point, where $x = 0$, and the end point, where $x = L$, both indicate the base station. We consider a vehicle making one circular trip to construct the model. The vehicle begins the trip at $x = 0$ and completes it at $x = L$.

As shown in Fig. 16.1, the starting and ending points of the ith power supply systems are denoted by x^o and x^f, respectively. These values indicate the distance measured from the base station. Suppose that N power supply systems have been allocated, then $i = 1, \ldots, N$. Note that N is one of our decision variables.

(1) *Decision variables*: For convenience, we define the spatial set X as the locations of the power supply systems along the route:

$$X = \left\{ x \middle| x_1^o \leq x \leq x_1^f, \ldots, x_N^o \leq x \leq x_N^o \right\}. \tag{16.8}$$

The goal of the optimization is to find the optimal values of the spatial set with the given battery capacity. We define t as a continuous variable indicating the vehicle travel time measured from the base station. Let $V(t)$ be the velocity projection of the bus at time t. The relationship between the displacement variable (x) and the temporal variable (t) is described as follows:

$$x = \int_{o}^{t} V(t)dt. \tag{16.9}$$

With the relationship expressed in (16.9), we also define the temporal set T as the time a vehicle travels over a power supply system:

$$T = \left\{ t \middle| t_1^o \leq t \leq t_1^f, \ldots, t_N^o \leq t \leq t_N^o \right\}. \tag{16.10}$$

Let us now represent Eq. (16.8) as function $x = Q(t)$. Note that this function is monotonically increasing. There therefore exists a unique solution for its inverse function, $t = Q^{-1}(x)$. We use this inverse function when we set up our optimization problem.

(2) *Cost function*: For the optimization model, we propose a cost function that consists of the total battery cost and the total cost of the power supply systems. With this cost function, the model is able to find economical design parameters by considering the trade-off between the size of the battery and the allocation of the power supply systems. We assume that the battery cost is a function of the battery capacity E^{capa}, which is reasonable because the battery packs widely used in commercial electric vehicles are composed of multiple battery cells. The capacity of the battery pack is determined by the number of cells within it. If the cost of each cell is known, the cost of the battery pack can easily be estimated. We denote F_{b}^{c} as the cost function of the battery. Suppose that k number of buses are operating. According to our rule, all buses operating on the route are equipped with the same-sized battery. Hence, the total battery cost is $k \cdot F_{\text{b}}^{\text{c}}(E^{\text{capa}})$.

The power supply system cost is divided into a fixed cost and a variable cost. The fixed cost is the cost incurred regardless of the length of the power supply system (specifically, the length is determined by the inductive cable). The major part of this cost is made up of the inverter cost and the labor cost to connect the power grid to the inverter. Note that each power supply system needs one inverter. The variable cost depends on the length of the power supply system: the longer the power supply system, the higher the cost. The functions of the fixed and variable costs are denoted by F_{f}^{c} and F_{v}^{c}, respectively. Because the fixed cost is identical across the power supply systems, the total fixed cost for N power supply system is $F_{\text{f}}^{\text{c}}(N)$. As the variable cost depends on the length of each power supply system, it is a function of the displacement of each power supply system's starting point x^o and end point x^f. The total cost is thus

$$k \cdot F_{\mathrm{b}}^{\mathrm{c}}(E^{\mathrm{capa}}) + F_{\mathrm{v}}^{\mathrm{c}}(X) + F_{\mathrm{v}}^{\mathrm{c}}(N). \tag{16.11}$$

(3) *Energy dynamics*: We denote $E(t)$ as the amount of energy in the battery at time t. The energy level should be within the lower and upper limits during travel:

$$E^{\mathrm{l}} \leq E(t) \leq E^{\mathrm{u}}, \tag{16.12}$$

where E^{l} and E^{u} are the lower and upper limits of the battery level, respectively. These values have the following relationship:

$$\begin{aligned} E^{\mathrm{u}} &= \alpha \cdot E^{\mathrm{capa}} \\ E^{\mathrm{l}} &= \beta \cdot E^{\mathrm{capa}} \\ 0 &< \alpha < \beta < 1. \end{aligned} \tag{16.13}$$

The upper and lower limit parameters, α and β, are usually given by the battery provider. For convenience, we introduce the variable E_0, which is

$$E_0 = E^{\mathrm{u}} - E^{\mathrm{l}} = (\alpha - \beta)E^{\mathrm{capa}}. \tag{16.14}$$

Value E_0 is the actual usable capacity of the battery. Note that in our optimization problem, we are interested in the cost of the battery relative to that of the power supply system. Therefore, it is more convenient to use the relative value in our battery cost, that is, $k \cdot F_{\mathrm{b}}^{\mathrm{c}}(E_0)$, rather than the absolute cost, $k \cdot F_{\mathrm{b}}^{\mathrm{c}}(E^{\mathrm{capa}})$. Given our assumption that the upper and lower limits of the battery are proportional to values α and β regardless of the battery size, the use of the relative value E^0 does not change the optimal solution. The degree of energy fluctuation depends on the rates of energy consumption and supply. We define $P_d(t)$ as the energy consumption rate. This quantity depends on the velocity trajectory, the road gradient, and the use of peripheral devices such as the bus air conditioner. We define $P_s(t)$ as the energy supply to the battery at time t. This quantity is actually the charging rate of the battery and depends on the allocation of the power supply systems.

Note that the state of charge level is assumed to be linearly proportional to the charging time, as long as the level of the charge is between E^{l} and E^{u}, as shown in Fig. 16.3. This assumption is based on the mechanism of the charging controller in the OLEV system. The charging limits are set to maximize the battery life. Therefore, if the vehicle is traveling on a route with a power supply system installed, the level of the charge in the battery will increase. Note that although the entire spectrum of the charging state is nonlinear, as shown in Fig. 16.3, the charging state within E^l and E^u, which is the actual state range in the OLEV system, is almost a straight line. Therefore, the charging rate is assumed to be constant.

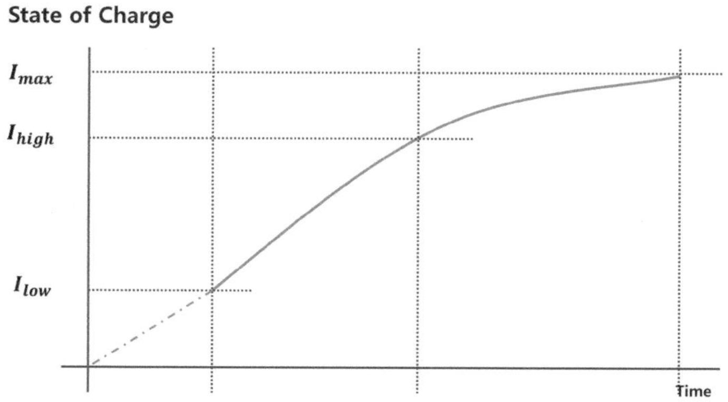

Fig. 16.3 State of charge profile

This assumption of linear energy charging and discharging is widely used in electric vehicle modeling [3].

The required power for the vehicles at a specific point x on the route, denoted by $P_{te}(x)$, is approximated with $F_{te}(x)$ and $V(x)$, which are the required force and velocity for point x, respectively. Again, $V(x)$ is the velocity profile, in which a vehicle is expected to run at $V(x)$ at point x on the route. The required force at each point is decomposed into the five applied forces: the rolling resistance force F_{rr}, air resistance force F_{ad}, hill climbing force F_{hc}, linear acceleration force F_{la}, and wheel acceleration force F_{wa}. In general, the magnitude values of both air resistance and wheel acceleration are relatively small; thus, we assume that these two forces can be neglected. As a consequence, the driving force can be described as

$$F_{te} = F_{rr} + F_{hc} + F_{la}. \tag{16.15}$$

Note that the acceleration force can be calculated from the velocity profile and that the hill climbing force can be calculated from the route information (hill profile). The power and force requirements, $P_{te}(x)$ and $F_{te}(x)$, respectively, are functions of the placement x. For convenience, we convert these variables into functions of time t, $P_{te}(t)$ and $F_{te}(t)$. Our primary focus in this modeling is to understand the battery charging behavior on the route. The charging rate is a function of time; hence, it is more convenient to describe the variables as a function of time rather than a function of displacement. This conversion can be achieved using the velocity profile equation $V(x)$ and the definition of the velocity $v(t) = \mathrm{d}x/\mathrm{d}t$; thus, $\mathrm{d}x/\mathrm{d}t = V(x)$. The required power at time t is then expressed as

$$P_{te}(t) = F_{te}(t) \cdot V(t). \tag{16.16}$$

We introduce the required battery power at time t, denoted by $P_d(t)$. In general, the electricity from the battery is served to the electric motor through several separate processes, such as gears and controllers, and there is further energy dissipation in the form of sound, light, and heat. As a result, some degree of efficiency loss is inevitable. We define the efficiency loss η and assume that it is a constant. Then, the battery power requirement is then described as

$$P_{\mathrm{d}}(t) = P_{\mathrm{te}}(t)/\eta. \tag{16.17}$$

Thus,

$$P_{\mathrm{s}}(t) = \begin{cases} 0, & \text{if } t \in T^{\mathrm{c}}, \\ p_{\mathrm{s}}, & \text{if } t \in T \quad \text{and} \quad E(t) < E^{\mathrm{u}}, \\ 0, & \text{if } t \in T \quad \text{and} \quad E(t) = E^{\mathrm{u}}. \end{cases} \tag{16.18}$$

The battery energy level in this case is described as

$$\frac{\mathrm{d}E(t)}{\mathrm{d}t} = -P_{\mathrm{d}}(t) + P_{\mathrm{s}}(t). \tag{16.19}$$

The general optimization problem with the cost function in (16.11) is constructed as

$$\min_{X,N,E_0} k \cdot F_{\mathrm{b}}^{\mathrm{c}}(E_0) + f_{\mathrm{c}}^{\mathrm{c}}(X) + f_{\mathrm{v}}^{\mathrm{c}}(N), \tag{16.20}$$

subject to Eqs. (16.9), (16.12), (16.13), (16.16)–(16.18).

16.3.3 Systems Optimization Category

The optimization problem described in the previous section is the generalized model. Although the general model illustrates well the underlying concept of the optimization, a more detailed equation setting is required to make use of the optimization model in the design of an actual OLEV system.

Here, we propose two optimization models, a closed environment system model and an open environment system model. For both models, the OLEV bus travels a service route with scheduled times for each stop. The major difference between the two environments is the degree of travel uncertainty. In the closed environment system, we assume that there are no traffic congestion and no unexpected traffic delays. The environment is separated from regular vehicular traffic and the drivers follow the velocity regulations. Good examples of an OLEV system that operates in a closed environment are the OLEV shuttle on the KAIST campus and the OLEV-based trolleys in Seoul Grand Park. In both systems, the drivers follow the speed regulations and travel the routes with scheduled stops. An open environment, in contrast, features traffic conditions that mirror those in a normal traffic environment on any city's streets. Compared with their counterparts in a closed environment, the drivers in an open environment system may face traffic congestion and other unexpected traffic events. From the optimization modeling perspective, the major difference between the two systems lies in the level of granularity in the energy demands on the route. As the velocity profile is estimated and the road gradient is known in the closed environment, we can evaluate the energy requirement at each point.

16.4 Closed Environment Model

The major characteristic of the closed environment model is our use of the route's velocity profile and road gradient. In a closed environment system, there are limited traffic congestion and few traffic lights. We expect the vehicles in this system to closely follow the timetabled schedule and velocity regulations. The velocity profile and road gradient allow us to estimate the energy demand at time t, denoted by $P_d(t)$. There are various methods to evaluate energy demand, and more detailed information can be found in [2, 3]. The objective function defined in the general optimization model is specified as follows:

$$k \cdot c_{\mathrm{b}} \cdot E_0 + c_{\mathrm{f}} \cdot N + \sum_{i=1}^{N} c_{\mathrm{v}} \cdot \left(x^f(i) - x^o(i)\right), \tag{16.21}$$

where c_{b} is the battery cost per unit of energy capacity, c_{f} is the fixed cost to install a power supply system, and c_{v} is the variable cost of each power supply system. The cost of the battery is linearly proportional to the unit capacity. This linear cost approximation is widely used in the industry [1–3]. We now construct the constraint equations on the basis of the general equation setting in the general optimization model. Suppose that a vehicle is about to move to the point at which the ith power supply system is located. With our notation, this point is indicated as $x^f(i)$. The time at which the vehicle is at this point is indicated by $t^f(i)$. Again, the mapping between the displacement variable (x) and temporal variable (t) is evaluated using Eq. (16.9). The level of energy at $t^f(i)$ is $E(t^f(i))$. The vehicle now departs from the ith power supply system and continuously travels along a route with no power supply systems until it reaches the next power supply system, i + 1, at point $x^0(i + 1)$. Again, the moment at which the vehicle arrives at the next power supply system is $t^0(i + 1)$. The following constraint must be satisfied:

$$E\left(t^f(i)\right) - \int_{t^f(i)}^{t^0(i+1)} P(t)\mathrm{d}t \geq E^{\mathrm{l}}. \tag{16.22}$$

The first term in Eq. (16.22) is the energy level when the vehicle leaves the ith power supply system, and the second term indicates the amount of energy consumed while the vehicle is traversing the area with no power supply system. Therefore, the term on the left-hand side represents the amount of energy in the battery when the vehicle arrives at the beginning of the $(i + 1)$th power supply system. The energy level needs to be greater than the lower energy limit of the battery. In other words, the power supply system needs to be installed in such a way that the energy level is maintained at about the lower limit. We now evaluate the amount of energy at the end of the ith power supply system, that is, the energy supply and consumption between $x^f(i)$ and $x^f(i + 1)$. The level of energy in the battery at $x^f(i)$ is $E(t^f(i))$. The level of energy at $x^f(i + 1)$ is thus described as

$$\min\left\{E\left(t^f(i)\right) - \int_{t^f(i)}^{t^f(i+1)} P_{\mathrm{d}}(t)\mathrm{d}t + p_{\mathrm{s}} \cdot \left(t^f(i+1) - t^o(i+1)\right), E^{\mathrm{u}}\right\}. \tag{16.23}$$

Note that the battery's energy level when the vehicle reaches the end of the $(i + 1)$th power supply system should be less than the upper limit of the battery E^{u}. If the charge amount is more than the upper limit, then the energy level would be E^{u} at $x^f(i + 1)$, owing to the power supply rule described in Eq. (16.18). Let T_1 be the time at which the vehicle reaches the base station and L^{m} be the maximum length of a power supply system. The closed environment optimization problem is then described as follows:

$$\min \quad k \cdot c_{\mathrm{b}} \cdot E_0 + c_{\mathrm{f}} \cdot N + \sum_{i=1}^{N} c_{\mathrm{v}} \cdot \left(x^f(i) - x^o(i)\right), \tag{16.24}$$

subject to

$$E(t^f(i)) - \int_{t^f(i)}^{t^0(i+1)} P_{\mathrm{d}}(t)\mathrm{d}t \geq E^{\mathrm{l}}, \quad \text{for } i = 0, \ldots, N, \tag{16.25}$$

$$E\left(t^f(i+1)\right) = \min\left\{E\left(t^f(i)\right) - \int_{t^f(i)}^{t^f(i+1)} P_{\mathrm{d}}(t)\mathrm{d}t + p_{\mathrm{s}} \cdot \left(t^f(i+1) - t^o(i+1)\right), E^{\mathrm{u}}\right\}, \quad \text{for } i = 0, \ldots, N-1, \tag{16.26}$$

$$t^f(0) = 0, \quad t^0(N+1) = T, \tag{16.27}$$

$$E(0) = E^{\mathrm{u}}, E(T_1) \geq 0, \tag{16.28}$$

$$x^f(i) - x^0(i) \leq L^m, \quad \text{and} \quad x^0(i) < x^f(i), \quad \text{for } i = 1, \ldots, N, \tag{16.29}$$

$$x^f(i) < x^0(i+1), \quad \text{for } i = 1, \ldots, N-1, \tag{16.30}$$

$$x^f(i), \quad x^0(i), \quad \text{and} \quad E_0 \geq R^+. \tag{16.31}$$

In the optimization model for the closed system, the inequality in Eq. (16.25) represents the minimum energy requirement constraint and Eq. (16.16) is the energy level when the bus leaves the end point of each power supply system. The boundary conditions are defined in Eqs. (16.27) and (16.28). The parameter relationships are defined in Eqs. (16.29) and (16.30). The decision variables and their bounds are defined in Eq. (16.31). The closed system optimization model can be solved using the metaheuristics method. Its solution algorithm is given in [1–3].

16.5 Open Environment Model

16.5.1 System Optimization Modeling

In an open environment, it is not easy to estimate the trajectory of the entire energy consumption over the route because of uncertainties that arise from traffic congestion and the traffic lights at intersections. As a result, a certain level of approximation is needed in the analysis. The open environment system is characterized by the two following major assumptions:

- The power supply systems are installed only in bus station areas.
- The energy demand between a pair of stations is known.

It is assumed that a vehicle stops at every station. The stations are thus good candidate locations for the installation of the power supply systems because the amount of charge in a battery is proportional to the length of time the vehicle spends above the power supply system. A OLEV bus is more likely to spend time over power supply systems installed in station areas over those along the road. However, the installation of power supply systems only at bus stations does not necessarily mean that advantage is not being taken of the OLEV system's dynamic charging ability; the battery is also charging while the bus is in motion.

As shown in Fig. 16.4, the bus station area includes not only a loading/offloading area but also a bus queueing area. Note that in the figure, the stopping zone is the location at which the bus makes a complete stop to load and offload passengers. When a bus approaches the station area, it slows down and moves slowly toward or waits in the station area. If a power supply system is installed in the station area, charging can be done as the bus approaches the stopping zone or after it reaches that zone. Therefore, although the power supply systems are installed in the station areas, the dynamic charging capability of the

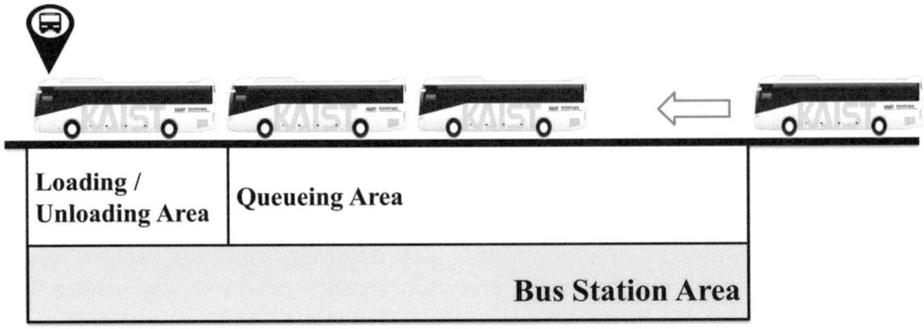

Fig. 16.4 Bus station

OLEV system can still be effectively used [5, 6]. Hereafter, *station* refers to the *station area*.

The other assumption is that the energy demand between the stations is known. It is assumed that information about the aggregate energy required to move from one station to the next is at least available. This assumption of known aggregate energy demand between points is commonly found in the charging allocation problems for electric and other alternative vehicles in the literature, including [7–9].

The charging infrastructure allocation problem is translated into the problem of identifying stations for the installation of the power supply systems and assigning an appropriate length of the power supply system to each station selected. To illustrate the problem, suppose that there is a single route with N_c candidate stations. As it is assumed that the base station will always have a power supply system, the candidate stations do not include the base station.

Let S_i be the notation for a station where $i = 0, \ldots, N_c$. The station index i is sequentially arranged such that $x_i < x_i + 1$, where x_i is the distance of the ith station from the base station. Station S_0 is the base station, which is the starting point of each trip. The bus circulates its route and comes back to the base station. For convenience, let the $(N_c + 1)$th station also be the base station. A continuous decision variable X_i is defined, which is the length of the power supply system installed at each station. The minimum and maximum lengths are denoted by $X^{\min}$ and $X^{\max}$, respectively. The minimum length is the length of the bus stopping zone (loading/unloading area), whereas the maximum length is the length of the entire station area.

The parameter p_s is the amount of energy supplied per unit length of a power supply system. The amount of energy supplied at station i, $p_s \cdot X_i$, is then approximated. In other words, the amount of energy charged to a battery at a station is linearly proportional to the length of the power supply system. The charge amount is actually proportional to the amount of time a bus spends over the power supply system. It is therefore reasonable to assume that the charge amount increases monotonically with an increase in the length of the power supply system. In the model, a linear approximation is used to express this monotonically increasing trend. The charge rate p_s is then estimated from a linear regression of the experimental results.

Note that the vehicle operates on a circular route that begins and ends at the base station. E_i is defined as the energy level of the battery just before the vehicle arrives at S_i. Let d_i be the energy demand between S_i and S_{i+1}. Note that E^u is the maximum energy level of the battery and that the battery is always charged to this level at the base station. Therefore, the energy level immediately before a bus reaches the first station is $E^u - d_0$. The maximum capacity is sufficiently large for it to travel to the first station, where it is supplied with $p_s \cdot X_1$ energy. The energy level when the vehicle leaves the station should then be $E_1 + p_s \cdot X_1$. This quantity should be less than the maximum capacity of the battery, E^u. It should also be large enough to allow the vehicle to travel to the next station, S_2. Hence, the following condition should be satisfied: $d_1 < E_1 + p_s \cdot X_1 < E_0$.

As noted, the minimum and maximum lengths of the power supply systems are denoted by X^{min} and X^{max}, respectively. Let y_i be a binary decision variable that indicates whether a power supply system is installed at S_i. If a power supply system is installed at S_i, then $y_i = 1$; otherwise, $y_i = 0$. If a power supply system is installed, a cost is incurred. Similar to the closed environment model, the installation cost comprises a fixed cost and a variable cost, denoted by c_f and c_v, respectively. The unit cost of the battery is denoted by c_b.

The notations, decision variables, and parameters for the open environment system are summarized in Table 16.1. If the energy dynamics described above are generalized, the optimization problem can be expressed as follows:

$$\min \quad k \cdot c_b \cdot E_0 + \sum_{i=1}^{N} c_v \cdot X_i + \sum_{i=1}^{N} c_f \cdot y_i, \tag{16.32}$$

subject to

Table 16.1 Notations for open environment OLEV system

Notation	Description
N_c	Number of candidate stations
S_i	Station i in the route
x_i	Distance to station S_i measured from the base station
X_i	Continuous decision variable indicating the length of the power supply system installed at S_i
X^{min}	Minimum length of the power supply system for S_i
X^{max}	Maximum length of the power supply system for S_i
p_s	Energy charge per unit length of the power supply system
E_i	Energy level in the battery just before the vehicle arrives at S_i
d_i	Energy demand between S_i and $S_i + 1$
y_i	Binary decision variable indicating whether a power supply system is installed at S_i

$$E_1 = E^{\mathrm{u}} - d_0, \tag{16.33}$$

$$E_i = E_{i-1} + P_{\mathrm{s}} \cdot X_{i-1} - d_{i-1}, \quad i = 2, \ldots, N+1, \tag{16.34}$$

$$E_i \geq E^{\mathrm{l}}, \quad i = 1, \ldots, N+1, \tag{16.35}$$

$$E_i + P_{\mathrm{s}} \cdot X_i \leq E^{\mathrm{u}}, \quad i = 1, \ldots, N, \tag{16.36}$$

$$y_i \cdot X^{\min} \leq X_i \leq y_i \cdot X^{\max}, \quad i = 1, \ldots, N, \tag{16.37}$$

$$y_i \in \{0, 1\}, \quad i = 1, \ldots, N-1, \tag{16.38}$$

$$X_i \in R^{+}, \quad i = 1, \ldots, N-1. \tag{16.39}$$

Equation (16.32) represents the cost function of the discrete model. Like the closed case, the cost consists of the total battery cost and the charging infrastructure cost. Equations (16.33) and (16.34) are the energy balance equations. These equations relate the energy supply and consumptions to the energy level in the battery. Equations (16.35) and (16.36) are the energy limits in the battery. The rest of the equations are auxiliary equations for the optimization problems.

16.5.2 Numerical Analysis

To demonstrate how the proposed optimization model can be used, we present a numerical case with a hypothetical dataset. The OLEV buses will circulate around a 30-km route. There are 12 candidate stations and a base station. Once a bus arrives at the base station, it stays there for a while and its battery is fully charged before it resumes service. The parameter values are listed in Table 16.2. Note that the relative cost instead of the absolute cost is used for convenience. The fixed and variable power supply system costs are the relative costs against the unit battery cost. For instance, $c_{\mathrm{f}} = 10$ means that the absolute value of the fixed cost of the power supply system is \$15,000 when the absolute battery cost is \$1000 per kWh.

Table 16.2 Parameters of the sample case

Parameters	Value
N_{c} (stations)	12
k (buses)	10
$X^{\min}$ (m)	10
$X^{\max}$ (m)	60
c_{b} (unit cost/kWh)	1
c_{f} (unit cost/m)	0.2
c_{v} (unit cost/station)	15
p_{s} (kWh/m)	0.1

Table 16.3 Stations (S_i) and energy demand (d)

S_i	0	1	2	3	4	5	6	7	8	9	10	11	12
d_i (kWh)	4	5	1.67	2	3.33	4	2.33	4	4.33	1.66	4	1	2.66

Table 16.4 Optimization result

Installed station	X
S_1	40 m
S_2	46.67 m
S_5	60 m
S_7	60 m
S_8	56.67 m
S_{10}	60 m
E_0	7.67 kWh

Table 16.3 shows the required energy from station S_i to S_{i+1}. The optimal solution for the case of p_s is 0.1 kWh/m, which is summarized in Table 16.4.

The results indicate that power supply systems need to be installed at stations S_1, S_2, S_5, S_7, S_8, and S_{10}, and that they need to be 40, 46.67, 60, 60, 56.67, and 60 m long, respectively. The battery capacity E_0 is 7.667 kWh. However, when the number of vehicles is reduced to four, the optimal battery capacity becomes 40 kWh, and no power supply system installation is recommended because there is no benefit to having power supply systems along the route. Instead, a bus should be equipped with a large battery to complete the route without a charge. Note that 40 kWh is sufficient energy to complete a loop. This result makes sense because the power supply systems constitute the charging infrastructure and there is a cost benefit only when there are enough buses in the OLEV system.

16.6 Conclusions

The OLEV system is an innovative electricity-powered transportation system that remotely receives electricity from power supply systems buried underground. In this chapter, the OLEV system structure is interpreted from the perspective of energy logistics. A mathematical optimization model for allocating the power supply systems and determining the battery size is then presented. Although the model considers a simple single circular route, the underlying concept can be extended to more complex models. It would be valuable to investigate the effects of randomness in the speed on the optimal parameter design model proposed herein. The development of a stochastic optimization model that considers random speed would also be useful. Another interesting research topic would be to consider more realistic cost values and to perform a cost–benefit analysis for an OLEV system. The OLEV system is a new technology. The existing OLEV systems were thus developed by custom-made manufacturing rather than by mass production. Although it may not be possible to obtain the exact cost figures, an approximated economic cost–benefit analysis could be performed. Providing a logical approach to the following question would be an example of such a cost–benefit analysis: What is the maximum cost of a power supply system that would render an OLEV system economically viable and allow it to compete with other purely electric vehicles? Answering this question would be a worthwhile future research topic.

References

1. Jang YJ, Jeong S, Ko YD (2015) System optimization of the on-line electric vehicle operating in a closed environment. Comput Ind Eng 80(February):222–235
2. Jang YJ, Suh ES, and Kim JW (2016) System architecture and mathematical models of electric transit bus system utilizing wireless power transfer technology. IEEE Syst J 10 (2):495–506
3. Ko YD, Jang YJ (2013) The optimal system design of the online electric vehicle utilizing wireless power transmission technology. IEEE Trans Intell Transp Syst 14(3):1255–1265
4. Falk T (2012) An electric bus that wirelessly charges at each stop. Last modified November 19, 2012 http://www.zdnet.com/article/an-electric-bus-that-wirelessly-charges-at-each-stop/
5. Mohrehkesh S, Nadeem T (2011) Toward a wireless charging for battery electric vehicles at traffic intersections. IEEE ITSC, 5–7 Oct 2011, Washington DC, USA, pp 113–118
6. Ulrich L (2012) State of charge. IEEE Spectr 49(1):56–59
7. Johnson J, Chowdhury M, He Y, Taiber J (2013) Utilizing real-time information transferring potentials to vehicles to improve the fast-charging process in electric vehicles. Transp Res C: Emerg Technol 26(January):352–366
8. Song YU, Park S, Kim W, Hong JS, Jeon D, Lee S, Park J (2013) A study on the validity of the infrastructure construction cost for the commercialization of online electric vehicles. J Soc e-Bus Stud 18(1):71–95
9. Wang YW, Lin CC (2009) Locating road-vehicle refueling stations. Transp Res E: Logistics Transp Rev 45(5):821–829

Part III
Other Applications for OLEV Technology

Chapter 17
Application of SMFIR to Trains

Byung Song Lee and Soon Man Hong

Abstract A 200 kW wireless power transfer system for a tram and 1 MW wireless power transfer system for high-speed trains developed at the Korea Railroad Research Institute in collaboration with KAIST are described. A review of the need and advantages of the wireless power transfer system for trains is also presented, including the wireless power transfer systems for railways developed in other countries (Transrapid 09, Bombardier PRIMOVE). It also describes the current barriers to the use of WPT system and future strategies for overcoming them.

17.1 Introduction

In Chap. 4, the axiomatic design of SMFIR (Shaped Magnetic Field in Resonance) was presented. SMFIR has many applications other than automobiles, including trains and coastal/river ships. This chapter presents an application of SMFIR, the wireless power transfer system, to railroads that use fixed rails. It is an ideal application of SMFIR because of the fixed rails and the close proximity between tracks and the bottom of the train. The speed of these high-speed trains will no longer be limited by the mechanical contact between a catenary and pantograph, and also, the tunnels can be much smaller when the overhead superstructure is no longer necessary [1]. In this chapter, the basic technology of wireless power transfer to trolleys and high-speed train is described in detail, including the power supply infrastructure and the power-pickup system. Most of the materials presented in the preceding chapters are equally applicable to trains. Potential advantages of

B.S. Lee (✉)
Korea Railroad Research Institute, 176, Cheoldobangmulgwan-ro,
Uiwang-si, Gyeonggi-do 16105, Korea
e-mail: bslee@krri.re.kr

S.M. Hong
Korea Railroad Corporation, 240 Jungang-ro, Dong-gu,
Daejeon 34618, South Korea
e-mail: thehongs@kaist.ac.kr

N.P. Suh and D.H. Cho (eds.), *The On-line Electric Vehicle*,
DOI 10.1007/978-3-319-51183-2_17

SMFIR-based train systems include lower installation and maintenance cost, compatibility with environment, high energy efficiency, and speed-independent electric power transfer. This chapter also outlines the factors that affect the reliability and safety. In addition, conventional applications of wireless power transfer system in railroad such as Transrapid 09 and PRIMOVE are introduced and assessed. Finally, application of 60 kHz magnetic resonance wireless power transfer to tram and high-speed train is presented and discussed.

17.2 Need for Wireless Power Transfer Systems for Railways

Many diesel-engine-based railways around the world have been replaced with electric-motor-based railways. In Korea, these electric systems account for 60% of the Korean railway system in 2010. The key decision when designing an electric railway is the method of feeding electrical power to the rolling stock. In general, the systems available for feeding electricity to a railway involve the use of either a catenary and a pantograph, or a third rail, but there are a number of difficulties associated with these contact-type power supply systems, such as high installation and maintenance costs, poor reliability, safety considerations, high levels of downtime because of accidents, and substantial environmental impacts. A new type of contact-free power supply system for electric railways would make future transport systems to be more effective.

Electric railway systems must operate punctually and safely, which are considered to be the main advantages of existing electric railway systems. However, there is a clear need for an electrical supply system that is better than the contact-type feeding systems that are currently available. Maintenance costs will be lower for any improved system that does not have a catenary. The risk of power loss caused by the disconnection of the electrical contact is also lower in a contactless electrical supply system. Wireless power transfer system requires very little maintenance compared to that required by the pantograph-and-catenary systems used in existing feeding systems, and there is little risk of disconnection, making it far safer.

The wireless power transfer system shows great promise from a technological viewpoint. First, the wireless power transfer system can reduce the construction cost of railway systems by removing the catenary poles and related infrastructures. The tunnel cross-sectional area also can be reduced since the poles and catenaries are removed. Second, the maintenance cost of the infrastructure can be reduced significantly because of the absence of mechanical contact and wear. Also, the wireless power transfer system is a more environment-friendly system because of the lack of dust produced by the wearing of contacts. The absence of pantograph reduces air resistance, which consumes a large amount of energy, allowing higher speeds. These advantages are critically important for high-speed and inter-city railways, which require large construction and maintenance cost. Third, the feeding and collection capacities of wireless power transfer system are not dependent on the speed and the external environment (temperature variations, rain, and snow), allowing the system to be less susceptible to failure. Fourth, the wireless power transfer system is more safe and reliable, since no high voltage and high current carrying wire are exposed to the general public and the outdoor environment. Lastly, the wireless power transfer system is helpful in achieving aesthetically improved cities because no overhead power cable modules are required, which is an important advantage for the light railways that serve urban areas.

17.3 Developed Wireless Power Transfer Systems for Railways in Korea

17.3.1 Wireless Low-Floor Tram

Railways using wireless power transfer system have been built in Korea. The wireless low-floor tram system operates by battery power alone, i.e., without a catenary. The battery is fully charged by bringing a pantograph into contact with a catenary at the stations. This system allows the aesthetic appearance of the city to be maintained because road-embedded power cable modules are not required in the downtown area, but it has the disadvantage of having to be equipped with a large-capacity battery. To overcome this, the Korea Railroad Research Institute and KAIST have conducted joint research on methods involving wireless power transfer system since 2011, and they succeeded in developing a working wireless power transfer system on the test line at Osong in June 2013. In the test, a round trip of about 800 m to and from a station can be made only using power supplied through wireless power transfer system, both when stopped and when moving [2].

The first model of a large-capacity (60 kHz) wireless power transfer system in a wireless low-floor tram was designed using a current collection voltage of 800 V, a 15-m * 2 ea feeding segment, a 30-mm air gap between the road surface and the current collection module, a 200-kW feeding capacity, and a power transfer efficiency of 86%. Compared to the 20-kHz wireless power transfer system used in an online electric bus developed by KAIST in 2011, the efficiency of the wireless low-floor tram was higher (because the air gap of tram was smaller than that of bus) and the costs of installing the feeding and current collection devices were lower because the sizes of the current collection modules were reduced by a third or the number of power conversion devices (inverters) on the track were reduced by up to a ninth [3]. In particular, the wireless low-floor tram used a resonant frequency of 60 kHz rather than 20 kHz (as used in the existing Transrapid and Bombardier Primove systems). Increasing the resonant frequency by a factor of three helped to decrease the weight of the system by allowing smaller feeding or current collection modules to be used, and increasing the power transfer rate. Using the frequency of

Fig. 17.1 Schematic diagram of a wireless power transfer system and its application to low-floor tram demonstration

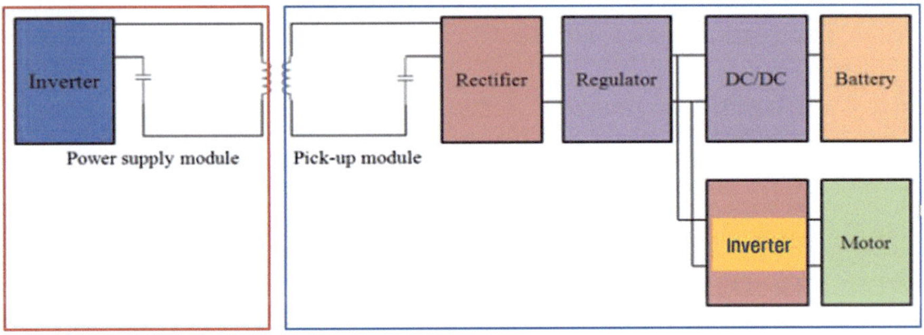

Fig. 17.2 Functional block diagram of a wireless transfer system applied to a low-floor tram

60 kHz could support large-capacity power conversion technology. Figure 17.1 shows a wireless low-floor tram using 60 kHz large-capacity wireless power transfer system technology. A concept diagram of the wireless power transfer system used in the wireless low-floor tram system is shown in Fig. 17.2. The wireless power transfer system is composed of a power supply that can provide a high power, an inverter that changes it to a frequency of 60 kHz, a feeding device that changes the high-frequency power into a magnetic field in the air, a pickup that changes the magnetic field in the air into high-frequency power, and a rectifier and regulator, which make the high-frequency power collected from the pantograph available as DC power. As can be seen in the schematic of the electrical power system shown in Fig. 17.3, the pickup module used in the wireless low-floor tram system changes the power collected from a pickup device into the constant voltage required to operate the tram through a power regulator, and charges the battery using the vehicle's DC/DC converter. The wireless power transfer system used in the wireless low-floor tram system is used not only to charge the batteries at a station but also to provide motor power. Figure 17.4 shows the improvements that would be achieved in the urban landscape when the catenary system is removed from a wireless low-floor tram system when the system uses wireless power transfer system.

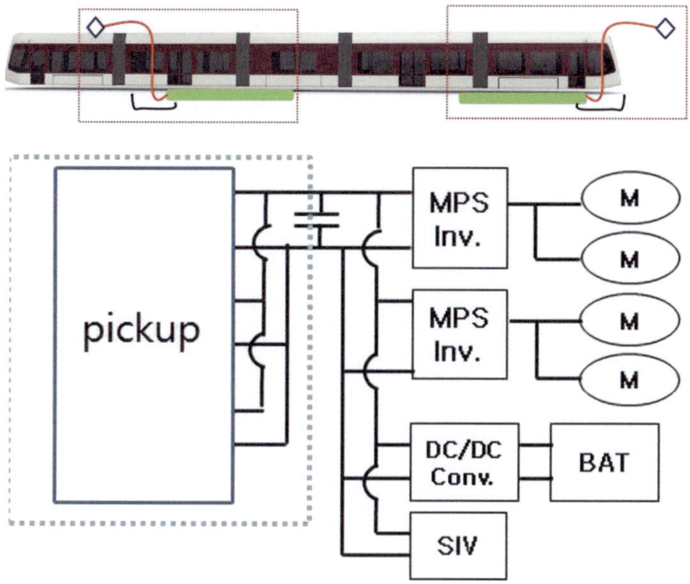

Fig. 17.3 Concept of wirelessly transferred power delivery to existing power collection system and block diagram of the power collector system

Fig. 17.4 Improved urban landscape by applying a wireless power transfer system to a low-floor tram

The benefits that can be achieved using wireless power transfer system technology in railway vehicles are summarized below.

- The power required by the vehicle is supplied from the bottom of the train through a non-contact method, so there is no need for maintenance or the replacement of parts such as catenary equipment. Maintenance costs could be

decreased by 52% (from 4,580,000 to 2,190,000 won/km) because maintenance work is at ground level rather than at an elevated level (also making the maintenance safer).
- Catenary equipment is not required, so the cross-sectional area of tunnels could be reduced by 20% where wireless power transfer system is used in subways, and construction costs could be reduced by 15% (a tunnel area of 56.0 m^2 costs 19.28 billion won/km, whereas a tunnel area of 45.05 m^2 costs 16.39 billion won/km).
- Catenary and pantograph systems that offer less efficient power transmission at high speeds are not used, so the development of high-speed trains would be expected to accelerate.
- The lack of a catenary on the ceiling of the train allows the height to be increased by up to 1.3 m, allowing multi-story carriage and wagon designs to be used. Accidents caused by feeder lines will also be prevented.
- Installing a 180-kW wireless power transfer system in a wireless low-floor tram operated using batteries would lead to the system being cheaper to operate because the capacities of the batteries mounted on the vehicles could be decreased by about three quarters over 10% road-embedded power cable module installation of the entire route (the current wireless low-floor tram battery capacity is 162 kWh, which could be decreased to 41.65 kWh).
- A battery capacity of 1.65 kWh would be sufficient to transport a tram 500 m using a 37-s charging period with 50 m of road-embedded power cable module segment installed.
- A battery capacity of 40 kWh would allow the tram to operate for 12 km in an emergency circumstance.
- Urban railways, including the wireless low-floor tram system in which wireless power transfer system technology is used, can share the roads with buses and automobiles, and are especially convenient because it is possible for other users to cross the tracks. Removing the need for a catenary on a system will make the urban environment more pleasant and allow complaints about overground urban railways damaging the urban landscape to be resolved.

17.3.2 High-Speed Train

For high-speed trains, developments of multi-MW wireless power transfer systems are required. As a first step of the research, a 1-MW, 5-cm air gap, 60-kHz, 128-m long, and 82.7% efficiency wireless power transfer system has been developed and demonstrated by Korea Railroad Research Institute (KRRI) with the collaboration of Korea Advanced Institute of Science and Technology (KAIST) in 2014 [4].

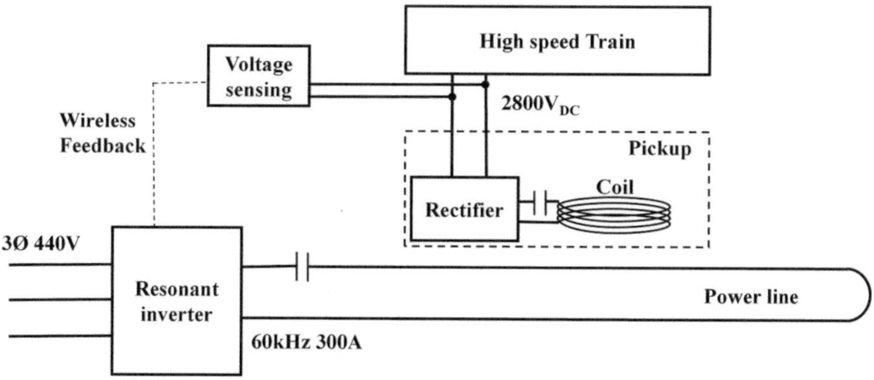

Fig. 17.5 Configuration of wireless power transfer system for high-speed train

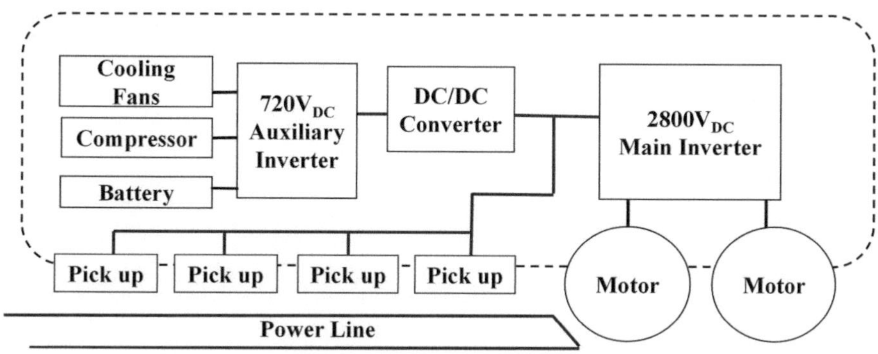

Fig. 17.6 Interface of wireless power transfer system with the target high-speed train

The configuration of the overall wireless power transfer system is shown in Fig. 17.5. A 1 MW resonant inverter supplied single-phase, 60 kHz, 300 A_{rms} current to the transmitter track. By magnetic induction, four pickups that installed under the high-speed train delivered power to the traction motors and auxiliary loads of the train. The resonant inverter controlled the output voltage of the pickup on the ground with wireless feedback of the measured output voltage. The output terminals of four pickups were connected to the traction inverter's DC link as shown in Fig. 17.6. Using an additional DC/DC converter, the auxiliary loads of the train was supplied by the wireless power transfer system as well. HEMU-430X (High-speed Electric Multiple Unit—430 km/h eXperiment) has been used as a target train. Figures 17.7, 17.8, 17.9, 17.10, and 17.11 show photos of the developed resonant inverter, transmitter track, and pickups. The power transfer efficiency of the developed system has been measured using a power analyzer and is

Fig. 17.7 Photograph of developed 1 MW, 60 kHz resonant inverter

Fig. 17.8 Photographs of developed 128-m transmitter track

summarized in Table 17.1. The efficiency dependence of the system on the load is shown in Fig. 17.12. At the rated load condition (load resistance = 10 Ω), the power transfer efficiency was over 82%. When the load decreased to 7% of the rated load (load resistance = 135 Ω), the efficiency decreased to 30%.

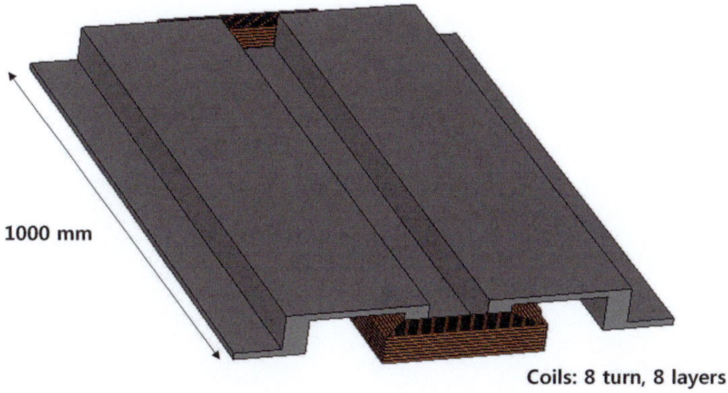

Fig. 17.9 Designed pickup configuration

Fig. 17.10 Photograph of target high-speed train

Fig. 17.11 Photograph of installed four pickup units

Table 17.1 Measured efficiency of developed system

Input power of the resonant inverter	Resonant inverter output current	Pickup output voltage	Pickup output power	Efficiency
989 kW	306 A	2810 V	818 kW	82.7%

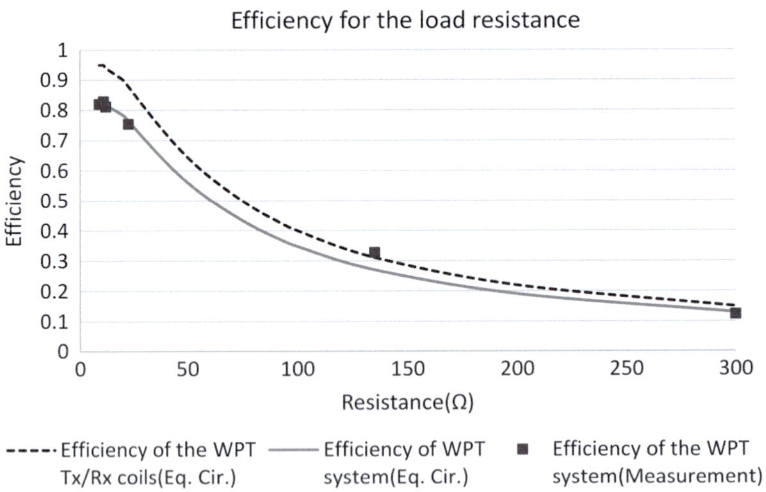

Fig. 17.12 Efficiency versus load resistance

17.4 Wireless Power Transfer Systems for Railways in Other Countries

Two railway systems currently use wireless power transfer system, the Transrapid 09 model high-speed maglev in Germany and the Bombardier Primove LRT in Canada.

17.4.1 *Transrapid 09*

The high-speed maglev system in use in Germany is known as Transrapid. It uses an electromagnetic suspension system with a linear synchronous motor. Electrical power to drive the trains is mostly supplied through an armature winding in a linear synchronous motor installed on a guide on the ground, so the system does not require a particularly high electrical power capacity, in contrast with existing high-speed train systems. However, power is supplied to the train to service the levitation electromagnet, as well as to control the train, the signaling, the air conditioning system, and an emergency battery.

Electrical power can be supplied to Transrapid trains in two ways. A contact-type power supply using a third rail can be used at low speeds (less than 100 km/h), and a non-contact power supply using a linear generator can be used at any speed, including high speeds (more than 100 km/h). Both power supply methods were used to supply power to trains in systems including the Transrapid 08 model, but the third rail system used at low speeds makes use of a mechanical contact method, and the following problems are therefore encountered:

- regular maintenance of the onboard current collector is required,
- sparks are generated when incomplete contact occurs during operation,
- power consumption increases at high speeds because of the increase in the air resistance,
- maintenance is required to keep the third rail within acceptable tolerances,
- the system is noisy because of wind effects,
- the system is sensitive to snow, freezing conditions, and contamination,
- the safety of the current collector can be compromised at high speeds.

The third rail method was replaced with a non-contact wireless power transfer system to provide power in the Transrapid 09 model, allowing the advantages of the maglev non-contact driving system while fundamentally solving the problems described above. The feeding module used in Transrapid 09 is of an air core type, with an air gap of 40 mm and without any magnetic material. The segment length is 100 m and there is a capacitor every 100 m. The frequency is 20 kHz, and insulated copper strand conductor wire is used because of the skin effect in high-frequency power transmission (Fig. 17.13).

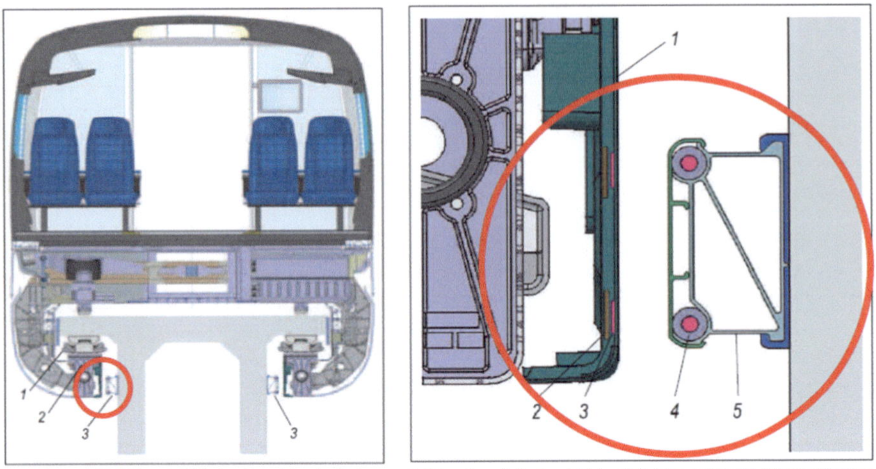

Fig. 17.13 Wireless power transfer system in the Transrapid 09 system

17.4.2 Bombardier PRIMOVE

The Bombardier Primove system, which is an LRT tram, also uses wireless power transfer system. Electrical power is supplied to a vehicle for traction, air conditioning, signaling, vehicle control, lighting, and to charge a battery, but wireless power transfer system is used to supply the power, thereby avoiding the need to install power cables in urban areas.

The initial prototype Primove model was designed with a 16.2-m feeding segment, a 65-mm air gap between the road surface and the current collection module, a 250-kW feeder capacity, a 250-kW battery, a power transmission efficiency of 93%, and a collected power of 750 V dc at a current of 800 A. A test run was completed on the Bautzen test track. A double-layer capacitor-based MITRAC energy saver was used in the prototype, and regenerative braking technology was used to accelerate and to run the vehicle, using the energy stored during braking, decreasing energy consumption by up to 30%. An unmanned automatic "EBI Drive 50" auxiliary train system has seen serviced in Primove for the first time anywhere in the world. Optimal acceleration and deceleration parameters are suggested in order to minimize energy consumption, ensure the vehicle travels smoothly, and reduce wear on the wheels, engine, and rails (Fig. 17.14).

The second Primove model had an 8.1-m feeding segment, 60–100 mm variable and controlled air gap between the road surface and the current collection module, a 200-kW feeder capacity, two 100-kW current collection modules, and a DC/DC conversion efficiency of 95%. The differences between the first and second models were that the onboard voltage in the second model was kept at a constant 750 V using variable pore control technology, which removed the need for an onboard regulator, and that three-phase air core winding was used in the feeding module to

Fig. 17.14 The Bombardier PRIMOVE system

Fig. 17.15 The second Bombardier PRIMOVE model

decrease the costs of manufacture and to obviate the need for an onboard air core current collection module. The second model was tested at Augsburg, near Munich. The total feeder was 275.4 m, and one inverter per segment was installed, meaning that 34 inverters were installed in total on the test track [5, 6] (Fig. 17.15).

17.5 Prospects for Wireless Power Transfer System in Railways

From the transport environment viewpoint, using wireless power transfer system, it may be possible to provide additional railway services in regions and over routes where this has not previously been possible because of geographical or spatial factors (such as sharp bends or the impossibility of installing an overhead catenary system). A wireless power transfer system for a tram could be installed in roads without the need to construct piers, so the tram system would have no negative impacts on the urban landscape, and the tram route would follow the same type of road everywhere, without negative aesthetic impacts.

From a political viewpoint, it is becoming a political necessity to construct railways that are less environmentally harmful and to implement policies to reduce greenhouse gas emissions, and investment in R&D in "environmentally friendly" electric railways using wireless power transfer system and battery-based technologies have been growing in advanced countries. The Korean government has focused its attention on the development of a transport system that is cost efficient and less environmentally harmful than current systems, so the prospect for wireless power transfers system technology is bright.

The barriers facing the development of wireless power transfer system and the strategies required to deal with the barriers are summarized below.

- Barriers to the use of wireless power transfer system in a railway system:
  - low power transfer efficiency;
  - high energy consumption;
  - a feeder system needs to be installed along the whole route;
  - excessive facilities are required because of the high level of reactive power.
- Strategies for using wireless power transfer system in a railway system:
  - achieve a high power transfer efficiency by using a high-frequency resonant circuit;
  - decrease energy consumption using a minimum loss core;
  - decrease construction costs by using concentrated feeding at stations and temporary feeding in acceleration sections;
  - minimize reactive power by minimizing the air gap size.

17.6 Conclusions

Application of SMFIR to railways is a logical next step in EGST. Developing an electric railway using wireless power transfer system to supply the power has been a global R&D theme at present. While Germany is at the stage of pre-commercialization for light train, Korea achieved the laboratory testing stage for tram and high-speed trains. When an electric railway system is modified to use wireless power transfer system, the existing overhead catenary feeding system in not needed, which should receive a positive response from the public. An electric railway that makes use of such a system would provide the users with an improved experience and encourage private transport users to switch to using public transport. A wireless power transfer system would not require an overhead catenary system, so the cross-sectional areas of tunnels could be reduced, lowering the construction costs. Such characteristics would bring great benefits to urban areas, where a diverse range of transport systems are required.

The lack of a catenary system for an electric railway using wireless power transfer system will make the urban landscape more harmonious and give a "high-tech" and new urban image. Using wireless power transfer system in combination with a battery charging system at stations would be expected to be more efficient than using a battery charging system alone because the wireless power transfer system would allow smaller batteries to be used.

A railway system, unlike a road transport system, would be suitable for the use of a limited-size wireless power transfer system because of the fixed rails and the small amount of variation in the air gap that would occur between the feeder and the current collector in a railway system. These characteristics and the steel wheels on trains will make it possible to design an efficient power transfer system, which should allow such a system to have a better energy efficiency than can be achieved in a road transport system.

Besides the decreased costs for tunnel and catenary construction and the lower maintenance costs that would be expected from using a wireless power transfer system, technical improvements, such as decreasing the battery weight and installation space by providing quick battery charging at a depot, a station, or a catenary section, may be expected. In traditional electric railways, the amount of regenerative energy afforded by braking depends on the operation pattern, and most of the regenerated energy is consumed by resistance. However, a separate power conversion system has recently been installed next to a railway to collect the regenerated energy. An electric railway using wireless power transfer system may be able to use regenerated energy directly without the need for an additional system either on the train or adjacent to the railway, implying that more of the regenerated energy would be used.

References

1. Land Transport and Maritime R&D report (2011) Research planning for developing the eco-friendly public transportation system based on the wireless power transfer technology, Korea
2. Lee B, Lee H, Park C (2010) Technology of inductive power transfer for railway vehicles. Auto J 32(3):36–40
3. www.olev.co.kr

4. Park C, Lee H, Lee B (2012) A study on the design parameters of inductive power transformers. Int J Appl Electromagn Mech 39:809–815
5. www.primove.bombardier.com
6. B. Transportation (2010) PRIMOVE contactless and catenary-free operation. EcoActive Technologies

Chapter 18
Electrification of Other Transportation Systems

Uooyeol Yoon

Abstract This chapter introduces the future green transportation systems based on SMFIR. The four systems considered are transportation systems in airport, harbor, ship transportation, and electric vehicles. In airports, cargos can be transported within the warehouse using electric carriers that use SMFIR. In harbors, containers are delivered to and from ships to ground-transportation stations, which can be done using OLEV-type transports. Also, the ships that operate in rivers and coastal areas can be propelled electrically with batteries that are charged through underwater SMFIR system. Also, various issues associated with wireless power transfer to electric vehicles such as health, safety, cost, and markets are discussed.

18.1 Introduction

SMFIR can be applied to other transportation systems in addition to buses and trains. This chapter presents four potential systems that can use SMFIR systems in order to reduce CO_2 emission: airport cargo transportation, container transport on port grounds, coastal and river ships, and passenger cars. In airports, cargos can be transported using electric carriers that are based on SMFIR within the warehouse. In harbors, containers are delivered to and from ships to ground-transportation stations (e.g., trucks), which can be done using special electric transports based on SMFIR. Also, the ships that operate in rivers and coastal areas can be propelled electrically using the SMFIR system with batteries. Finally, smaller electric passenger cars can use wireless charging systems that are based on SMFIR. Various issues associated with wireless power transfer to electric vehicles such as health, safety, cost, and markets are discussed.

U. Yoon (✉)
Wireless Power Transfer Research Center KAIST, Korea Advanced Institute of Science and Technology(KAIST), 303 Truth Hall 193 Munji-Ro, Yuseong-Gu, Daejeon 305-732, South Korea
e-mail: uyoon@kaist.ac.kr

N.P. Suh and D.H. Cho (eds.), *The On-line Electric Vehicle*,
DOI 10.1007/978-3-319-51183-2_18

18.2 Cargo Transportation Within Airports

Wireless power transfer technology can be applied to electric cargo carriers in airports. Figure 18.1 shows an electric carrier in airport warehouse to store and carry cargo.

Figure 18.2 shows the concept of wireless power transfer of an electric carrier for airport application. The electric carrier does not have to travel long distances, and thus, long running time is unnecessary. A wireless power transfer system for an electric carrier is thus suitable for aerospace transportation. Figure 18.3 shows the topology of wireless power transfer for the electric carrier.

18.3 Harbor Transportation

A shuttle carrier in harbor is an application of SMFIR. Figure 18.4 illustrates transportation at a harbor. The shuttle carriers have moved the container box from ship to container terminal. With the shuttle carrier, handover does not require the simultaneous presence of a crane and shuttle, thereby avoiding waiting and therefore increasing the utilization rates of the equipment. The working space has several randomly accessible transfer positions, ensuring there is always a position to locate

Fig. 18.1 Electric cargo carrier in an airport warehouse (source/ref http://www.supplychaindigital.com/warehousing/3573/Four-Things-to-Consider-When-Choosing-Warehouse-Locations)

Fig. 18.2 Concept of wireless power transfer electric carrier for airport application (*source* drawing)

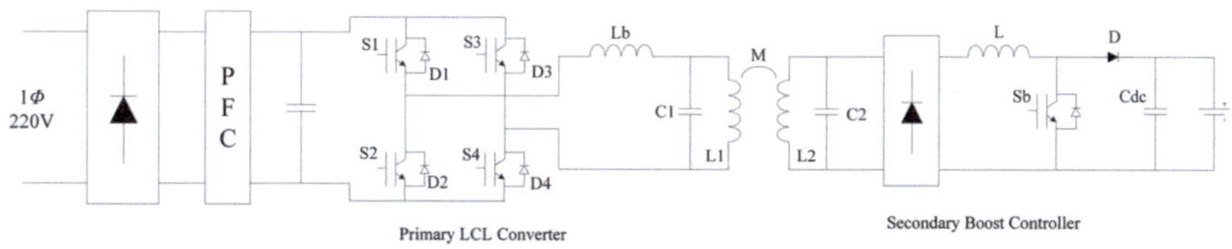

Fig. 18.3 Topology of wireless power transfer system for electric carrier application

Fig. 18.4 Example of the transportation in harbor (https://www.kalmar.com.au/globalassets/newsroom/images/hires/kalmar-history-2000s_print_45920.jpg)

Fig. 18.5 Shuttle carrier in harbor (https://www.kalmar.com.au/globalassets/newsroom/images/hires/trapac_2-13-66_print_49438.jpg)

the next container. By means of this setup, each set of the apparatus can operate at its optimum pace and with optimal performance [1].

Figure 18.5 shows a shuttle carrier in a harbor. It is an ideal application for SMFIR. By eliminating the diesel engines currently used with SMFIR-based electric drives, it can eliminate the pollution problem that exists in busy harbors. The shuttle carrier does not have to travel a long distance, and the operation time is relatively short. Therefore, the wireless power transfer system may be relatively short in length for dynamic charging. An optimum length of the underground power system and the size of the battery of the shuttle carrier can be chosen for optimum cost. The wireless power transfer for the shuttle carrier can be designed as shown in Fig. 18.6.

18.4 Wireless Electric Power Transfer to Ship Transportation

In Norway, a battery-powered electric car ferry was developed. It is capable of carrying 120 cars and 360 passengers, and it can fully recharge in just 10 min. Ampere—formerly ZeroCat—is a groundbreaking ferry constructed for Norled by the Norwegian Shipyard Fjellstrand. It is the world's first electric-powered car ferry

Fig. 18.6 Wireless power transfer system for shuttle carrier in harbor

and generates zero emissions and minimum sound. The ferry was delivered in October 2014, and commercial operations began in May 2015. The ferry's electric power train was designed by Norwegian shipyard Fjellstrand with battery technology from Siemens with an 800-kW battery that weighs 11 tons and drives two screws [2]. Although the battery is quite heavy, the ship only weighs half as much as a conventional catamaran ferry, thanks to twin hulls made of aluminum. Those hulls have a slim design, which further increases efficiency.

This electric ferry is suitable for applying wireless power transfer technology because the ferry is docked at the wharf for relatively long periods. The docking time of the ferry is sufficient to charge the battery of the electric ferry. Furthermore, the sailing distance of the ferry is short. Figure 18.7 shows the concept of the electric ferry with wireless power transfer technology.

It is not only its drive system that makes the new ferry so environmentally friendly. Its electric motors are nearly silent and do not burn any fossil fuels. They also do not produce any pollutants. By contrast, a conventional ferry traveling the same route consumes around one million liters of diesel fuel and emits 2680 tons of carbon dioxide and 37 tons of nitrogen oxide each year. Project specialists have also adopted a new approach for the ferry's design. Unlike most electric cars, this ship was developed from the ground up as an electrically powered vessel. This has had a noticeable effect on its weight in particular. Despite its ten-ton batteries and capacity for 360 passengers and 120 vehicles, the ship will be only half as heavy as

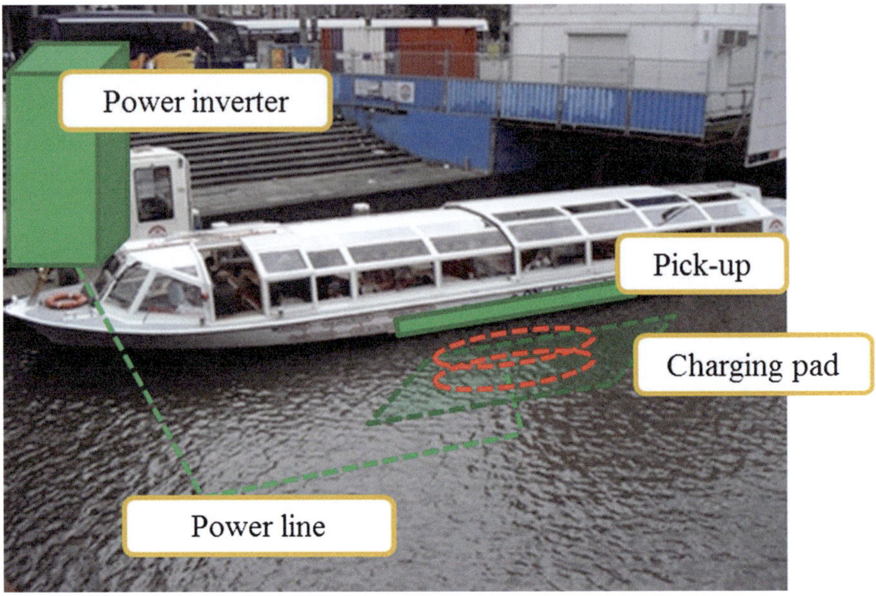

Fig. 18.7 Concept of electric ferry with wireless power transfer technology

a conventional ferry. That is because it will be made exclusively of light aluminum rather than the steel normally used in shipbuilding. The ship's corrosion-resistant aluminum structure also means it does not require a special coat of paint used to protect steel ships against rust. This also lowers the ferry's operating costs. In addition, the ship's designers searched for the most energy-efficient systems available.

One of the companies in the Netherlands was interested in applying SMFIR to their river ferries by transmitting magnetic power to the vessel under water. The idea was to install the underwater charging facility to charge batteries on ships while they are at the pier. It is conceptually possible, but was not implemented.

18.5 Electric Vehicle (EV) Transportation

18.5.1 Application of Wireless Power Transfer System to Electric Vehicles

Wireless power transfer is about to take a huge leap in scale, as kilowatts of electric power can now be transferred over an air gap of hundreds of millimeters while still maintaining high-energy transfer efficiency. As shown in Fig. 18.8, wireless power transfer technology for electric vehicles is a no fuss, simple solution for charging

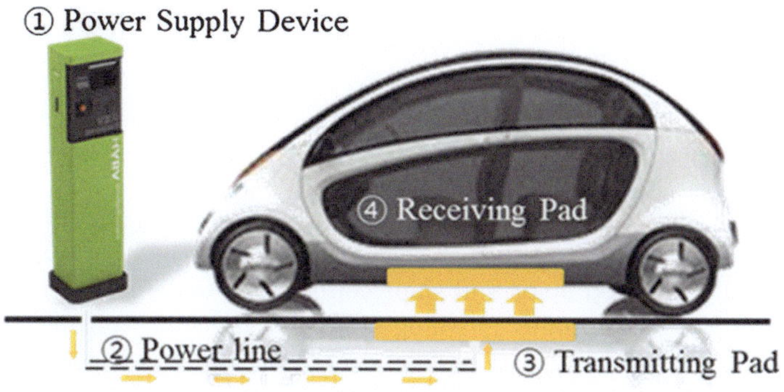

Fig. 18.8 Commercial case of wireless power transfer for EV

electric vehicles and hybrid electric vehicles. While the wireless power transfer technology is ideally suited to stationary wireless charging, it also opens up the possibility of dynamic wireless charging for the electric vehicle battery while the vehicle is being driven. The wireless power transfer technology will lead to a shift in charging behavior. Drivers will charge their electric vehicle often and potentially by using dynamic charging to complement local stationary charging, removing range anxiety. This means that batteries could be smaller, which would result in reduction in electric vehicle cost and vehicle weight. With further spread of shared vehicles and self-driving cars, the elimination of the need for stationary charging for a long time would make these vehicles more functional.

18.5.2 The Need for Wireless Power Transfer of Electric Vehicles

Many users are likely to have easy access to home charging of the batteries in their garage. The market will support these needs with different charging methods. Many EV users will have to use street parking due to a lack of dedicated private parking lots, in city apartments for instance. Public charging demand will therefore increase as electric vehicle ownership expands. Utilities consequently have two options: to deploy plug-in charging with an increase in street charging facilities, which carries the risk of vandalism and loosening and oxidization of electric connections; or to deploy underground wireless charging. The latter removes the health and safety risk to the general public of tripping and falling over high-voltage cables. There is an economical benefit to wireless charging as well: Keeping the battery charge at between 40 and 80% will optimize the life of the electric vehicle battery. Wireless charging lends itself perfectly to this kind of user behavior.

18.6 Conclusions

This chapter introduced the possibility of wireless power transfer technology for future green transportation systems. Several transportation systems are described: airport transportation, harbor transportation, ship transportation, and electric vehicles. In airport and harbor transportation, it was shown that the wireless power transfer technology can be applied to load-carrying transports. In ship transportation, electrically powered boats and ships can be used to carry cargo and passengers. Finally, SMFIR and other charging systems for electric vehicles can replace the present IC engine-based automobile fleets. The wireless charging system is likely to be more convenient and safe than plug-in-type charging systems.

References

1. Kalmar Corporation(2014) http://www.kalmarglobal.com/equipment/shuttle-carriers/
2. Siemens Corporation(2014) http://www.siemens.com/innovation/en/home/pictures-of-the-future/mobility-and-motors/electromobility-electric-ferries.html

Chapter 19
Other Applications of SMFIR

Gu Ho Jung

Abstract This chapter presents other applications of Shaped Magnetic Field In Resonance (SMFIR). In addition to EE-type SMFIR, several different kinds of wireless power transfer systems that are being promoted are reviewed. Finally, new applications such as bikes and motorbikes are introduced.

19.1 Introduction

Wireless electromagnetic power transfer has been a fascinating subject ever since the late nineteenth century thanks to the pioneering work of Nicola Tesla (1856–1943), a Serbian American engineer, who invented the transformer and AC induction motor. The subject of wireless power transfer system has received much attention in the early twenty-first century with ubiquitous use of handheld devices such as laptops, cell phones, and portable tools. The heavy power transfer by wireless means has gained much attention because of the emergence of electric vehicles (EVs) [1]. The basic idea was to replace the plug-in type charging of batteries with wireless transfer of electric power in garages and parking lots. In this chapter, we review many different kinds of wireless power transfer systems currently being promoted.

19.2 Special Features of SMFIR

KAIST is the world's first to develop the Shaped Magnetic Field In Resonance (SMFIR) technology that safely transfers a large amount of energy to electric vehicles either in motion or stationary [5]. SMFIR is a unique technology that can

G.H. Jung (✉)
Korea Advanced Institute of Science and Technology (KAIST),
303 Truth Hall 193 Munji-Ro, Yuseong-gu, Daejeon 305-732, South Korea
e-mail: ghjung9595@kaist.ac.kr

N.P. Suh and D.H. Cho (eds.), *The On-line Electric Vehicle*,
DOI 10.1007/978-3-319-51183-2_19

transmit a large amount of power over a distance of up to a few meters, depending on the application. The unique characteristic of SMFIR is the ability to transmit high power, more than 100 kW, over a large distance by shaping the magnetic field. The SMFIR design makes it possible to increase the distance for power transfer "indefinitely" since the distance between the poles of the magnetic field determines the shape of the magnetic field. Furthermore, the two-dimensional magnetic field of SMFIR enables power transfer to moving vehicles [2, 3].

Based on the SMFIR technology, our OLEV transport system is wirelessly powered through a power grid embedded under the road. A pickup module installed under the vehicle works to collect the magnetic field efficiently from under-road power cables and convert it into electric energy for vehicle operation, as extensively presented in previous chapters, including the basic design of the system based on Axiomatic Design in Chap. 2.

Figure 19.1 shows OLEV system with EE-type structure using SMFIR and its magnetic loop shapes. Two road embedded power cable modules with opposite current directions under the road induce magnetic flux around each one. These induced magnetic fluxes are added at the location between the road embedded power cable modules and are cancelled outside of the road embedded power cable modules. A pickup module installed under the vehicle works to collect the magnetic field efficiently from under-road power cables and convert it into electric energy for vehicle operation. In this case, it is possible to get higher power efficiency and power for electric vehicle by designing shape of the ferrite core and optimizing its magnetic field.

The developed SMFIR technologies can be extended to other applications such as high-speed railroad systems and E-bike, where we need to improve power efficiency and increase output power with compact and light-weight module. The easiest solution is to increase the switching frequency from the current 20–60 kHz,

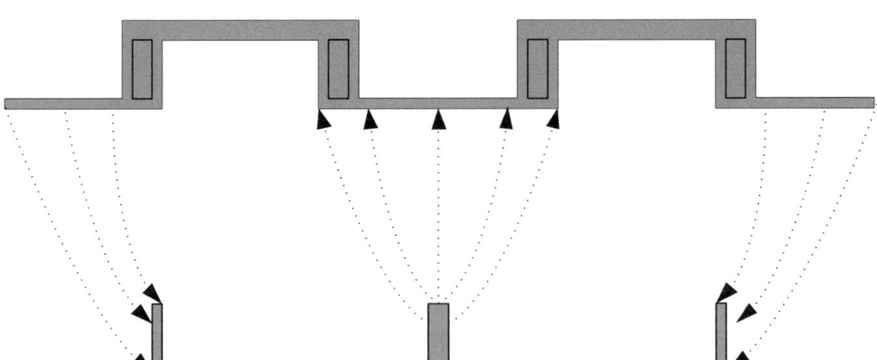

Fig. 19.1 OLEV system with EE-type structure using SMFIR and its magnetic flux loop shapes

etc., from which power loss of power cable and ferrite core will increase significantly with higher switching frequency. Therefore, we have to develop new power cable and ferrite core with lower power loss at higher frequencies. Also, we have to adopt a new power semiconductor such as SiC-FET to develop the inverter and regulator with higher power efficiency.

19.3 Portable Device

Hanlim POSTECH co. of Korea has developed a wireless charger for mobile phones, and it plans to release all kinds of wireless charging products through continuous advancement of related technology. Currently, an air gap of less than 2–3 cm must be maintained for wireless charging. However, wireless charging with a larger air gap is expected to be possible in the future with advancements in wireless charging technology, enabling a desk can serve as a wireless charger and a mobile phone can be charged by simply placing it on the desk. Similarly, a cup-holder can act as a wireless charger and a mobile phone can be charged inside the holder.

Wireless charging technology is being utilized in many portable devices, especially smartphones. For example, using a desk at school or the office as the charger, devices such as smartphones and laptop computers, can be charged wirelessly. This approach is already being applied in everyday life. Starbucks in the USA provides wireless charging services in the tables of some stores, allowing customers to charge their smartphones while they enjoy their cup of coffee. It has already installed these wireless rechargers in 12 locations in Boston, Massachusetts, and nine in San Jose, California. Powermat Spot that Starbucks has installed as wireless recharger is compatible with standards established by Power Matters Alliance, an industry group that includes Samsung, Blackberry, Microsoft, LG and others.

The Dell also applied this wireless charging technology to the stand of its laptop computer, Latitude Z, which can charge the laptop without the use of cables or any visible plug. It used the adaptive coupling technology of Michigan-based eCoupled (a brand of Fulton Innovation). This base station has a permanently embedded coil with 35–45 mm in diameter, and the laptop is charged wirelessly when it is placed near the base station. Members of the Wireless Power Consortium including Dell, Palm, and Sony are already working on products that involve inductive charging. Dell has adopted the wireless charging function as the first of PC manufactures.

For wireless charging for various portable devices, the Alliance for Wireless Power (A4WP) WPT system is initiated and Fig. 19.2 shows the basic WPT system configuration between a PTU and a PTU for A4WP. The PTU consists of three main functional units with a TX resonator, matching circuit, and a power amp and the PRU includes an RX resonator, a rectifier, and a DC-DC converter.

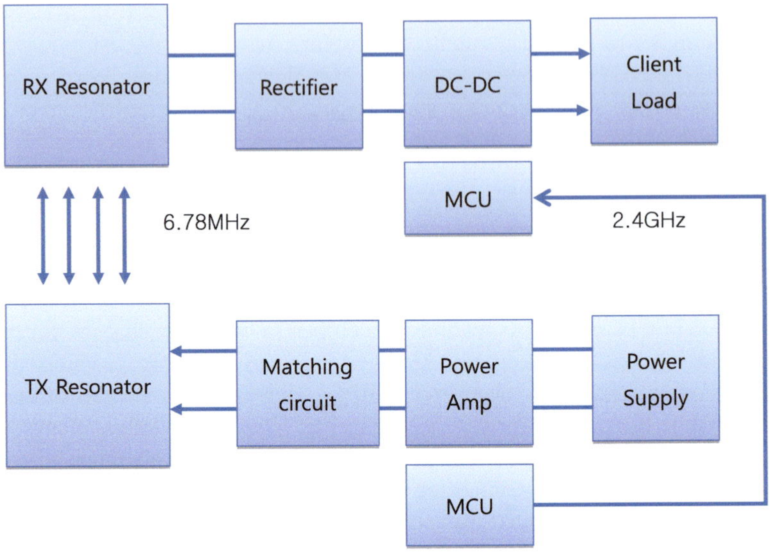

Fig. 19.2 Wireless Power Transfer System configuration for A4WP

19.4 Home Appliances

WiTricity's wireless power TV used a standard Colpitts oscillator whose inductive element consists of a single loop of copper wire 25 cm in radius, as shown in Fig. 19.3. Single copper driving loop A was used as a part of the driving circuit, which outputs a sine wave with frequency of 9.9 MHz. Load was attached to a loop of wire B. The various K_s, K and K_D represent direct couplings between the objects. Driving loop A is inductively coupled to the source coil S. Load loop B with its own loop of insulated wire is attached to a calibrated light bulb, which is inductively coupled to D. WiTricity calculated the power transfer efficiency q using the equation q = $Pw/(Ps + P_D + Pw)$ between the source coil S and the load B, where Ps, P_D, and Pw are the power loss dissipated in the source coil, device coil, and load. From the experimental results of maximum efficiency according to wireless power transfer system distance, they were able to fully light up a 60-W light bulb from distances more than 2 m away [4].

As shown in Fig. 19.4, Hirer of China also developed a wireless power TV using both MIT's WiTricity and WHDI wireless video technology. A large power supply unit exists in the back of the TV and emits RF to the backside of the TV through a totally harmless method. A coil inside the TV receives current. The TV can thereby receive enough current if it is placed in a straight line with the power supply.

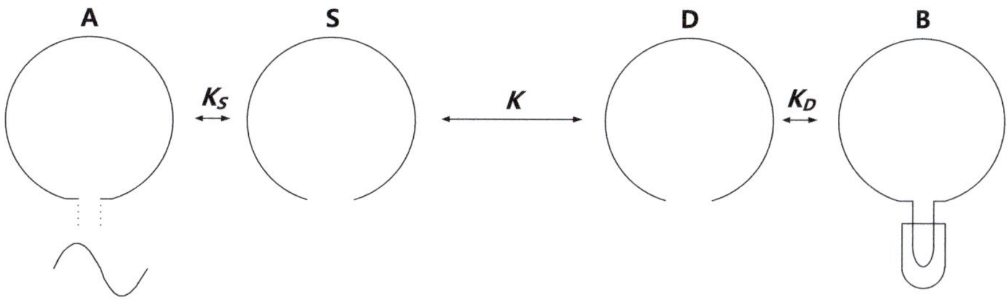

Fig. 19.3 Schematic configuration of experimental setup by WiTricity. D and loop A are adjusted to ensure that their direct coupling is zero, while coils S and D are aligned coaxially. The direct couplings between B and A and between B and S are negligible

Fig. 19.4 Concept of Haier's wireless power TV

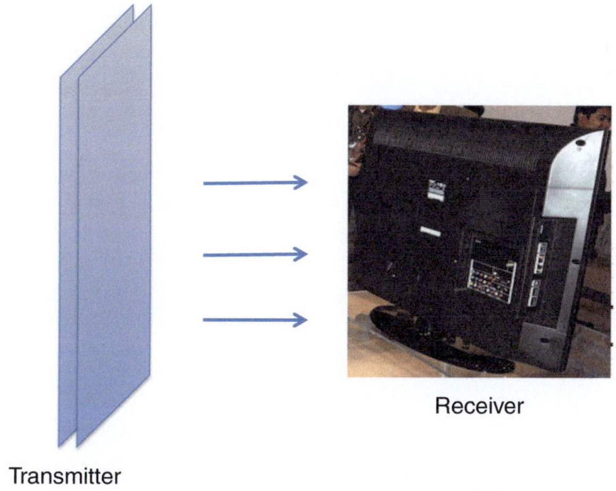

19.5 Wireless Power Distribution

Nikola Tesla worked on his Wardenclyffe Tower for long-range wireless energy transfer in the nineteenth century. After that, MIT's Marin Soljacic, Aristeidis Karalis, and John Joannopoulos have invented WiTricity technology that could deliver power wirelessly by harnessing the properties of resonance. The distance between the power source and capture device ranged from a few centimeters to over 2.5 m, where a 60-W light bulb was illuminated and power efficiency according to distances was proven experimentally. Although this technology has received much attention and many companies acquired licenses for the technology from WiTricity, it is not easy to find any commercial applications.

As shown in Fig. 19.5, NASA has been working on an ambitious project named Solar Power Satellite via Arbitrarily Large Phased Array (SPS-ALPHA) to harness solar power in space. If successful, this project will make it possible the

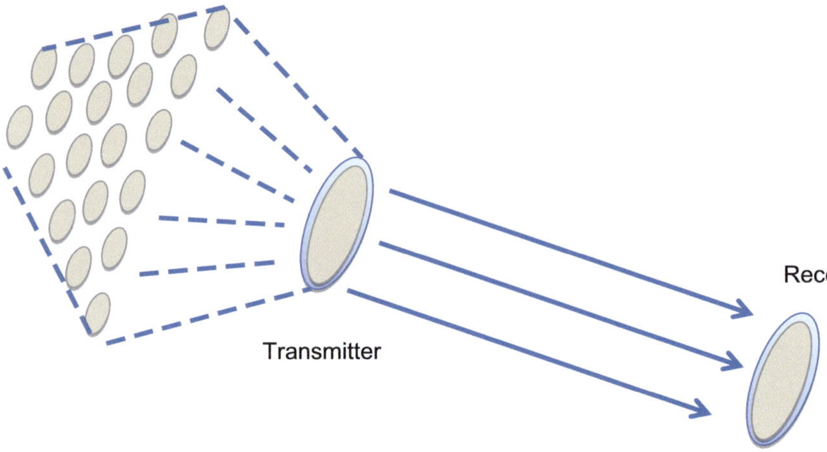

Fig. 19.5 Concept of SPS-ALPHA project of NASA

construction of huge platforms with tens of thousands of small elements that can deliver remotely 10–1000 s of megawatts using wireless power transfer system to markets on Earth and missions in space.

19.6 Bikes and Motorbikes

The demand for electric motorbikes in many countries has been increasing to improve environment by reducing harmful gases. To accommodate the ever-increasing use of these electric bikes, large-scale parking lots for motorbikes have been installed in southeast Asian nations such as Indonesia and Vietnam. A wireless power transfer system that can charge numerous motorbikes simultaneously would be a good solution for these parking lots.

Figure 19.6 shows an example of a wireless power transfer system for an electric bike. It consists of an power inverter, power line, pickup module, and regulator (or DC/DC converter). In this case, the inverter is located inside the bike stand and it generates high-frequency current. This alternating current creates a high-frequency magnetic field around the power cable inside the power module, and then, this magnetic field is converted into high-frequency AC voltage through a pickup module on the bicycle. Finally, the regulator or DC/DC converter charges the battery with appropriate voltage and current. The gap between the power module and pickup module is a few centimeters. The battery is charged with a wireless charging method using a magnetic field.

In the case of bikes and motorbikes, the rated power of the pickup module is below 500 W and varies according to the type of bike and motorbike.

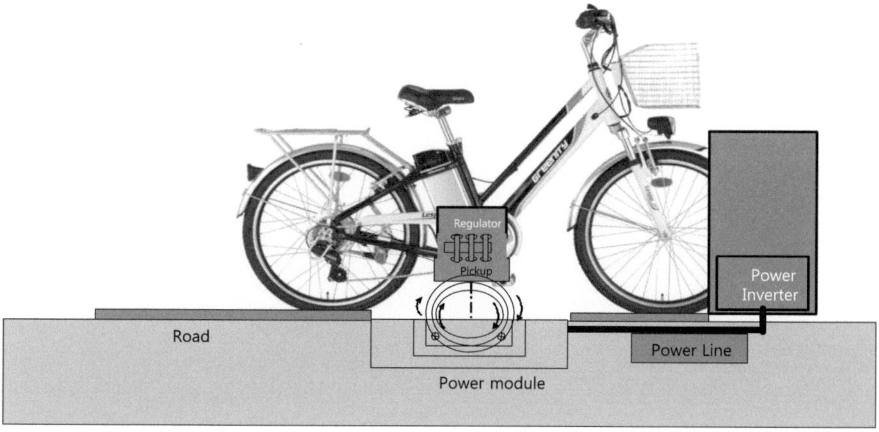

Fig. 19.6 Example of wireless power transfer system for a bike

19.7 Conclusions

This chapter introduced the other applications of Wireless power transfer system, including the SMFIR technology proposed by KAIST and future improvements including core, cable, and new semiconductor devices were discussed.

First of all, special features of SMFIR were explained with EE-type structure using SMFIR and its magnetic loop shapes. Also, wireless power technology applications for portable devices and home appliances including cell phone, laptop computer were mentioned. Finally NASA's ambitious project named Solar Power Satellite via Arbitrarily Large Phased Array (SPS-ALPHA) for wireless power distribution was introduced with its applications for Bikes and Motorbikes.

References

1. California partners for advanced transit and highways. http://www.path.berkeley.edu/. Accessed 2009
2. Cho DH, Jung GH, Yoon UY, Lee BS (2014) Development & implementation of electric tram system with wireless charging technology. ICT Express J 1(1):34–38
3. Jung GH, Shin BS, Lee SH, Shin JG, Kim YS, Jeon SJ (2012) High efficient inductive power supply and pickup system for On-Line Electric Bus. Paper presented at the 2012 IEEE Electric Vehicle Conference (IEVC), Greenville, SC:1–5
4. Kurs A (2007) Power transfer through strongly coupled resonances. Master of Science in Physics at the MIT
5. Shin JG, Shin SY, Kim YS, Ahn SY, Lee SH, Jung GH, Jeon SJ, Cho DH (2014) Design and implementation of shaped magnetic-resonance-based wireless power transfer system for roadway-powered moving electric vehicles. IEEE Trans Ind Electron J 61(3):1179–1192

Part IV
Performance, Cost, Regulatory, and Safety Considerations

Chapter 20
Electrified Transportation System Performance: Conventional Versus Online Electric Vehicles

Amro M. Farid

Abstract In recent years, the electrification of ground transportation has emerged as a trend to support energy efficiency and CO_2 emissions reduction targets. The true success, however, of this trend depends on the successful integration of electric vehicles into the infrastructure systems that support them. Left unmanaged, conventional electric vehicles may suffer from delays due to charging or cause destabilizing charging loads on the electrical grid. Online electric vehicles have emerged to remediate the need for stationary charging and its effects. This chapter seeks to objectively compare the systemic impacts of these two electric vehicle concepts on the combined electrical power grid and road transportation system. It applies a recently developed hybrid dynamic system model of the transportation-electricity nexus that holistically incorporates vehicle dispatch, route choice, charging station queues, coordinated charging, and vehicle-to-grid stabilization. It draws upon Axiomatic Design for Large Flexible Engineering System Theory to superimpose a marked petri net model layer on a continuous-time kinematic and electrical state evolution. The results show that online electric vehicles, unlike their conventional vehicle counterparts, are able to avoid charging station queues and thus are able to meet the needs of a greater variety of transportation uses cases including commercial and public fleets. Their impacts on the power system also differ. While conventional electric vehicles are likely to require greater investment to expand power system generation capacity, online electric vehicles are likely to incur greater operating costs to manage their charging loads. The chapter concludes with several directions for future work in the development of intelligent transportation-energy systems which can serve to reduce both costs for both vehicle concepts.

A.M. Farid (✉)
Thayer School of Engineering, Dartmouth College,
14 Engineering Drive, Hanover, NH 03755, USA
e-mail: amfarid@dartmouth.edu; amfarid@mit.edu

A.M. Farid
MIT Mechanical Engineering Department, 77 Massachusetts Ave,
Cambridge, MA 02139, USA

N.P. Suh and D.H. Cho (eds.), *The On-line Electric Vehicle*,
DOI 10.1007/978-3-319-51183-2_20

20.1 Introduction

The previous chapters have devoted themselves to the design of the online electric vehicle (OLEV). This chapter now turns its focus to the context of this newly designed artifact to study the performance of OLEV relative to plug-in electric vehicles. More specifically, it draws a rigorous comparison of the two technologies with respect to transportation availability and charging loads to the electric grid, and the associated investment and operating costs to mitigate them. In such a way, this chapter completes the assessment of the OLEVE not just as an individual technology but also as one that exists within a larger integrated transportation-electricity infrastructure.

This infrastructure, in recent years, has emerged as a trend to support energy efficiency and CO_2 emissions reduction targets [1–5]. Relative to their internal combustion vehicle (ICV) counterparts, electric vehicles (EV), be they trains, buses, or cars, have a greater "well-to-wheel" energy efficiency [5, 6]. They also have the added benefit of not emitting any carbon dioxide in operation and rather shift their emissions to the existing local fleet of power generation technology [7].

The success of electric vehicles depends on their successful integration with the infrastructure systems that support them. From a transportation perspective, conventional electric cars typically only have a short range of 150 km [8] but may still require several hours to charge [9]. This affects when a vehicle can begin its journey and the route it intends to take. From an electricity perspective, the charging loads can draw large power demands which may exceed transformer ratings, and cause undesirable line congestion or voltage deviations [10–13]. These loads may be further exacerbated temporally by similar charging patterns driven by similar work and travel lifestyles or geographically by the relative sparsity of charging infrastructure in high-demand areas [12]. In effect, the electric vehicles and their supporting charging infrastructure couple the transportation and electrical systems into a nexus.

Definition 1 Transportation-Electricity Nexus (TEN) [14]: A system-of-systems composed of a system with the artifacts necessary to describe at least one mode of transport united with an interdependent system composed of the artifacts necessary to generate, transmit, distribute, and consume electricity.

As a result, the performance in the transportation domain cannot be studied independently of the performance in the electrical domain. Furthermore, efforts to operate and control the performance in either domain require an assessment model whose scope includes the functionality of both systems. Consider an EV taxi fleet operator [10–12]. They must dispatch their vehicles like any other conventional fleet operator, but with the added constraint that the vehicles are available after the required charging time. Once en route, these vehicles must choose a route subject to the nearby online (wireless) and conventional (plug-in) charging facilities. In real-time, however, much like gas stations, these charging facilities may not be available due to the development of queues. Instead, the EV taxi driver may opt to charge elsewhere. Once a set of EV taxis arrive at a conventional charging station, the EV taxi fleet operator may wish to implement a coordinated charging scheme [15–26] so as to limit the charging loads on the electrical grid. The local electric utility may even incentivize this EV taxi operator to implement a "vehicle-to-grid" scheme [27–29] to stabilize variability in grid conditions. The aforementioned five transportation-electric nexus operations management decisions are summarized in 1. While these decisions are coupled, the degree to which they can be coordinated ultimately depends on the presence of a well-designed intelligent transportation-energy system (ITES) [10].

Designing such an intelligent transportation-energy system requires a careful assessment of the couplings between the kinematic and electrical states in a TEN. Thus far, to our knowledge, only two works have been able to assess this coupling. A simplified study based on the city of Berlin was implemented on MATSIM [30]. Meanwhile, the first full-scale study was completed in the city of Abu Dhabi [10–12] using the Clean Mobility Simulator [31]. The former assumed a home charging (i.e., always available) use case and thus neglected the impacts of charging station capacity on the transportation system as well as on the power system. The latter study sought a more holistic approach to system performance measurement. "Quality of Service" (QOS) [11, 12] was introduced as a transportation performance measure to address the availability concerns expressed in EV adoption public attitude surveys. Meanwhile, power system line and bus safety criteria were introduced on the basis of IEEE reliability standards [10–12]. The methodologies for the performance assessment of a TEN are still very much in the course of development. As part of its methodological contribution, this book chapter also quantifies the required additional power system operating reserves and generation capacity.

This chapter's primary contribution is to objectively compare the systemic impacts of both electric vehicle concepts on the transportation-electricity nexus. This requires several notable contributions to the existing literature. Of the two previously mentioned studies, neither considered online electric vehicles, either on their own or relative to their conventional electric vehicle counterparts. Thus, this chapter makes a methodological contribution in how to conduct such an assessment. To that end, it uses the recently developed hybrid dynamic model (HDM) for a transportation-electricity nexus [14]. It draws upon Axiomatic Design for Large Flexible Engineering System Theory [14, 32–45] to superimpose a marked petri net model layer on a continuous-time kinematic and electrical state evolution. This work, therefore, also represents the first application of the model to a moderately sized system. This system is a newly developed transportation-electrification test case which objectively draws out the differences between online and conventional electric vehicles. The model also serves as the basis for a newly developed transportation-electricity nexus simulations. Together, these contributions enable an objective comparison of the two transportation-electrification concepts. The results show that online electric vehicles, unlike their conventional vehicle counterparts, are able to avoid charging station queues and thus able to meet the needs of a greater variety of transportation use cases including commercial and public fleets. Their impacts on the power system also differ. While conventional electric vehicles are likely to require greater investment to expand power system generation capacity, online electric vehicles are likely to incur greater operating costs to manage their charging loads.

The book chapter, therefore, proceeds as follows. Section 20.2 presents the hybrid dynamic model for a transportation-electricity nexus. Section 20.3 then details the test case used to make the comparison. Section 20.4 provides the rationalization for the new simulator. The results and discussion of the work are then presented in Sect. 20.5. The work is concluded in Sect. 20.6. Finally, Sect. 20.7 provides some insights for future work within the context of intelligent transportation-energy systems so as to reduce costs for both vehicle concepts.

20.2 Transportation-Electricity Nexus Hybrid Dynamic Model

This section describes the recently developed transportation-electricity nexus hybrid dynamic model [14] in five steps. Prior to proceeding, it assumes a sufficient formal background in graph theory [46–48] and petri nets [49–51], which is otherwise obtained in Appendix. The section begins with a rationale for such a model and then describes its development. Next, it applies Axiomatic Design for Large Flexible Engineering Systems Theory [14, 32–45]. The resulting system knowledge base is then used to construct a timed petri net model [14]. This gives the added advantage of describing the system state in a distributed fashion that more intuitively represents the microscopic flow of vehicles within a transportation system. The events of fixed duration are then replaced with continuous-time differential equations that describe the kinematic and electrical state evolution of each vehicle. The section concludes with a description of model outputs of greatest interest.

20.2.1 The Need for a Hybrid Dynamic Model

The assessment of holistic system performance for a transportation-electricity nexus requires an equally holistic system model. As has been previously argued in the case of the energy-water nexus [52, 53], the main challenge in transportation-electrification is that engineers are typically trained within disciplines (e.g., mechanical, electrical, chemical, civil) rather than broad-scoped problem areas such as transportation-electrification. This often leads to *silo* thinking that generates piece-meal technical solutions that are restricted by the boundaries, competences, and methods of the respective engineering field. Nevertheless, if many of the traditional methods from multiple disciplines are combined into a single analytical model, then new effective solutions can be developed that target the main technical barriers at the heart of problem. That said, a transportation-electricity nexus model is not simply a traffic simulation model stitched to a power system model. The whole, in form and function, is very much likely to be greater than the sum of the parts. Instead, integrated technical modeling frameworks that transcend the traditional boundaries of the various engineering disciplines are required.

Graph theory is often proposed as such a modeling framework in many application domains [46–48]. Indeed, many models of transportation [54–58] and power systems [59, 60] rely upon the graph theoretic definitions provided in Appendix. In transportation systems, the nodes often physically represent intersections and stations, while edges/arcs represent roads, rails, or transportation routes. In power systems, the nodes often physically represent generators, substations, and loads, while the edges represent the road embedded power cable modules. This is an abstracted model of a system's form, neglecting an *explicit* description of the system's function [61]. Thus, in detailed engineering, it becomes difficult to link nodes and edges to physical variables of engineering physics. Furthermore, graph theoretic application domains are restricted to large flexible *homo-functional* engineering systems where artifacts (of some kind) are transported between physical locations. While transportation functions from one location to another are fundamentally different, ultimately they are of the same type. Thus, it is less than clear how graph theory may be applied to systems like the transportation-electricity nexus that span multiple energy domains and are thus fundamentally *hetero-functional*. In summary, graph theory does not do the following [14, 36–40, 45].

1. Explicitly differentiate between heterogeneous modes of transport with fundamentally different graphs for each
2. Explicitly describe the system function, especially functions that do not involve a type of transportation.
3. Facilitate further detailed engineering design in terms of system function and form.
4. Facilitate the description of faulted, or intentionally offline functionality to support reconfigurable operation.

20.2.2 Axiomatic Design for Large Flexible Engineering Systems

To overcome the limitations described above, the hybrid dynamic model builds upon Axiomatic Design for Large Flexible Systems Theory where the allocation of system function to system to system form is explicitly described.

Definition 2 Large Flexible Engineering System (LFES) [36–40, 44]: an engineering system with many functional requirements that not only evolve over time, but also can be fulfilled by one or more design parameters.

In the context of this work, the *functional requirements* and *design parameters* mentioned in Definition 2 are understood to be *mutually exclusive and collectively exhaustive* sets of the system's processes (P) and resources (R), respectively. This change of terminology is applied consistently for the rest of the chapter.

The system resources $R = B \cup H$ may be classified into a set of stations $B = \{b_1 \ldots b_{\sigma(B)}\}$, and transporting resources $H = \{h_1 \ldots h_{\sigma(H)}\}$ [14, 32–37, 41, 42, 45]. Depending on the nature of the design problem, previous work has interpreted these resources as vehicles [34, 35, 42], stations, or even whole modes of transport [32, 39, 45]. In (traditional) transportation systems, the system processes are taken as the set of transportation processes.

Definition 3 Transportation process [42]: Given an arbitrary origin station b_{y_1} and an arbitrary destination b_{y_2} within a set of stations B, a transportation-resource-independent process $p_\tau \in P_T$ transports individuals between b_{y_1} and b_{y_2}. A convention is adopted between the indices of stations and transportation processes such that:

$$\tau = \sigma(B)(y_1 - 1) + y_2 \tag{20.1}$$

where the $\sigma()$ gives the size of a set.

There are $\sigma^2(B)$ such transportation processes of which $\sigma(B)$ are "null processes" where no motion (i.e., the parking process) occurs. Note that these transportation processes are analogous to the *potential* edges between nodes and the self-loops in a traditional graph. Nevertheless, they are *formally* different because the analog of an edge in Axiomatic Design requires that an edge represent a feasible combination of process *and* resource.

It is also important to note that the inclusion of the $\sigma(B)$ parking processes is essential to the TEN model. Many traffic simulation packages, perhaps due to their historical focus on road congestion management, define vehicle trips between *distinct* origins and destinations [62, 63]. The parking processes, at best, have little to add in that regard and can be left out of the scope of simulation. At worst, they dramatically expand the number of vehicles participating in the evolution of the traffic system's state. In a transportation-electricity nexus, however, many of these parking functions are also associated with charging and thus cannot be neglected. Thus, transportation-electricity nexus modeling must be done from the perspective of the traveler and thus is consonant with ongoing multi-agent system trends in future urban transportation systems [64, 65].

The essential Axiomatic Design activity of mapping function to form is then completed via a LFES knowledge base.

Definition 4 LFES Knowledge Base [14, 32–37, 41, 42, 45]: A binary matrix J_S of size $\sigma(P) \times \sigma(R)$ whose element $J_S(w, v) \in \{0, 1\}$ is equal to one when action (in the SysML sense) e_{wv} exists as a system process $p_w \in P$ being executed by a resource r_v in R.

The Axiomatic Design equation for large flexible engineering systems then mutually relates the system processes and resources [14, 32–37, 41, 42, 45]

$$P = J_S \odot R \tag{20.2}$$

where $\odot$ is the Boolean equivalent of matrix multiplication [14, 32–37, 41, 42, 45]. It intuitively means that a given system process p_w can be fulfilled by a system resource r_v ***or*** any other resource as long as $J_S(w, v) = 1$. This is a fundamentally different design equation than the one used in Axiomatic Design for large fixed systems described extensively in the previous chapters of this book because the mapping between function and form is based upon an ***OR*** relationship rather than an ***AND*** relationship. Consequently, the independence axiom remains fulfilled between process p_w and resource r_v even if additional processes and resources are added. As a result, there is no requirement that $\sigma(P) = \sigma(R)$ in LFESs.

From a graph theory perspective, the system knowledge base itself forms a bipartite graph which maps the set of system processes to their resources. Unlike traditional graph theory, however, the explicit treatment of system processes in Axiomatic Design allows the study of hetero-functional large flexible engineering systems such as production systems [33–38, 40–43, 66–68], power systems [69, 70], water distribution systems [37, 38, 67], their nexus [71], transportation systems [14, 32, 45], and transportation-electrification [14]. Furthermore, since there are no conditions placed on the heterogeneity of the system resources, the system knowledge base also succinctly suffices to describe multi-modal transportation systems with fundamentally different graphs (e.g., rail and road networks).

The LFES Knowledge Base in Definition 4 is then sufficient to define a knowledge base for (traditional) transportation systems.

Definition 5 Transportation System Knowledge Base [14, 32, 45]: Given a set of transportation processes P_T and a set of transportation resources R, an action $\epsilon_{\tau v} \in \mathcal{E}_T$ can be defined for each feasible combination of transportation process $p_{t\tau}$ being realized by resource r_v. The Transportation System Knowledge Base J_T is a binary matrix of size $\sigma(P_T) \times \sigma(R)$ where element $J_T(\tau, v) \in \{0, 1\}$ is equal to one when action $\epsilon_{\tau v}$ exists.

A careful inspection of Definitions 4 and 5 shows that traditional transportation systems assume that $P = P_T$. In the transportation-electricity nexus, however, this assumption is not true because charging functionality is not considered. The transportation processes alone do not form a *collectively exhaustive* set of the system processes.

The derivation of the TEN knowledge base, therefore, requires the introduction of a set of charging processes.

Definition 6 Charging Process [14, 39]: A resource-independent process $p_c \in P_c$ that positively or negatively affects an electric vehicle's state of charge (SOC). These processes may draw or inject the required energy into the interdependent electricity grid.

In the context of this work, $P_C = \{p_{c1}, \ldots, p_{c4}\}$ where.

- p_{c1} null charging does not change the electric vehicles state of charge
- p_{c2} discharge the EV SOC to the electric vehicle's propulsion system
- p_{c3} charge the EV SOC by wire
- p_{c4} charge the EV SOC wirelessly

At this high level of design, no assumption is made on sign (or directionality) of the power transfer. These processes may be further differentiated depending on the need for different rates of SOC change. These processes may be realized by the set of transportation resources R. Conventional (non-electrified) stations effectively implement p_{c1}, while conventional roads implement p_{c2}. Charging stations and electrified rails are capable of p_{c3} regardless of whether they are simply charging or implementing more advanced "vehicle-to-grid" technology [27–29]. Finally, the electrified roads associated with OLEVs [72–75] described throughout much of this book are capable of p_{c4}.

From the set of charging processes, the associated charging system knowledge base is defined.

Definition 7 Charging System Knowledge Base [14, 39]: Given a set of charging processes P_c and a set of transportation resources R, an action $\epsilon_{\kappa v} \in \mathcal{E}_C$ can be defined for each feasible combination of charging process $p_{c\kappa}$ being realized by resource r_v. The Charging System Knowledge Base J_C is a binary matrix of size $\sigma(P_C) \times \sigma(R)$ where element $J_C(\kappa, v) \in \{0, 1\}$ is equal to one when action $\epsilon_{\kappa v}$ exists.

The derivation of the TEN knowledge base then requires that the *collectively exhaustive and mutually exclusive* system processes be defined from among the feasible combinations of transportation and charging processes. For example, these can include "staying in place at b_{y1} while charging by wire" or "moving from b_{y1} to b_{y2} while charging wirelessly." While the system knowledge base can be constructed manually by inspection, such an approach is ultimately tedious given its size. Instead, it is more readily obtained from the much smaller transportation and charging knowledge bases.

$$J_S = \left[J_C \otimes \mathbf{1}^{\sigma(P_T)}\right] \cdot \left[\mathbf{1}^{\sigma(P_C)} \otimes J_H\right] \tag{20.3}$$

where $\otimes$ is the Kronecker tensor product and $\mathbf{1}^n$ is a ones vector of length n.

Interestingly, the LFES knowledge base has an additional property in that it defines the system's scleronomic (i.e., sequence-independent) degrees of freedom [42].

Definition 8 Scleronomic Transportation Degrees of Freedom [14, 32, 45]: The set of independent transportation actions $\mathcal{E}_S$ that completely defines the available transportation processes in a transportation system. Their number is given by:

$$DOF_S = \sigma(\mathcal{E}_S) = \sum_{w}^{\sigma(P)} \sum_{v}^{\sigma(R)} J_S(w, v) \tag{20.4}$$

Ultimately, the Axiomatic Design for Large Flexible Engineering Systems Theory is a sufficiently rich foundation to describe the structure of a transportation-electricity nexus. The next subsection turns to defining its discrete-event system dynamics.

20.2.3 A Timed Petri Net Model

In this subsection, the TEN's scleronomic degrees of freedom are used to establish a timed petri net model. As with the previous discussion on graph theory, a formal background in petri nets is assumed and is otherwise obtained in Appendix. In developing the timed petri net model, the emphasis was placed on maintaining its *intuitive* link to the physical reality. Therefore, Definition 24 of a timed petri net is given the following physical meaning:

Definition 9 *Transportation-Electricity Nexus Timed Petri Net*: A timed petri net where

- B, as the set of places, represents transportation stations.
- $\mathcal{E}$, as the set of discrete events, represents the scleronomic degrees of freedom (as defined in the previous section).
- M, as the set of arcs, represents the logical relationship from the events to the places and from the places to the events.
- $W : M \rightarrow \{0, 1\}$ is the weighting function on the arcs.
- Q_B, as the place marking (or discrete state) vector, represents the queue of vehicles at a given transportation station awaiting an event.
- $Q_{\mathcal{E}}$, as the event marking vector, represents the number of vehicles undergoing the events (e.g., parking, charging, moving from one place to another).
- D, as the event durations, represent the duration of time that each event requires for completion.

The physical meanings of Q_B and $Q_{\mathcal{E}}$ are particularly important and are subtly different from many other petri net models. In this model, parking is explicitly modeled as a transition with its associated timing, duration, and required capacity. In the case of conventional parking, this is assumed to occur immediately upon arrival. The leftover tokens in Q_B represent the electric vehicles that have arrived at a charging station and are awaiting their place in a queue for the associated charging event. It has a direct impact on an electric vehicle's utilization. The tokens in $Q_{\mathcal{E}}$ directly affect the utilization of parking lots and charging stations and the congestion in conventional and electrified roads.

The discrete-event dynamic evolution of the timed petri net is then given by Definition 25:

$$Q_B[k+1] = Q_B[k] + M_{\mathcal{N}}^{+} U_k^{+} - M_{\mathcal{N}}^{-} U_k^{-} \tag{20.5}$$

$$Q_{\mathcal{E}}[k+1] = Q_{\mathcal{E}}[k] - U_k^{+} + U_k^{-} \tag{20.6}$$

where the input firing vectors U_k^- are derived from the traffic demand data, which provide the following information:

- when a given vehicle will begin a trip from origin to destination
- which route (i.e., sequence of roads) it will take.
- how long it will remain at its destination
- when it will charge along the way (while moving or parked)

Naturally, a given vehicle $l \in L$ will take several trips over the course of the day, and a transportation-electricity nexus model must equally consider when the vehicle is parked as when it is moving.

These traffic demand data are most easily captured in a vehicle firing matrix $\mathcal{U}_k$.

Definition 10 *Vehicle Firing Matrix*: a binary vehicle firing matrix $\mathcal{U}_k$ of size $\sigma(\mathcal{E}_S) \times \sigma(L)$ whose element $\mathcal{U}_k(\psi, l) = 1$ when the kth firing timing triggers a vehicle l to take scleronomic degree of freedom ψ for action.

Consequently, the input firing vector for the timed petri net U_k^- is easily calculated.

$$U_k^- = \mathcal{U}_k \mathbf{1}^{\sigma(H)T} \tag{20.7}$$

The calculation of the output firing vectors is explained in Appendix.

20.2.4 Refinement to a Hybrid Dynamic Model

In this subsection, the timed petri net model is replaced with a hybrid dynamic model. While the timed petri net model has significant advantages in terms of its simplicity, it assumes fixed duration events which are better replaced by variable durations determined by a set of continuous-time differential equations that describe the evolution of each vehicle. The development of the hybrid dynamic model gains its inspiration from hybrid automata [49]. To that effect, it is defined as follows:

Definition 11 Transportation-Electricity Nexus Hybrid Dynamic Model: A 10-tuple $\mathcal{H} = (B, \mathcal{E}, M, W, Q, \Phi, U, X, F, domain)$ where

- $(B, \mathcal{E}, A, W, Q)$ is the underlying marked petri net (Definition 9).
- Φ is the discrete state petri net transition function (Definition 23),
- $\mathcal{U}$ is a binary vehicle firing matrix (Definition 10).

- $X = [x_1, \ldots, x_{\sigma(L)}]$ is a continuous-time vector representing the kinematic and electric state of each vehicle in a fleet of size $\sigma(L)$.
- f is a vector field. $f : Q \times X \times U \rightarrow X$. It describes the continuous-time evolution of these vehicles.
- Domain is a set of invariant conditions [49] which associates a discrete state Q to an interval of X and U within which X and U must remain in order to also remain in the discrete state Q.

The state x_l must represent the kinematic and electrical state of a given vehicle $l \in L$. While the model can easily accommodate an elaborate description of the vehicle's internal dynamics, it is important to recall that doing so might be practically infeasible given the shear number of vehicles being simulation within the TEN. To that effect, a minimalistic model is chosen. $x_l = [z_l, \dot{z}_l, s_l]^T$ where.

- z_l is the distance of the vehicle along a road segment in relative coordinates
- $\dot{z}_l$ is the speed of the vehicle along the road segment
- s_l is the vehicle's state of charge.

The vector field f is implemented as a state space differential equation of the form:

$$\dot{X} = f(Q, X, U_k) \tag{20.8}$$

In free-driving conditions, the dynamics of each vehicle become entirely uncoupled and the state of the vehicle becomes purely a function of the vehicle firing matrix.

$$\begin{bmatrix} \dot{z}_l \\ \ddot{z}_l \\ \dot{s}_l \end{bmatrix} = \begin{bmatrix} \beta_v \\ 0 \\ \alpha_v \end{bmatrix} U_{kl} \tag{20.9}$$

The vehicle speed is set to a constant speed β_1 whether it is moving along a road or parked at a station. Additionally, the charging rate α_1 is sufficient to describe all four types of charging processes. One advantage of the free-driving model is that it retains the timings of the underlying timed petri net model. In contrast, under more normal driving conditions with some congestion, a car following model is typically used [63, 76, 77]. Consequently, the state of charge is often modeled to change with the vehicle speed and vehicle firing matrix.

$$\begin{aligned} \dot{z}_l &= \dot{z}_l \\ \ddot{z}_l &= \alpha \dot{z}_l^{\beta}(t) \frac{(\dot{z}_{l-1}(t-T) - \dot{z}_l(t-T))}{(z_{l-1}(t-T) - z_l(t-T))^{\gamma}} \\ \dot{s}_l &= f(\dot{z}_l) + \alpha_v U_{kl} \end{aligned} \tag{20.10}$$

where $\ddot{z}_l$, $\dot{z}_l$, z_l are the acceleration, speed, and position of the lth vehicle which follows the $l-1$ vehicle. $\alpha > 0$, β, and γ are model parameters that control the proportionalities, and T is reaction time [76, 77].

Finally, the *domain* describes a set of invariant conditions upon which a given discrete state remains valid. In the context of the HDM, these conditions are useful for constraining the vehicles distance along the road segment and its state of charge within limits. For example,

$$\begin{aligned} 0 \leq z_h \leq z_{vmax} \\ 0 \leq s_h \leq S_{hmax} \end{aligned} \tag{20.11}$$

where D_v may be the road length and S_{hmax} may be the vehicle's battery capacity.

20.2.5 Hybrid Dynamic Model Outputs

To conclude the definition of the hybrid dynamic model, it is important to identify relevant model outputs for subsequent discussion. As previously mentioned, the model was developed to maintain a highly intuitive nature. The place marking vector Q_B represents the queue of vehicles at a given station. It is important as a model output to measure the number of electric vehicles waiting to be charged. The event marking vector itself $Q_{\mathcal{E}}$ represents the number of vehicles undergoing the events. It can be divided into the number of vehicles flowing in roads and the number of parked vehicles. Finally, the total charging load f_c— the net load required by the TEN from the power system—is easily found as a linear combination of the elements in the event marking vector.

$$f_c = C_\alpha^T Q_{\mathcal{E}} \tag{20.12}$$

where C_α is a vector of size $\sigma(\mathcal{E}) \times 1$ whose elements $C_\alpha(v) = \alpha_v$, the charging rate in kilowatts for the given event.

20.3 Transportation-Electrification Test Case

With a hybrid dynamic model for the transportation-electricity nexus in place, the chapter shifts focus to the comparison of electrified transportation system performance in the cases of conventional and online electric vehicles. To that end, it is necessary to use a transportation-electrification test case that does not create undue bias between the scenarios. Indeed, one critique of the traffic simulation literature is that it focuses on specific traffic topologies rather than investigating the fundamental dynamics of traffic behavior [78]. Furthermore, a test case must be of a sufficiently moderate size to predict how the scenarios may scale up to a full deployment. Finally, the electrification topology should be regular enough to facilitate the interpretation of how the system performance emerges from the system behavior in the respective scenarios. Furthermore, it must be of a sufficiently moderate size to predict how the scenarios may scale up to a full deployment. Finally, the electrification topology should be regular enough to facilitate the interpretation of how the system performance emerges from the system behavior in the respective scenarios. With these considerations in mind, and in the absence of a real-life test case with the same attributes, a hypothetical test case aptly named "Symmetrica" was developed [79]. While its specific characteristics may differ from the reality of specific regions, its characteristics do offer much in developing insight and intuition into the dynamics of transportation-electricity nexus. For the sake of simplicity, and without loss of generality, the roads are assumed to be free of congestion and the free-driving model will then be applied.

20.3.1 Road Topology

The Symmetrica road topology is shown in Fig. 20.1. It is a suburban 12 $\times$ 12 km grid with intersections at every kilometer. Each road segment has a free speed of 60 km/hr.

20.3.2 Electrification Topologies

Two electrification topologies were designed for Symmetrica. Both were designed to be able to deliver a maximum of 46.8 MW of charging load—the size of a medium capacity generator. The equivalence in the peak charging rate between the two scenarios was introduced to avoid biasing effects on vehicle travel patterns and power system balance. For simplicity, the charging infrastructure was also assumed to come with the necessary upgrades in line, transformer, and substation capacities to support the charging. Electric vehicle charging in private residences do pose such a challenge, but it is specifically neglected in this work. Consequently, this work does not address the local line and voltage safety criterions found in recent work [10–13]. Furthermore, this work assumes that the two electrification scenarios have equivalent investment cost and instead addresses the additional costs required to mitigate the effects of charging on the power system. The details of each scenario are as follows.

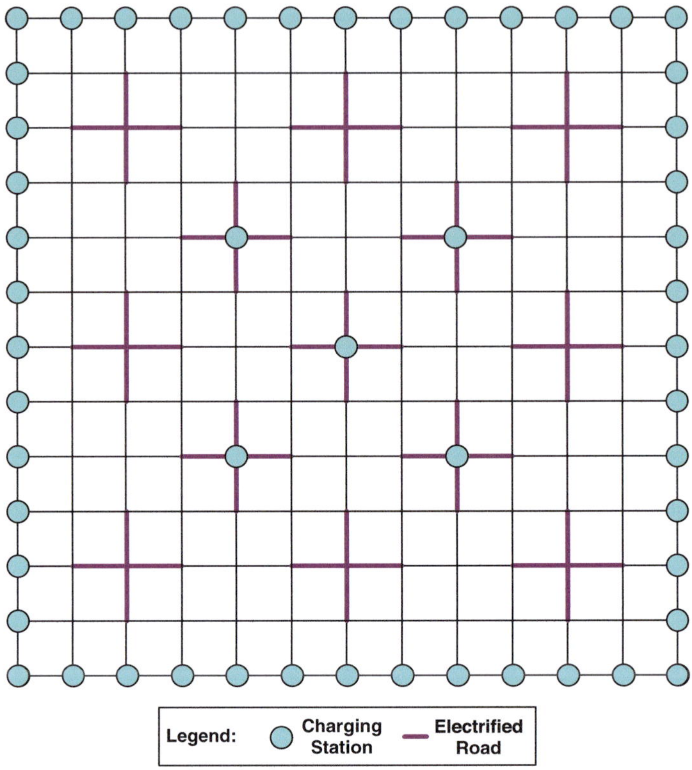

Fig. 20.1 Topology of the Symmetrica transportation-electrification nexus

20.3.2.1 Conventional Electrification Topology

The conventional electrification topology consists of two groups of charging stations. There are 5 charging station in the city center at coordinates (4,4), (4,8), (8,4), (8,8), and (6,6). These are marked as cyan circles in Fig. 20.1. Each of these can deliver 30 kW to each of 25 vehicles at a time. In order to include the potential for home charging, a charging station was placed at every intersection along the periphery, each delivering 18 kW to each of 50 vehicles at a time. It is important to note that a given intersection here does not represent an individual home where there may only be 1 or 2 spaces to charge in a garage. Instead, the intersections along the periphery represent all of the homes associated with the vehicles that exited Symmetrica through that intersection. Thus, it is a centralized representation of a distributed charging capacity outside of the city. In theory, full electric vehicle deployment suggests ubiquitous home charging, and thus, this capacity is infinite. (Every vehicle owner should be able to return home to charge). However, a fair comparison with the online electric vehicle case requires that the installed charging capacity of scenarios be equal. Here, the total installed charging capacity is (5)(0.03)(25) + (48)(0.018)(50) = 46.8 MW.

20.3.2.2 Online Electric Vehicle Topology

The online electric vehicle topology consists of 13 groups of electrified road segments which appear in Fig. 20.1 as magenta-colored 2 km × 2 km road crosses. Each road segment is able to deliver 30 kW to each of 28 vehicles at a time. In order to clearly distinguish the differences in system performance between the two electrification scenarios, this scenario assumes no stationary charging. This electrification scenario is also capable of delivering up to (4)(13)(0.03)(30) = 46.8 MW of power.

20.3.3 Traffic Demand

The traffic demand represents a simplification of an average work day. Symmetrica starts the day empty of any vehicles. Vehicles enter from any of the intersections along the symmetric periphery and go to five work locations which coincide with the five conventional charging stations depicted as cyan circles. In such a regular topology, there are many "shortest route" choices between a given origin and destination. For example, there are 6 such routes just from (0,0) to (2,2). The traffic demand consists of all of these shortest routes with the added constraint that such routes must pass through the centers of the electrified road segment crosses depicted in magenta. This ensures that the same traffic demand can be applied in both electrification scenarios without bias. It also serves to more evenly distribute the traffic and not place undue congestion on the electrified roads. Note that the number of routes thus follows an exponential distribution with the required distance.

The traffic demand makes use of this exponential distribution to generate the timing and congestion in the morning and evening rush hour commutes. For a given origin–destination pair, there are many possible routes. Each of these is initiated every minute, one vehicle at a time, in such a way that they are centered around 7:30 am. A total of 8252 vehicles are simulated. In all, the first vehicles enter Symmetrica at 5:00 am and the end of the morning commute is marked with the last vehicle at 10 am. Upon arriving to the five work locations, the vehicles remain there for 8 h and then return "home" to the Symmetrica periphery intersection from which they entered along the route that they took in the morning. No further assumption is made on the transportation use case (e.g., private usage, taxi, car sharing, etc.). Instead, the potential implications on use cases are discussed in connection with the results described in Sect. 20.5.

20.3.4 *Charging Demand*

All of the electric vehicles begin the day at a full charge. The online electric vehicles are assumed to discharge power at a rate of 30 kW when moving at the road free speed of 60 km/hr. The conventional electric vehicles have larger batteries and thus weigh more. They are assumed to consume 20% or 36 kW. The online electric vehicle battery size is a modest 2.5 kWhr which is sufficient in this case to not require any conventional charging. The conventional electric vehicle battery size is a manageable size of 12 kWhr. These values are within the physical limits of current technology although the dissipation rates are exaggerated given the relatively small distances in Symmetrica.

20.4 MATLAB Simulation for Urban Mobility Electrification

A performance comparison was conducted by simulation in MATLAB. This intentional methodological departure from many available commercial as well as open-source microscopic traffic simulation software packages was deeply considered. Several reviews contrast their respective features and performance of these packages [62, 64, 65, 80, 81]. No commercial packages consider electrification of roads as well as parking stations and are thus deemed inadequate. The open-source packages can potentially be retrofitted with electrification functionality. This is, however, a non-trivial *post hoc* endeavor. As mentioned in the Axiomatic Design for Large Flexible Engineering Systems discussion in Sect. 2.2, microscopic traffic simulators were meant to simulate vehicles *in motion* between *distinct* origins and destinations. This is adequate as long as the vehicles do not change state while parked but very much inadequate otherwise, as in the case of electrified transportation. Furthermore, online electric vehicles simulation requires that a given road have the potential to provide electrified *and* non-electrified transport. Finally, microscopic traffic simulation packages only keep track of the state of *moving* vehicles. In contrast, transportation-electrification requires a state for all vehicles regardless of whether they are moving. These differences represent dramatic expansions in simulation functionality—ones that are not likely to be fulfilled by incremental tweaks but rather large-scale architectural changes to the software. In contrast, the simulations presented are built directly upon the Axiomatic Design model presented in Sect. 20.2. All of the required functionality is considered early in the simulator design in agreement with Axiomatic Design principles. As a result, the entire simulation straightforwardly falls upon three highly intuitive matrix-based equations of motion (Eqs. 20.5, 20.6, and 20.8). That two of these are linear further facilitates the design.

In a "Big Data" application such as microscopic transportation-electrification, it is also essential to consider the computational speed of the simulation. Consider the size of the vehicle firing matrix alone: It is $\sigma(\mathcal{E}) \times \sigma(L) \times \sigma(T)$. A full-scale simulation, for the traditional one day, at the traditional resolution of one second, in a moderately sized city such as Abu Dhabi, would require approximately $15e3 * 1e6 * 8.64e4 = 1.2960e15$ elements. Therefore, a well-considered simulation must employ (1) matrix representations rather than scalar approaches; (2) sparse matrix techniques and advanced numerical methods; (3) parallel processing wherever possible; and (4) efficiently compiled code.

In contrast to other simulation packages implemented in JAVA and C++, MATLAB was chosen as an implementation platform. Software development in MATLAB is known to be as easy as other high-level languages such as JAVA. Furthermore, it is purposefully built to handle matrix operations. It also has advanced libraries to support sparse matrices, the solution of differential equations and other advanced numerical methods. Recent versions of MATLAB also offer significant parallel processing functionality to leverage modern multi-core processors and graphical processing units. Finally, although MATLAB is an interpreted language to facilitate development, final version code can be easily compiled into C or GPU code for further performance enhancements. These features have been repeatedly tested and demonstrated across several application domains and thus represent proven technology. In contrast, simulation in JAVA and C++ would likely require custom and expert development of many of these advanced functionalities, thus expanding development time and cost.

Finally, the development of simulations that are founded upon a rigorous dynamic model opens the door for the development of control and optimization methods as part of an intelligent transportation-energy system.

20.5 Results and Discussion

The results of the Symmetrica test case are now presented. The discussion addresses four areas of transportation-electricity nexus performance. The first two address the transportation system: (1) the traffic behavior of moving and parked vehicles and (2) the charging behavior as a queue and electrical load. The last two address the implications on the power system assuming no corrective action was taken in the transportation system: (3) the implications on power system generation capacity and (4) the implication on the required power system operating reserves. Each of these is now addressed in turn below. Figure 20.2 summarizes many of the simulation results within one holistic framework.

20.5.1 Traffic Behavior: Moving and Parked Vehicles

The top two subplots of Fig. 20.2 show the Symmetrica traffic behavior between 5 am and just past 6:30 pm. The number of driving vehicles on any given road

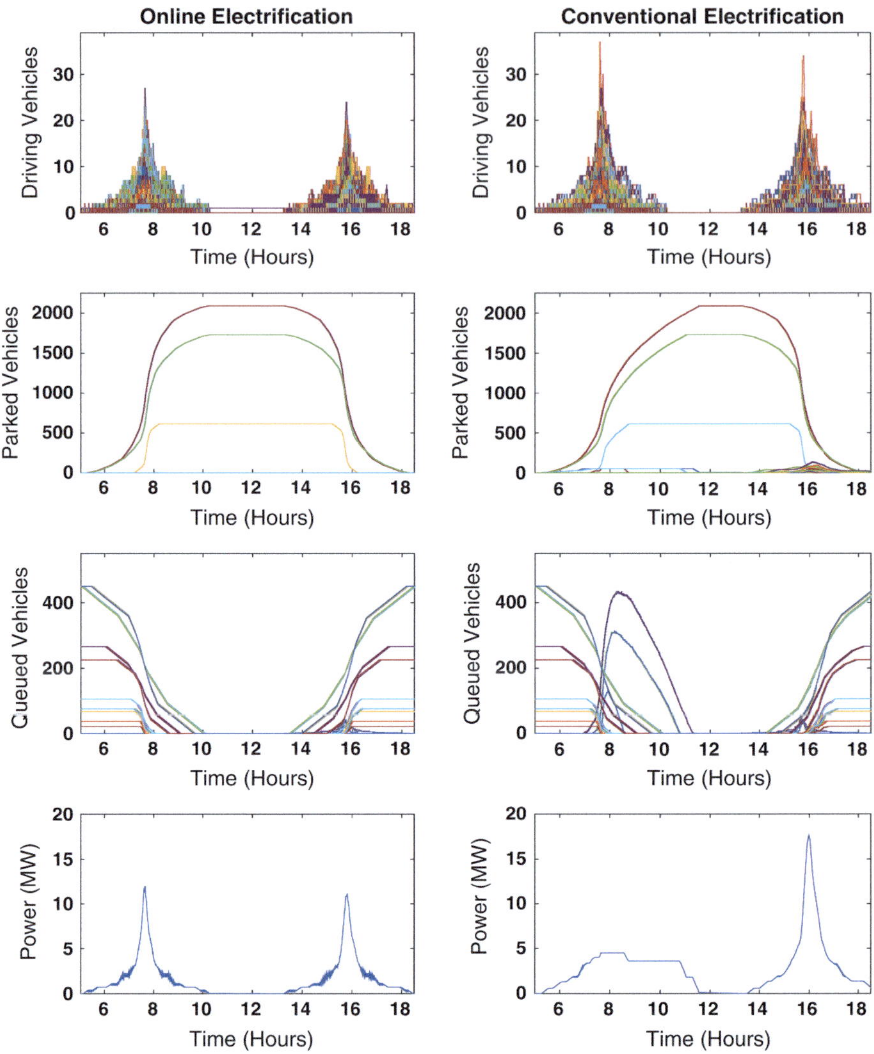

Fig. 20.2 Simulation results of the Symmetrica test case for online and conventional transportation-electrification

segment is shown with a distinct color. In the hybrid dynamic system model, this corresponds to the rows of Q_E that are associated with events along road segments. As expected, both the online and conventional electrification scenarios reflect the traffic demand which has two exponentially distributed peaks of traffic congestion corresponding to the morning and afternoon commutes. These are qualitatively the same but numerically different. Note that the conventional electrification scenario has peaks of 37 and 34 vehicles, while the online electrification scenario has peaks of 27 and 24. This difference is caused by the online electrification scenario having two events along electrified roads and not just one. As a result, the associated traffic is divided among them. In actuality, the physical road still carries the same number of vehicles. It is also worth noting that the two scenarios demonstrate the same timing for the afternoon commute. This suggests that the 8 h devoted to work was sufficient to allow all electric vehicles to depart without delay. Therefore, both electrification scenarios demonstrate a 100% quality of service [10–12] for this specific use case.

The next two subplots in Fig. 20.2 show the quantity of parked vehicles in Symmetrica's parking lots. In the hybrid dynamic system model, this corresponds to the rows of Q_E that are associated with parking events at the five work locations. As expected, the number of parked vehicles rises sharply shortly after the peak congestion times and eventually stabilizes as moving vehicles reach their workplace. Note that the number of parked vehicles in the conventional electrification scenario appears delayed despite the equivalent traffic behavior timing in the previous two plots. This is because the electric vehicles either begin or await charging upon arrival to the workplace and therefore are not counted as "non-electrified" parking. Thus, these parked vehicles come to mean "vehicles-ready-for-usage." The equivalence of the peaks of the two plots confirms that, in this use case, the same quality of service of 100% is achieved. Also note that the numerical scale of parked vehicles is much greater than those in motion. Here, the peak is just over 2000 vehicles—two orders of magnitude greater. From a systems perspective, roads are associated with distributed quantities, while parking lots are associated with centralized quantities. While this is perhaps a straightforward insight in transportation behavior, the implications on electrification infrastructure appear profoundly and are dramatic in the subsequent plots.

The next two subplots in Fig. 20.2 show the quantity of queued vehicles in Symmetrica. In the hybrid dynamic model, this corresponds to Q_B. In the morning, vehicles in both electrification scenarios are queued at the periphery (as a representation of homes). These queues dramatically shorten over the course of the morning commute hours until they disappear. They only begin to form again in the late afternoon hours as vehicles return home. This is another way of viewing the "distributed-to-centralized-to-distributed" state evolution of the vehicle fleet over the course of the day. The obvious difference between the two electrification scenarios is the presence of queues with conventional charging stations. Online electric vehicles charge as they move. They do not form queues in the stationary sense. If the electrified road does not have sufficient capacity, then some vehicles do not receive the full amount of charge that they require. In contrast, conventional charging only completes charging once a full state of charge has been achieved. Consequently, charging queues inevitably form. It is here that the centralized state of the vehicle fleet means that more than 400 vehicles can be waiting to charge at a time. Furthermore, this queue does not clear until nearly the afternoon hours. Had this queue been longer, with perhaps an increase in the traffic demand or electric vehicle penetration rate, it could have meant that some vehicles would not be available on time for the afternoon commute home. This would represent a reduced quality of service. Alternatively, had this electrification scenario been applied to a taxi use case, then many taxis would have to wait several hours after their morning customer just to take on another fare. Such a low vehicle utilization rate could prove to be a non-starter for EV taxi adoption. In contrast, the online electric vehicle taxis could continue to operate in a seemingly perpetual way.

Thus, the intelligent transportation-energy system premise posed at the beginning of the chapter manifests itself. Conventional charging of electric vehicles means that electric vehicle dispatch and route choice have immediate implications on the formation of queues and the associated quality of service. A planning centric solution would be to either expand the charging rate of the conventional chargers or increase the number of charging slots. While this requires an added investment cost, it is well justified if it directly improves the quality of service of the desired set of electric vehicle use cases. Alternatively, an intelligent transportation-energy system can serve to dispatch vehicles at better times, choose the best routes, and potentially defer vehicles to charging stations with shorter queues. Of course, none of these solutions are required in the case of online electric vehicles.

The next two subplots in Fig. 20.2 show the aggregate charging load of electric vehicles in Symmetrica. In the hybrid dynamic model, this corresponds to the output function Eq. 20.12. It is here that the differences between the two electrification concepts become most apparent. In the online electric vehicle scenario, charging occurs literally "on-the-go," as required, in an entirely unmitigated way, as a result of road congestion. Thus, two sharp charging peaks emerge in time with the road congestion. Consequently, this traffic demand does not cause any charging load in the late morning and early afternoon hours. In contrast, the conventional charging scenario is associated with a "stock" of vehicles and not their flow. Charging is associated with two time periods: during work and at home after the commute. For charging at work, the number of charging vehicles climbs rapidly shortly after the peak congestion. Here, the centralization of electric vehicles means that charging capacity is limited, saturates, and causes charging queues. The charging load eventually falls in a stepwise fashion as the queues are cleared in the late morning hours. For home charging, the vehicle fleet is very much distributed. As a result, the charging load does not saturate and more closely resembles the sharp peak of online electric vehicles. Note that the online charging scenario has a fleeting peak charging power of 12.5 MW, well below the installed capacity of 46.8 MW. In contrast, the conventional charging shows two very different characteristics: an enduring level near 5 MW and a sharp peak near 17.5 MW. As an aside, it should be noted that the area under the charging load curve represents the total energy consumed. It is numerically greater in the case of conventional electric vehicles given the assumption of larger and heavier batteries. Managing the magnitude and shape of these charging load profiles is of ultimate importance for successful integration with future electricity grids.

Again, the premise of intelligent transportation-energy system posed at the beginning of the chapter manifests itself. Transportation-electrification requires electric vehicles that, at worst, do not destabilize power grid operation and, at best, are active players in its stabilization. While many vehicle-to-grid stabilization schemes have thought to make use of the parked vehicles for the provision of parked vehicles, the reality is that the potential is much greater and much more dynamic. In the conventional charging scenario, for example, it is now clear that charging station selection can serve to manage queues and thus improve the performance of coordinated charging and vehicle-to-grid stabilization schemes. That said, both control decisions only serve to flatten and elongate the charging load profile. This occurs under the assumption that the transportation use case does not in actuality impose a binding time constraint on how quickly the vehicles need to be available. An immediate trade-off appears: The charging profile can be flattened for grid stabilization or sharpened for transportation system performance. For online electric vehicles, the management of the associated charging load peaks is critical to power grid operation. Furthermore, the absence of charging stations and their queues eliminates many of the control levers found in the conventional charging scenario. Instead, vehicle dispatch and route choice can be deployed as mitigation measures. Small delays of vehicle dispatch can effectively dull a sharp charging load. Equally possible, the online electric vehicle may take a different route, travel in non-electrified lanes, or simply shield itself from the wireless charging - especially in the presence of a larger backup battery. A combination of all of these options can be easily achieved with greater coordination through an intelligent transportation-energy system.

20.5.2 Required Power System Generation Capacity

The last two subplots of Fig. 20.2 show that electrified transportation causes a non-negligible charging load on the power grid. As previously discussed, this charging load can be ameliorated with an intelligent transportation-energy system primarily in the transportation domain. Alternatively, the unaltered charging load can be passed to the power grid for mitigation in the electrical domain. There, it is necessary to assess whether the power grid requires revised planning and operations management. This subsection addresses the first in terms of the need for additional power system generation capacity.

Traditionally speaking, the first tool in power system generation capacity planning is the load duration curve. The load profile from the previous year is taken as a time series and then sorted from high to low [82]. Figure 20.3 shows the load duration curve for the Bonneville Power Administration in 2013 in black. Base load units are required for the entire year, followed by "shoulder" load units for most of the year, followed by peak load units for only tens of days in the year. Generally speaking, these generation units have increasing marginal costs in the order of mention to form an upward sloping power generation supply curve. Power system generation capacity planning thus asks two important questions: (1) Will the peak load increase enough to require the installation of expensive peak load generation capacity? (2) Will the peak load broaden enough to merit a replacement of peak

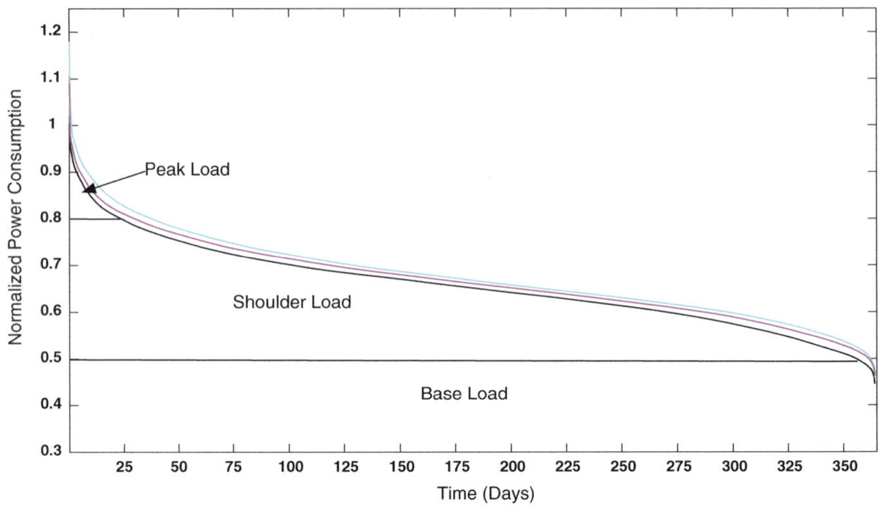

Fig. 20.3 Load duration curves: (a) base case w/o electrification (*black*), (b) conventional electrification (*cyan*), and (c) online electrification (*magenta*)

load capacity with shoulder load capacity? These two questions are now addressed in the case of the conventional and online electrification scenarios.

To assess these impacts, the two charging load curves were zero-padded so as to form a time series for a full day. Then, as a reasonable assumption of a worst case scenario, it was assumed that all days in the year are equivalent to the work day that had been previously simulated. This yielded two full year charging load time series. The base case was then scaled so as to be 4 times larger than the magnitude of the home charging load. This 25% value was chosen as a reasonable approximation of the percentage of all energy consumed by the transportation sector in the United States [83]. The two full year charging load time series were then added to the scaled base case. The three resulting load duration curves are shown in Fig. 20.3: (1) a base case without electrification (black), (2) a base case with conventional electrification (cyan), and (3) a base case with online electrification (magenta).

Simply speaking, the results show that when it comes to electric vehicle charging, timing is critical. Table 20.1 summarizes the numerical figures. Although the magnitude of the charging loads in the online electrification scenario is greater, they come during the morning and afternoon commute hours, times not traditionally associated with peak loads. Thus, the peak load expands by 8.96% (6.31 MW)—a relatively small number when compared with the maximum value of 25%. In the meantime, the midday and evening charging of the conventional electrification scenario more closely coincides with the traditional peaks of the base case load profile. Thus, the peak load expands by 17.9% (12.6 MW) even though the magnitude of the charging load was much smaller for much of the day. In 2013, the cost of a peak load advanced gas turbine was $676/kW [84]. Thus, for the provided test case, and all else held equal, the conventional charging scenario requires 4.23 M$

Table 20.1 Effects of transportation-electrification on power system peak load capacity and investment cost

	Base case	Conventional electrification	Online electrification
Peak load in MW (% increase)	70.4 (0%)	83.0 (17.9%)	76.7 (8.96%)
Required (M$)	Investment 0	8.5	4.27

more capital investment. It is also worth noting that in this test case, neither electrification scenario was capable of flattening the load duration curve. Therefore, there is no incentive to shift the generation portfolio from expensive peak load plants to more competitive shoulder load plants. In both cases, an intelligent transportation-energy system can serve to reduce peak loads and expand shoulder loads for even greater capital cost savings.

20.5.3 Required Power System Operating Reserves

While it is important to assess the long-term impact of charging load curves, it is equally important to recall that they present an immediate challenge to present day power grid operations. Power grid balance must be achieved at all times and at all frequencies. In the meantime, load curves arising from the electrification of transportation have frequency characteristics ranging from seconds to several hours. Therefore, the techno-economic impacts of transportation-electrification on power grid operations must be assessed in a holistic manner that spans multiple timescales. Recently, the concept of power grid enterprise control has been advanced to conduct such an assessment [85–88]. This concept was successfully used to improve the methodologies used to conduct integration studies on renewable energy [89–92], energy storage [93, 94], and demand side resources [95]. Such an enterprise control approach should also be applied to transportation-electrification.

While a complete enterprise control methodology is outside the scope of this work, the provided data test case is sufficient to draw accurate conclusions on the quantity and cost of additional power system operating reserves that are required for reliable operation. Recent research in power grid enterprise control has provided *closed-form* derivations for the required power system operating reserves given the characteristics of the net load [96–99].

The required quantities power system load following, ramping, and regulation reserves for the two electrification scenarios are presented in Table 20.2. These values assume a power grid enterprise control structure similar to that of PJM-ISO with a day-ahead market time step of 1 h, a real-time market time step of 5 min, and a sampling rate of 1 min. The cost of load following reserves and regulation reserves is taken as \$15.42/MW and \$49.77/MW, respectively [100].

Table 20.2 Effects of transportation-electrification on power system operating reserves and costs

	Base case	Conventional electrification	Online electrification
Load following reserves in MW (% increase)	1.005	2.543 (153%)	1.863 (85%)
Load following reserve additional cost (k$)	0	208	116
Ramping reserves in kW (% increase)	103	182 (77.0%)	196 (91.4%)
Regulation reserves in kW (% increase)	167	247 (48.1%)	297 (78.0%)
Regulation reserve additional cost (k$)	0	420	681
Total additional operating cost (k$)	0	628	797

For simplicity, the forecast errors on the conventional and charging loads are assumed to be zero and uncorrelated.

The results showed that conventional electrification required on average 0.7 MW more of load following reserves leading to an difference of 92 k$ in additional cost. Meanwhile, the sharp peaks associated with the online electrification scenario, respectively, lead to a difference of 14 and 50 kW in additional ramping and regulation reserves. This accounts for 261 k$ per year. In all, the online electrification scenario incurs 169 k$ more in annual operating costs. In both cases, an intelligent transportation-energy system can serve to reduce the variability of charging loads at the timescale of all three types of reserves for even greater savings in operational cost.

20.6 Conclusion: Conventional Versus Online Electric Vehicle System Performance

In conclusion, this chapter has provided a systematic techno-economic performance assessment of conventional and online transportation-electrification scenarios. It was completed with the recognition that electrified transportation effectively couples the transportation system to the electrical power grid in a "transportation-electricity nexus." To conduct the study, a hybrid dynamic model of the nexus was presented as an objective basis of comparing the two scenarios. This model was then simulated in MATLAB using a newly developed test case called Symmetrica. It featured equivalent traffic demand with sufficient charging capacity in both scenarios to avoid reduced quality of service, voltage deviations, or line outages. It did, however, specifically address the geographical distribution and physical capacity of the charging infrastructures. In competing the fully analysis, the chapter has also provided a methodological contribution in addition to the provided results.

The simulation results show several findings:

- The simulation of a transportation-electricity nexus requires the consideration of parking and charging states in addition to the vehicle motion models found in traditional microscopic traffic simulators.
- Charging capacity limits in conventional transportation-electrification causes charging queues that limit the variety of use cases to which electric vehicles are suitable. While 100% quality of service is achievable for private use cases that include charging at home and work, it is less so for commercial and public use cases.
- Online electric vehicles are well suited for all uses, particularly commercial and public uses where vehicle utilization is paramount.
- Online charging loads closely follow vehicle congestion and are likely characterized by sharp peaks.
- Conventional charging loads exhibit a plateau shape in the afternoon when vehicles are centralized at workplaces and charging capacity is limited. They exhibit sharp peaks in the evening when vehicles are distributed at residences and charging capacity is nearly ubiquitous.
- The timing of conventional charging coincides with peak conventional loads and thus is more likely to require power generation capacity expansion. In contrast, online charging is much less synchronized and is less likely to require as much of an increase.
- The sharp peaks that characterize online transportation-electrification require substantial operating reserves for reliable power grid operation. These incur additional operating costs relative to a conventional charging scenario.

20.7 Future Work: Intelligent Transportation-Energy Systems

A consistent theme in this book chapter has been the potential role of an intelligent transportation-energy system, as first mentioned in [10]. The physical coupling of the transportation and power systems brings about the coupled operations management decision shown in Table 20.3. The first step to designing such an ITES begins with a deep understanding of the nature of the coupling in the form of a rigorous quantitative model such as the one presented in this work. This model must then be simulated to demonstrate these couplings. Test cases play an important role in that regard as they offer insight and intuition into fundamental dynamics rather than moving directly to case study results. Ultimately, the control and optimization methods developed within an intelligent transportation-energy system must demonstrate their benefits on arbitrary topologies and not just a specific region.

As this chapter has demonstrated, the potential benefits of an ITES are great. In conventional charging, an ITES can serve to modify vehicle dispatch and route

Table 20.3 Intelligent transportation-energy system operations decisions in the transportation-electricity nexus

• **Vehicle Dispatch**: When a given EV should undertake a trip (from origin to destination)
• **Route Choice**: Which set of roads and intersections it should take along the way
• **Charging Station Queue Management**: When and where it should charge in light of real-time development of queues
• **Coordinated Charging**: At a given charging station, when the EVs should charge to meet customer departure times and power grid constraints
• **Vehicle-to-Grid Stabilization**: Given the dynamics of the power grid, how can the EVs be used as energy storage for stabilization

choice and manage queues so as to minimize the waiting time for charging. This leads to direct improvements in the quality of service and expands the use cases for which electric vehicles can be adopted. Charging queue management, coordinated charging, and vehicle-to-grid stabilization also serve to reshape the charging load profile for both technical and economic benefits on the power grid. In the case of online electric vehicles, quality of service has been effectively guaranteed. However, an ITES can still modify vehicle dispatch and route choice and also implement coordinated charging and vehicle-to-grid stabilization to smoothen charging loads that will directly follow traffic congestion. In both cases, the implementation of an ITES leads to direct savings in investment and operating costs. Shifting the timing of charging peaks avoids the installation of peak load capacity. Meanwhile, flattening the shape of charging loads means that fewer operating reserves will ultimately be required. An intelligent transportation-energy system, therefore, presents a valuable opportunity to improve the techno-economic case for a sustainable and electrified transportation system. And while multi-modal transportation-electrification systems present formidable challenges for future work, it is likely that they will ultimately be facilitated by recent advances in connected vehicle technology and multi-agent systems.

Appendix

The hybrid dynamic system model presented in Sect. 20.2 is based upon graph theory, petri nets, and Axiomatic Design for large flexible systems. As the first two of these subjects are from different disciplines, a brief introduction to their fundamental concepts is provided for the potentially uninitiated reader.

Graph Theory

Graph theory is a long-established field of mathematics with applications in many fields of science and engineering where artifacts are transported between physical locations [46–48]. A number of definitions are introduced for later use in the discussion.

Definition 12 *Graph* [48]: $G = \{B, E\}$ consists of a collection of nodes B and collection of edges E. Each edge $e \in E$ is said to join two nodes, which are called its end points. If e joins $b_1, b_2 \in B$, we write $e = \langle b_1, b_2 \rangle$. Nodes b_1 and b_2 in this case are said to be adjacent. Edge e is said to be incident with nodes b_1 and b_2, respectively.

Definition 13 *Bipartite Graph* [48]: Graph G is bipartite if $B(G)$ can be partitioned into two disjoint subsets B_1 and B_2 such that each edge $e \in E(G)$ has one end point in B_1 and the other in $B_2 : E(G) \subseteq \{e = \langle b_1, b_2 \rangle | b_1 \in B_1, \text{ and } b_2 \in B_2\}$.

Definition 14 *Directed Graph (digraph)* [48]: D, consists of a collection nodes B, and a collection of arcs E, for which we write $D = (B, E)$. Each arc $e = \langle b_1, b_2 \rangle$, is said to join node $b_1 \in B$ to another (not necessarily distinct) node b_2. Vertex b_1 is called the tail of e, whereas b_2 is its head.

Definition 15 *Incidence In Matrix* [48]: M^+ of size $\sigma(B) \times \sigma(E)$ is given by:

$$M^+(i,j) = \begin{cases} 1 & \textit{if } b_y \textit{ is the tail of arc } e_j \\ 0 & \textit{otherwise} \end{cases} \tag{20.13}$$

where the operator $\sigma()$ gives the size of a set.

Definition 16 *Incidence Out Matrix* [48]: M^- of size $\sigma(B) \times \sigma(E)$ is given by:

$$M^-(i,j) = \begin{cases} 1 & \textit{if } b_y \textit{ is the head of arc } e_j \\ 0 & \textit{otherwise} \end{cases} \tag{20.14}$$

Definition 17 *Incidence matrix* [48]: M of size $\sigma(B) \times \sigma(E)$ is given by:

$$M = M^+ - M^- \tag{20.15}$$

Definition 18 *Adjacency matrix* [48]: A, is binary and of size $\sigma(B) \times \sigma(B)$ and its elements are given by:

$$A(y_1, y_2) = \begin{cases} 1 & \textit{if} \langle b_{y_1}, b_{y_2} \rangle \textit{exists} \\ 0 & \textit{otherwise} \end{cases} \tag{20.16}$$

Petri Nets

Petri nets offer a long-established method for modeling and simulating the discrete-event dynamics of a system. Their usage is described by the following definitions.

Definition 19 *Marked Petri Net (Graph)* [49]: A bipartite directed graph represented as a 5-tuple $\mathcal{N} = (B, \mathcal{E}, A, W, Q_B)$ where:

- B is a finite set of places of size $\sigma(B)$.
- $\mathcal{E}$ is a finite set of (instantaneous) transitions/events of size $\sigma(E)$.
- $M \subseteq (B \times \mathcal{E}) \cup (\mathcal{E} \times B)$ is a set of arcs of size $\sigma(M)$ from places to transitions and from transitions to places in the graph.
- $W : A \rightarrow \{0, 1\}$ is the weighting function on arcs.
- Q_B is a marking (or discrete state) vector of size $\sigma(B) \times 1 \in \mathbb{N}^{\sigma(B)}$.

As with graphs, three incidence matrices are defined for petri nets.

Definition 20 *Petri Net Incidence In Matrix* [49]: $M^+_{\mathcal{N}}$ of size $\sigma(B) \times \sigma(\mathcal{E})$ is given by:

$$M^+_{\mathcal{N}}(i,j) = \begin{cases} 1 & \textit{if } b_y \textit{ is the tail of event } \epsilon_j \\ 0 & \textit{otherwise} \end{cases} \tag{20.17}$$

Definition 21 *Petri Net Incidence Out Matrix* [49]: $M^-_{\mathcal{N}}$ of size $\sigma(B) \times \sigma(\mathcal{E})$ is given by:

$$M^-_{\mathcal{N}}(i,j) = \begin{cases} 1 & \textit{if } b_y \textit{ is the head of event } \epsilon_j \\ 0 & \textit{otherwise} \end{cases} \tag{20.18}$$

Definition 22 *Petri Net Incidence matrix* [49]: $M_{\mathcal{N}}$ of size $\sigma(B) \times \sigma(\mathcal{E})$ is given by:

$$M_{\mathcal{N}} = M^+_{\mathcal{N}} - M^-_{\mathcal{N}} \tag{20.19}$$

Definition 23 *Petri Net (Discrete-Event) Dynamics* [49]: Given a binary firing vector U_{Dk} of size $\sigma(\mathcal{E}) \times 1$ and a petri net incidence matrix $M_{\mathcal{N}}$ of size $\sigma(B) \times \sigma(\mathcal{E})$, the evolution of the marking vector Q_B is given by the state transition function $\Phi(Q_B, U_{Dk})$:

$$Q_B[k+1] = \Phi(Q_B, U_{Dk}) = Q_B[k] + M_{\mathcal{N}} U_k \tag{20.20}$$

While marked petri nets are sufficient for discrete-event dynamics, they do assume events of infinitesimal duration. In the case of transportation systems, it is necessary to associate a duration to each of these events which may be either deterministic [50] or stochastic [101]. The former is defined as follows:

Definition 24 *Timed Petri Net (Graph)* [50]: A 6-tuple $\mathcal{N}_T = (B, \mathcal{E}, A, W, Q, D)$ where $(B, \mathcal{E}, A, W, Q)$ is a marked petri net where Q is marking vector of size $[\sigma(B) + \sigma(\mathcal{E})] \times 1 \in \mathbb{N}^{[\sigma(B)+\sigma(\mathcal{E})]}$ that includes marking of events $Q_{\mathcal{E}}$ in addition to the marking of places Q_B. $Q = [Q_B; Q_{\mathcal{E}}]$. D is a duration vector of size $\sigma(\mathcal{E}) \times 1$ representing the finite duration required to fire the event.

The following petri net dynamics are used in the context of this work.

Definition 25 *Timed Petri Net (Discrete-Event) Dynamics*: The evolution of the marking vector $Q = [Q_B; Q_{\mathcal{E}}]$ is given by the state transition function $Q[k+1] = \Phi_T(Q[k], U_k^+, U_k^-)$.

$$Q_B[k+1] = Q_B[k] + M_{\mathcal{N}}^+ U_k^+ - M_{\mathcal{N}}^- U_k^- \tag{20.21}$$

$$Q_{\mathcal{E}}[k+1] = Q_{\mathcal{E}}[k] - U_k^+ + U_k^- \tag{20.22}$$

where U_k^- is the kth input firing vector and U_k^+ is k^{th} output firing vector.

The input firing vector U_{Dk}^- is taken as exogenous, while the output firing vector U_{Dk}^+ is calculated from the event durations D by means of a scheduled event list.

The state transition function in Definition 25 has a minor modification from the one commonly used elsewhere in the petri net literature [50]. Normally, tokens remain in place until the event duration has passed. Here, the tokens are taken from the place marking vector and appear instead in the transition marking vector. They reappear in the marking vector after the event duration has passed. This modification is made so that the petri net dynamics more closely represent the physical reality as described in Sect. 20.2. The rules of timed petri net operation, including when transitions are enabled, remain otherwise the same.

Definition 26 *Scheduled Event List* [49]: A tuple $S = (u_{vk}, t_k)$ consisting of all elements u_{vk} in firing vectors U_k and their associated times t_k. For every element, $u_{vk}^- \in U_k^-$, there exists another element $u_{vk}^+ \in U_k^+$ which occurs at $t = t_k + d_v$.

The output firing vectors U_{Dk}^+ are then calculated from their elements for all the unique times $t = t_k + d_v$.

References

1. Anair D, Mahmassani A (2012) State of charge: electric vehicles' global warming emissions and fuel-cost savings across the United States. Tech. Rep. april, union of concerned scientists: citizens and scientists for environmental solutions, Cambridge, MA, USA
2. Karabasoglu O, Michalek J (2013) Influence of driving patterns on life cycle cost and emissions of hybrid and plug-in electric vehicle powertrains. Energy Policy 60(0):445–461. doi:10.1016/j.enpol.2013.03.047. URL http://www.sciencedirect.com/science/article/pii/S0301421513002255
3. Pasaoglu G, Honselaar M, Thiel C (2012) Potential vehicle fleet CO2 reductions and cost implications for various vehicle technology deployment scenarios in Europe. Energy Policy 40:404–421. doi:10.1016/j.enpol.2011.10.025

4. Raykin L, Roorda MJ, MacLean HL (2012) Impacts of driving patterns on tank-to-wheel energy use of plug-in hybrid electric vehicles. Trans Res Part D: Trans Environ 17(3):243–250. doi:10.1016/j.trd.2011.12.002. URL http://www.sciencedirect.com/science/article/pii/S1361920911001568
5. Yang Z, Wu Y (2012) Projection of automobile energy consumption and CO2 emissions with different propulsion/fuel system scenarios in Beijing. In Remote Sensing, Environment and Transportation Engineering (RSETE), 2012 2nd international conference on, pp 2012–2015. doi:10.1109/RSETE.2012.6260668
6. Soylu S (2011) Electric vehicles—the benefits and barriers. Intech Open Access Publisher, Rijeka, Croatia
7. Litman T (2013) Comprehensive evaluation of transport energy conservation and emission reduction policies. Trans Res Part: Policy Pract 47:1–23
8. Skippon S, Garwood M (2011) Responses to battery electric vehicles: UK consumer attitudes and attributions of symbolic meaning following direct experience to reduce psychological distance. Trans Res Part D: Trans Environ 16(7):525–531. doi:10.1016/j.trd.2011.05.005
9. Pointon J (2012) The multi-unit dwelling vehicle charging challenge. In Electric vehicles virtual summit 2012, vol 69. The Smart Grid Observer
10. Al Junaibi R, Viswanath A, Farid AM (2013) Technical feasibility assessment of electric vehicles: an Abu Dhabi example. In 2nd IEEE International conference on connected vehicles and expo. Las Vegas, NV, USA, pp 1–8
11. Junaibi RA (2013) Technical feasibility assessment of electric vehicles in Abu Dhabi. Master's thesis, Wasdar Institute of Science and Technology Engineering Systems and Management Department
12. Junaibi RA, Farid AM (2013) A method for the technical feasibility assessment of electrical vehicle penetration. In 7th annual IEEE systems conference, Orlando, FL, United states, pp 1–6
13. Kassakian JG, Schmalensee R, Desgroseilliers G, Heidel TD, Afridi K, Farid AM, Grochow JM, Hogan WW, Jacoby HD, Kirtley JL, Michaels HG, Perez-Arriaga I, Perreault DJ, Rose NL, Wilson, GL, Abudaldah N, Chen M, Donohoo PE, Gunter SJ, Kwok PJ, Sakhrani VA, Wang J, Whitaker A, Yap XL, Zhang RY, of Technology MI (2011) The future of the electric grid: an interdisciplinary MIT Study. MIT Press, Cambridge, MA. URL http://web.mit.edu/mitei/research/studies/documents/electric-grid-2011/Electric_Grid_Full_Report.pdf
14. Viswanath A, Farid AM (2014) A hybrid dynamic system model for the assessment of transportation-electrification. In American control conference, 2014, Portland, Oregon, pp 1–7
15. Clement-Nyns K, Haesen E, Driesen J (2010) The impact of charging plug-in hybrid electric vehicles on a residential distribution grid. Power Syst IEEE Trans 25(1):371–380. doi:10.1109/TPWRS.2009.2036481
16. Dyke KJ, Schofield N, Barnes M (2010) The impact of transport-electrification on electrical networks. Ind Electron IEEE Trans 57(12):3917–3926. doi:10.1109/TIE.2010.2040563
17. Erol-Kantarci M, Sarker JH, Mouftah HT (2012) Quality of service in plug-in electric vehicle charging infrastructure. In electric vehicle conference (IEVC), 2012 IEEE international, pp 1–5. doi:10.1109/IEVC.2012.6183227
18. Gan L, Topcu U, Low S (2011) Optimal decentralized protocol for electric vehicle charging. In IEEE conference on decision and control and european control conference, Ieee (2011), pp 5798–5804. doi:10.1109/CDC.2011.6161220
19. Gong Q., Midlam-Mohler S, Serra E, Marano V, Rizzoni G (2013) PEV charging control for a parking lot based on queuing theory. In 2013 American control conference. Washington, D.C., pp 1126–1131
20. Lopes JAP, Soares FJ, Almeida PMR (2011) Integration of electric vehicles in the electric power system. Proc IEEE 99(1):168–183. doi:10.1109/JPROC.2010.2066250

21. Ma Z, Callaway D, Hiskens I (2012) Optimal charging control for plug-in electric vehicles. In Control and optimization methods for electric smart grids, Springer Berlin Heidelberg, Berlin, Germany, pp 259–273
22. Palensky P, Dietrich D (2011) Demand side management: demand response, intelligent energy systems, and smart loads. Ind Inf IEEE Trans 7(3):381–388. doi:10.1109/TII.2011.2158841
23. Pieltain Fernandez L, Roman TGS, Cossent R, Domingo CM, Fria As P (2011) Assessment of the impact of plug-in electric vehicles on distribution networks. Power Sys IEEE Trans 26 (1):206–213. doi:10.1109/TPWRS.2010.2049133
24. Qian K, Zhou C, Allan M, Yuan Y (2011) Modeling of load demand due to ev battery charging in distribution systems. Power Syst IEEE Trans 26(2):802–810. doi:10.1109/TPWRS.2010.2057456
25. Saber AY, Venayagamoorthy GK (2011) Plug-in vehicles and renewable energy sources for cost and emission reductions. Ind Electron IEEE Trans 58(4):1229–1238. doi:10.1109/TIE.2010.2047828
26. Sortomme E, Hindi MM, MacPherson SDJ, Venkata SS (2011) Coordinated charging of plug-in hybrid electric vehicles to minimize distribution system losses. Smart Grid, IEEE Trans 2(1):198–205. doi:10.1109/TSG.2010.2090913
27. Kempton W, Tomić J (2005) Vehicle-to-grid power implementation: From stabilizing the grid to supporting large-scale renewable energy. J Power Sources 144(1):280–294. doi:10.1016/j.jpowsour.2004.12.022
28. Sovacool BK, Hirsh RF (2009) Beyond batteries: an examination of the benefits and barriers to plug-in hybrid electric vehicles (PHEVs) and a vehicle-to-grid (V2G) transition. Energy Policy 37(3):1095–1103
29. Su W, Rahimi-eichi H, Zeng W, Chow MY (2012) A survey on the electrification of transportation in a smart grid environment. IEEE Trans Industr Inf 8(1):1–10
30. Galus MD, Waraich RA, Noembrini F, Steurs K, Georges G, Boulouchos K, Axhausen KW, Andersson G (2012) Integrating power systems, transport systems and vehicle technology for electric mobility impact assessment and efficient control. IEEE Trans Smart Grid 3 (2):934–949. doi:10.1109/TSG.2012.2190628. URL http://ieeexplore.ieee.org/lpdocs/epic03/wrapper.htm?arnumber=6204241
31. Sonoda T, Kawaguchi K, Kamino Y, Koyanagi Y, Ogawa H, Ono H (2012) Environment-Conscious urban design simulator "Clean Mobility Simulator"—traffic simulator that includes electric vehicles. Mitsubishi Heavy Ind Tech Rev 49(1):78–83
32. Baca EES, Farid AM, Tsai IT (2013) An Axiomatic Design approach to passenger itinerary enumeration in reconfigurable transportation systems. In: Proceedings of ICAD2013 the seventh international conference on axiomatic design, Worcester, MA, USA, pp 138–145
33. Farid AM (2007) Reconfigurability measurement in automated manufacturing systems. Ph. D. thesis, University of Cambridge Engineering Department Institute for manufacturing
34. Farid AM (2008) Product degrees of freedom as manufacturing system reconfiguration potential measures. International transactions on systems science and applications 4(3):227–242. URL http://siwn.org.uk/press/sai/itssa0004.htm
35. Farid AM (2013) An axiomatic design approach to production path enumeration in reconfigurable manufacturing systems. In 2013 IEEE international conference on systems man and cybernetics, pp 1–8
36. Farid AM (2014) Static resilience of large flexible engineering systems: Part I—Axiomatic design model. In 4th international engineering systems symposium, Hoboken, N.J., pp 1–8 URL http://amfarid.scripts.mit.edu/resources/CESUN1.pdf
37. Farid AM (2014) Static resilience of large flexible engineering systems: Part II—Axiomatic design measures. In 4th international engineering systems symposium, Hoboken, N.J., pp 1–8. URL http://amfarid.scripts.mit.edu/resources/RMS-C10.pdf
38. Farid AM (2015) Static resilience of large flexible engineering systems: axiomatic design model and measures. IEEE sys J Systems(99):1–12. doi:10.1109/JSYST.2015.2428284

39. Farid AM (2016) A hybrid dynamic system model for multi-modal transportation-electrification. IEEE transactions on control system technology (in press) 1(1):1–12. doi:10.1109/TCST.2016.2579602
40. Farid AM (2016) An engineering systems introduction to axiomatic design. In Farid AM, Suh NP (eds) Axiomatic design in large systems: complex products, buildings and manufacturing systems, Chap. 1, Springer, Berlin, Heidelberg, pp 1–47. doi:10.1007/978-3-319-32388-6
41. Farid AM, McFarlane DC (2006) A development of degrees of freedom for manufacturing systems. In IMS'2006: 5th international symposium on intelligent manufacturing systems: agents and virtual worlds, pp 1–6. Sakarya, Turkey
42. Farid AM, McFarlane DC (2008) Production degrees of freedom as manufacturing system reconfiguration potential measures. Proceedings of the institution of mechanical engineers, Part B (Journal of Engineering Manufacture) 222(B10):1301–1314. doi:10.1243/09544054JEM1056
43. Schoonenberg WC, Farid AM (2015)A dynamic production model for industrial systems energy management. In 2015 IEEE International conference on systems man and cybernetics, Hong Kong, pp 1–7
44. Suh NP (2001) Axiomatic design: advances and applications. Oxford University Press
45. Viswanath A, Baca EES, Farid AM (2014) An Axiomatic design approach to passenger itinerary enumeration in reconfigurable transportation systems. IEEE Trans Intell Trans Syst 15(3):915–924. doi:10.1109/TITS.2013.2293340
46. Lewis TG (2011) Network science: theory and applications. Wiley, Hoboken, N.J
47. Newman M (2009) Networks: an introduction. Oxford University Press, Oxford, United Kingdom
48. van Steen M (2010) Graph theory and complex networks: an introduction. Maarten van Steen
49. Cassandras CG (2007) Introduction to discrete event systems, 2nd edn. Springer, New York
50. Popova-Zeugmann L (2013) Time petri nets. Springer
51. Reisig W (2013) Understanding petri nets. Springer
52. Farid AM, Lubega WN (2013) Powering and watering agriculture: application of energy-water nexus planning. In: GHTC 2013: IEEE global humanitarian technology conference, Silicon Valley, CA, USA, pp 1–6
53. Lubega WN, Santhosh A, Farid AM, Youcef-Toumi K (2013) Opportunities for integrated energy and water management in the gcc—a keynote paper. In EU-GCC renewable energy policy experts' workshop, december, Masdar Institute, Abu Dhabi, UAE, pp 1–33
54. Hame L, Hakula H (2013) Dynamic journeying in scheduled networks. Intell Trans Syst IEEE Trans 14(1):360–369. doi:10.1109/TITS.2012.2213817
55. Ip, W.H., Wang, D.: Resilience and Friability of Transportation Networks: Evaluation, Analysis and Optimization. IEEE Systems Journal **5**(2), 189–198 (2011). doi:10.1109/JSYST.2010.2096670
56. Pillac V, Gendreau M, Guéret C, Medaglia AL (2013) A review of dynamic vehicle routing problems. European Journal of Operational Research 225(1):1–11 . doi:10.1016/j.ejor.2012.08.015. URL http://www.sciencedirect.com/science/article/pii/S0377221712006388
57. Zografos KG, Androutsopoulos KN (2008) Algorithms for itinerary planning in multimodal transportation networks. Intell Trans Syst IEEE Trans 9(1):175–184. doi:10.1109/TITS.2008.915650
58. Zografos KG, Androutsopoulos KN, Spitadakis V (2009) Design and assessment of an online passenger information system for integrated multimodal trip planning. Intell Trans Syst IEEE Trans 10(2):311–323. doi:10.1109/TITS.2009.2020198
59. Ash J, Newth D (2007) Optimizing complex networks for resilience against cascading failure. Physica a: statistical mechanics and its applications 380:673–683. doi:10.1016/j.physa.2006.12.058. URL http://www.sciencedirect.com/science/article/pii/S0378437107002543
60. Zimmerman RD, Murillo-Sanchez CE, Thomas RJ (2011) MATPOWER: Steady-State operations, planning, and analysis tools for power systems research and education. IEEE Trans Power Sys 26(1):12–19. doi:10.1109/TPWRS.2010.2051168

61. De Weck OL, Roos D, Magee CL (2011) Engineering systems: meeting human needs in a complex technological world. MIT Press, Cambridge, Mass. . URL http://www.knovel.com/knovel2/Toc.jsp?BookID=4611 http://mitpress-ebooks.mit.edu/product/engineering-systems
62. Barcelo J (2010) Fundamentals of traffic simulation. Springer, New York
63. Treiber M, Kesting A (2013) Traffic flow dynamics: data, models and simulation. Springer, Heidelberg; New York
64. Kokkinogenis Z, Passos LS, Rossetti R, Gabriel J (2011) Towards the next-generation traffic simulation tools: a first evaluation. In 6th iberian conference on information systems and technologies, pp 15–18
65. Passos L, Rossetti R, Kokkinogenis Z (2011) Towards the next-generation traffic simulation tools: a first appraisal. In Information systems and technologies (CISTI), 2011 6th Iberian conference on, pp 1–6
66. Farid AM (2014) Axiomatic design & design structure matrix measures for reconfigurability & its key characteristics in automated manufacturing systems. In International conference on axiomatic design, Campus de Caparica, Portugal, pp 1–8
67. Farid AM (2014) Intelligent transportation-energy systems: Abu dhabi feasibility study. In Gulf traffic conference, Dubai, UAE, pp 1–32
68. Farid AM (2014) Measures of reconfigurability and its key characteristics in intelligent manufacturing systems and its key characteristics in intelligent manufacturing systems. J Intell Manufact pp 1–17
69. Farid AM (2014) Multi-agent system design principles for resilient operation of future power systems. In IEEE international workshop on intelligent energy systems, San Diego, CA, pp 1–7
70. Farid AM (2015). Multi-agent system design principles for resilient coordination and control of future power systems. Intell Ind Syst 1(3):255–269. doi:10.1007/s40903-015-0013-x
71. Lubega WN, Farid AM (2014) A reference system architecture for the energy-water nexus. IEEE systems J PP(99):1–11
72. Ahn S, Kim J, Dong-Ho Cho (2012) Wireless power transfer in on-line electric vehicle. In Wireless power transfer, 2nd edn. River Publishers, p 416
73. Jang YJ, Ko YD, Jeong S (2012) Creating innovation with systems integration: road and vehicle integrated electric transportation system. In Systems conference (SysCon), 2012 IEEE international, pp 1–4. doi:10.1109/SysCon.2012.6189531
74. Jang YJ, Ko YD, Jeong S (2012) Optimal design of the wireless charging electric vehicle. In Electric vehicle conference (IEVC), 2012 IEEE international, pp 1–5. doi:10.1109/IEVC.2012.6183294
75. Rakouth H, Absmeier J, Brown Jr A, Suh IS, John M, Sumner R, Henderson R (2013) EV charging through wireless power transfer: analysis of efficiency optimization and technology trends. In BT - FISITA 2012 World automotive congress, november 27, 2012—november 30, 2012 (Lecture Notes in Electrical Engineering), vol 192 LNEE, Springer Verlag, Delphi Automotive Systems, MI, United States, pp 871–884
76. Brackstone M, McDonald M (1999) Car-following: a historical review. Transportation research. Part F: traffic psychology and behavior pp 138–145:181–196. URL http://www.sciencedirect.com/science/article/pii/S136984780000005X
77. Olstam JJ, Tapani A (2004) Comparison of Car-following models. VTI meddelande 960A, pp 1–45
78. Treiber M, Kesting A (2010) An open-source microscopic traffic simulator. Intell Transp Syst Mag IEEE 2(3):6–13. doi:10.1109/MITS.2010.939208
79. Farid AM (2015) Symmetrica: test case for transportation-electrification research. Infrastruct Complexity (2015)(9):1–10. doi:10.1186/s40551-015-0012-9
80. Alecsandru CD (2006) A stochastic mesoscopic cell-transmission model for operational analysis of large-scale transportation networks. Ph.D. thesis, Louisiana
81. Allan DF, Farid AM (2015) A benchmark analysis of open source transportation-electrification simulation tools. In 2015 IEEE conference on intelligent transportation systems. Las Palmas de Gran Canaria Canary Islands, Spain, pp 1–7

82. Wood AJ, Wollenberg BF (2014) Power generation, operation, and control, 3rd edn. Wiley, Hoboken, NJ, USA
83. Committee on Americas Energy Future (2009) Americas energy future: technology and transformation. National Academies Press
84. United States Energy Information Administration (2013) Updated capital cost estimates for utility scale electricity generating plants. Tech. rep., U.s. Department of Energy
85. Farid AM (2015) Evolution of electricity grid. In Wind energy 2050: On the shape of near 100% RE grid, Chap. 5, World Wind Energy Association, Bonn, Germany, pp 45–48. URL http://amfarid.scripts.mit.edu/resources/Books/SPG-BC03.pdf
86. Farid AM (2015) Evolution of the physical power grid. In wind energy 2050: on the shape of near 100% RE grid, Chap. 6, World Wind Energy Association, Bonn, Germany, pp 49–56. URL http://amfarid.scripts.mit.edu/resources/Books/SPG-BC04.pdf
87. Farid AM, Jiang B, Muzhikyan A, Youcef-Toumi, K (2015) The need for holistic enterprise control assessment methods for the future electricity grid. Renew Sustain Energ Rev 56 (1):669–685. doi:10.1016/j.rser.2015.11.007
88. Farid AM, Muzhikyan A (2013) The need for holistic assessment methods for the future electricity grid. In GCC CIGRE power 2013, Abu Dhabi, UAE, pp 1–12
89. Muzhikyan A, Farid AM, Youcef-Toumi K (2013) Variable energy resource induced power system imbalances: a generalized assessment approach. In IEEE conference on technologies for sustainability, Portland, Oregon, pp 1–8. URL http://dx.doi.org.libproxy.mit.edu/10.1109/SusTech.2013.6617329
90. Muzhikyan A, Farid AM, Youcef-Toumi K (2013) Variable energy resource induced power system imbalances: mitigation by increased system flexibility, spinning reserves and regulation. In IEEE conference on technologies for sustainability, Portland, Oregon, pp 1–7. URL http://dx.doi.org.libproxy.mit.edu/10.1109/SusTech.2013.6617292
91. Muzhikyan A, Farid AM, Youcef-Toumi K (2015) An enterprise control assessment method for variable energy resource induced power system imbalances Part 1: Methodology. IEEE Trans Ind Electron 62(4):2448–2458. doi:10.1109/TIE.2015.2395391
92. Muzhikyan A, Farid AM, Youcef-Toumi K (2015) An enterprise control assessment method for variable energy resource induced power system imbalances Part 2: Results. IEEE Trans Ind Electron 62(4):2459–2467 .doi:10.1109/TIE.2015.2395380
93. Muzhikyan A, Farid AM, Youcef-Toumi K (2014) A power grid enterprise control method for energy storage system integration. In IEEE innovative smart grid technologies conference Europe, Istanbul, Turkey, pp 1–6. URL http://amfarid.scripts.mit.edu/resources/Conferences/SPG-C43.pdf
94. Muzhikyan A, Farid AM, Youcef-Toumi K (2016) Relative merits of load following reserves and energy storage market integration towards power system imbalances. Int J Electr Power Energy Syst 74(1):222–229. doi:10.1016/j.ijepes.2015.07.013
95. Jiang B, Muzhikyan A, Farid AM, Youcef-Toumi K (2015) Impacts of industrial baseline errors in demand side management enabled enterprise control. In IECON 2015—41st annual conference of the IEEE industrial electronics society, pp 1–6. Yokohama, Japan
96. Muzhikyan A, Farid AM, Youcef-Toumi K (2014) An enhanced method for the determination of load following reserves. In: American Control Conference, 2014, pp 1–8. Portland, Oregon. doi:10.1109/ACC.2014.6859254
97. Muzhikyan A, Farid AM, Youcef-Toumi K: An enhanced method for determination of the ramping reserves. In IEEE american control conference, Los Angeles, CA, USA, pp 1–8. (2015). URL http://amfarid.scripts.mit.edu/resources/Conferences/SPG-C46.pdf
98. Muzhikyan A, Farid AM, Youcef-Toumi K (2015) An enhanced method for determination of the regulation reserves. In IEEE american control conference, Los Angeles, CA, USA, pp 1–8. URL http://amfarid.scripts.mit.edu/resources/Conferences/SPG-C47.pdf

99. Muzhikyan A, Farid AM, Youcef-Toumi K (2016) An a priori analytical method for determination of operating reserves requirements. Int J Energy Power Syst (provisionally accepted with minor revisions) 1(1):1–11
100. Monitoring Analytics (2014) State of the market report for pjm. Tech. rep., PJM-ISO
101. Marsan M (1990) Stochastic petri nets: an elementary introduction. In Rozenberg G (ed) Advances in petri nets 1989 SE—1 (Lecture notes in computer science) vol 424. Springer Berlin Heidelberg, pp 1–29

Chapter 21
Energy Efficiency Consideration of an OLEV Bus System

In-Soo Suh

Abstract A detailed comparison of several vehicle systems is presented with emphasis on public transit applications. A plug-in series CNG-electric hybrid (SHEV), a battery-powered electric vehicle (BEV), and an electric vehicle (EV) with OLEV bus are compared and analyzed in terms of their overall operational energy efficiency and competitiveness.

21.1 Introduction

An OLEV can be categorized as an electric vehicle because it is driven by an electric power train with one or more electric motors and a storage medium for electric energy—usually a battery. The key difference between an OLEV and a general electric vehicle is that an OLEV has a set of pickup devices installed under the vehicle to collect the electromagnetic field energy. A set of electrical devices including rectifiers and regulators, which convert and deliver the electricity in the required form inside the vehicle, must also be installed within the OLEV [1–3]. An additional power control and management system, known as a Power Distribution Unit (PDU), must also be incorporated in the vehicle to control the power flow from the different electrical energy sources and loads.

In order to show the differences in the system architecture of an OLEV vehicle, a detailed comparison between a series internal-combustion-electric hybrid vehicle (SHEV), powered by compressed natural gas (CNG), and an OLEV system is presented [4–7]. Schematic diagrams of the compared systems are presented in Fig. 21.1.

I.-S. Suh (✉)
Korea Advanced Institute of Science and Technology (KAIST),
5338 Capri Dr., Troy, MI 48098, USA
e-mail: insoo.suh@kaist.ac.kr

Present Address:
I.-S. Suh
General Motors, 30500 Mound Rd., Warren, MI 48092, USA

N.P. Suh and D.H. Cho (eds.), *The On-line Electric Vehicle*,
DOI 10.1007/978-3-319-51183-2_21

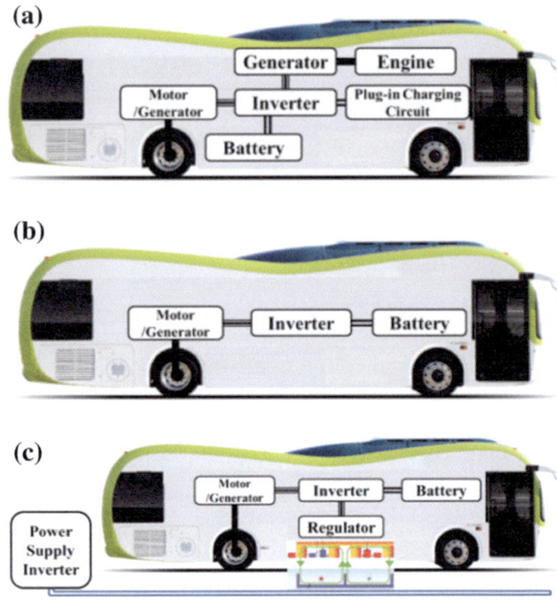

Fig. 21.1 Block diagram of three power train system **a** Plug-in series hybrid electric vehicle (SHEV), **b** Pure electric vehicle (PEV), and **c** OLEV

A plug-in series CNG-electric hybrid (SHEV), a purely electric vehicle (BEV), and an electric vehicle (EV) with dynamic wireless charging capability (OLEV) for a typical transit bus are compared and analyzed in terms of their overall operational energy efficiency and competitiveness, with an emphasis on public transit applications [8]. The three different types of vehicles for efficiency and cost of ownership comparison purpose are shown in Fig. 21.1. The power train alternatives are compared over three different standard test drive cycles: the New York City Cycle (NY), the High-Speed Heavy Duty Diesel Truck Cycle (HHDT), and finally a simulated test drive cycle corresponding to a real implementation of an OLEV track at a university campus (on-campus), of which test speed profiles are summarized in Fig. 21.2.

21.2 Series ICE-Electric Hybrid Vehicle (SHEV)

The SHEV consumes electrical energy stored in batteries; the energy is supplied by a generator that is powered by a compressed natural gas (CNG) burning engine in this discussion. Since the battery bus and OLEV are powered directly by electricity, it is necessary to convert the energy content and cost of CNG to make a meaningful comparison. An SOC (state of charge) fraction in kWh can be converted to CNG energy consumption in grams or MJ. The electrical energy from SOC variation can be converted to consumed energy via the equations below, using CNG's low heating value (QLHV = 40,000 kJ/kg) and a typical averaged CNG engine thermal conversion heat efficiency, η, of 30%:

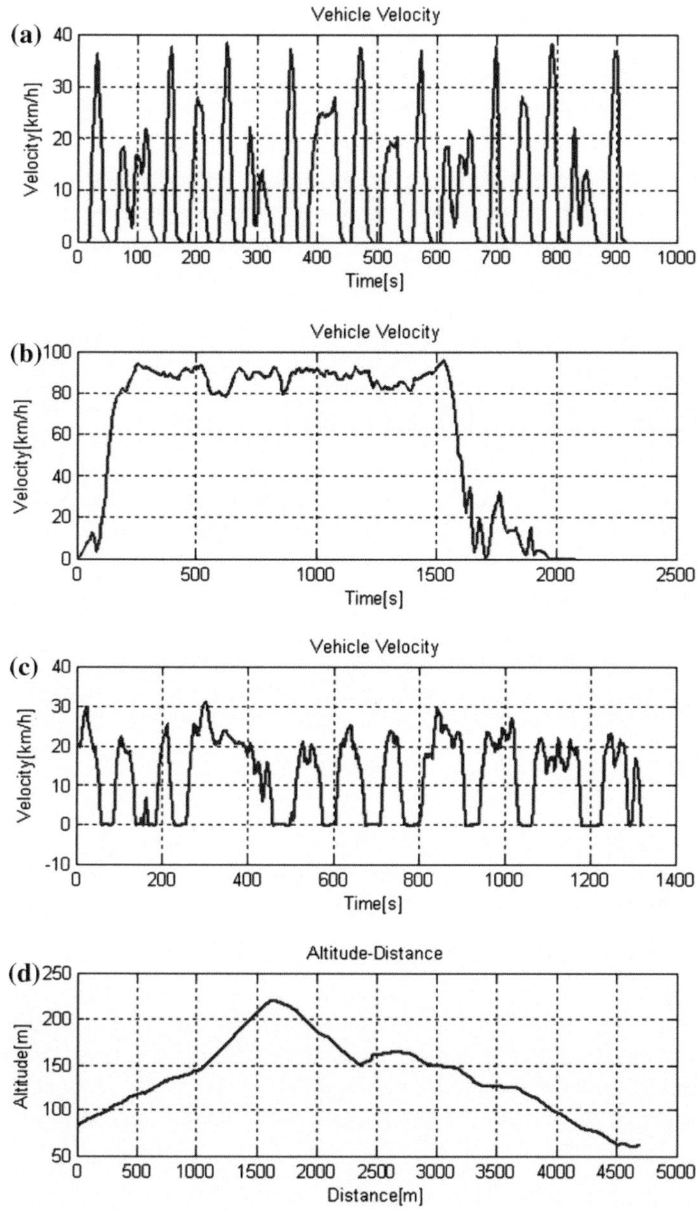

Fig. 21.2 Test drive cycle for simulation **a** NY Manhattan (NY), **b** High-Speed Heavy Duty Diesel Truck (HHDT), and **c** on-campus drive cycle recorded, and **d** on-campus drive cycle recorded altitude

$$\text{Fuel Consumption} = \text{Fuel}_{\text{eng}} + \text{Fuel}_{\text{bat}} \tag{21.1}$$

$$\text{Fuel}_{\text{bat}} = 1200\ \text{kJ}/\% \times \Delta\text{SOC}\,\% \times \frac{1}{40,000\ \text{KJ}/\text{kg}} \times \frac{1}{0.61\ \text{kg}/\text{L}} \times \frac{1}{\eta} \tag{21.2}$$

The results over three different drive cycles are summarized in Table 21.1. For typical city driving, represented by the NY drive cycle, the fuel economy of an SHEV is about 2.41 km/l, which has been improved from that of a baseline (non-hybrid) CNG bus, 2 km/L. For a typical highway driving mode, represented by the HHDT drive cycle, the fuel economy has been calculated as 2.7 km/L.

21.3 Battery Electric Vehicle (BEV)

The driving range and efficiency of a pure battery electric vehicle are summarized in Table 21.2, when the bus is continuously operated over the respective cycles of New York, HHDT, and on-campus. The measure of operational efficiency here is the driving distance per consumed energy as measured at the battery. With the increased numbers of battery pack from 21 to 20, the efficiency decreases because of the increased weight of the batteries.

21.4 OLEV Bus System

A simplified dynamic power flow model for SHEV and OLEV are shown in Fig. 21.3.

NY drive cycle analysis on OLEV

(1) Only stationary charging at stops
In the simulation of the NY drive cycle for OLEV driving case, it is assumed to have five stops in every 500 m distance while driving 2.7 km of distance, and the duration of each stop is assumed as 25 s considering typical public transit bus operation. The red circles in Fig. 21.4 indicate the stationary wireless charging spot during the NY drive cycle in the simulation model.

Table 21.1 Simulated efficiency summary—SHEV

	NY	HHDT	On-campus
Driving distance (km)	2.7	37.2	4.67
Fuel consumption (L)	1.43	13.8	2.56
SOC variation	0.05	0.02	0.07
Efficiency (km/L)	2.41	2.70	2.19

Table 21.2 Summary of calculated energy efficiency—BEV

		NY	HHDT	On-campus
Driving distance (km)		2.7	37	4.6
Battery packs	(units)	14–20	14–20	14–20
Energy consumption (MJ)	14	12.61	162.2	20.26
	16	13.34	166.7	21.6
	18	14.09	171.0	22.94
	20	14.85	175.5	24.34
SOC variation	14	−0.080	−0.1	−0.014
	16	−0.073	−0.09	−0.0128
	18	−0.068	−0.082	0.0122
	20	−0.065	−0.076	−0.0106
Driving range (km)	14	236	260	252
	16	258	288	268
	18	278	314	287
	20	291	341	302
Efficiency (km/kWh)	14	0.771	0.821	0.817
	16	0.729	0.799	0.767
	18	0.690	0.779	0.722
	20	0.655	0.759	0.680

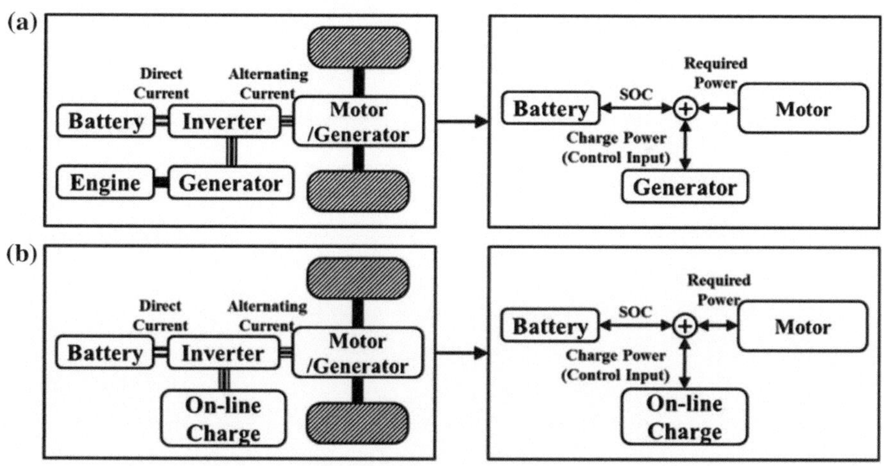

Fig. 21.3 Dynamic power flow model for **a** SHEV **b** OLEV

(2) Partial range of powered track over NY cycle
The partial range of powered track installed for simulation purpose over the low speed ranges is indicated with a dotted rectangle box in Fig. 21.5 for dynamic wireless charging, in addition to stationary charging at stops as earlier. The simulated results are shown in Fig. 21.6.

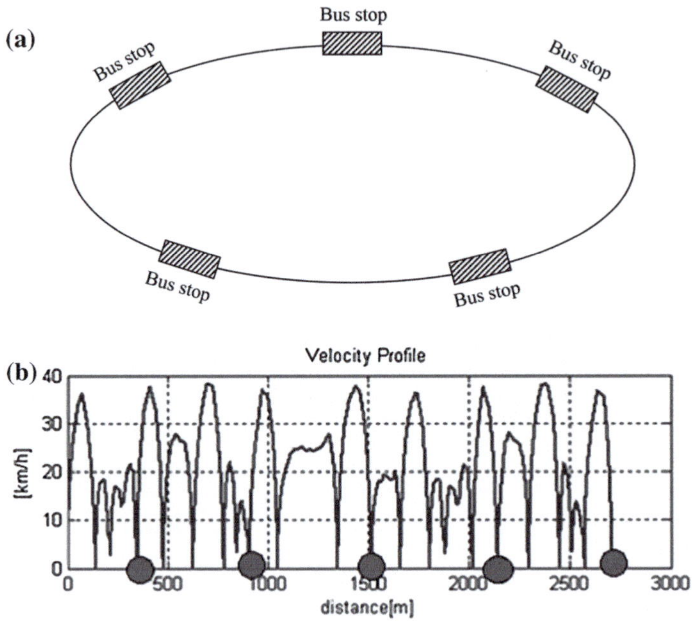

Fig. 21.4 Stationary charging only at stops over NY cycle. **a** Stationary charging spots, **b** stationary charging spots over NY cycle of vehicle speed over time

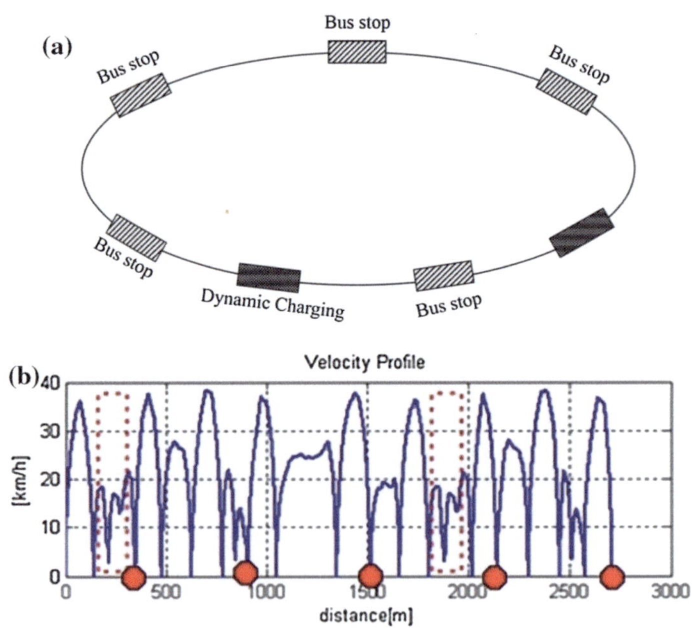

Fig. 21.5 Partial range of powered track for dynamic charging in addition to the stationary charging at stops over NY cycle. **a** Stationary and dynamic wireless charging spots, **b** stationary and dynamic charging period over NY cycle

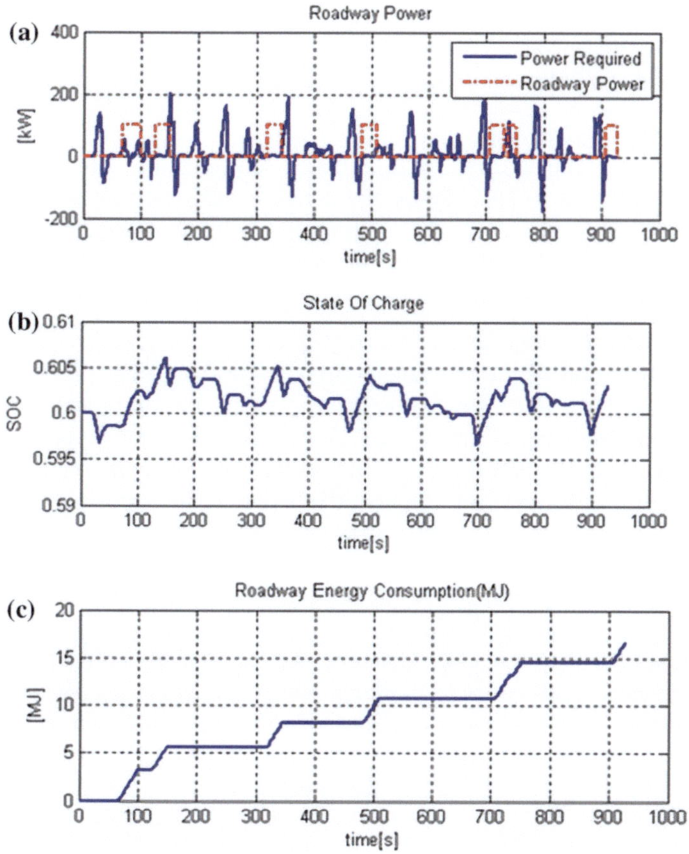

Fig. 21.6 Simulation results of OLEV over NY drive cycle with a partial range of wireless power supply **a** required power for driving and wirelessly delivered power from on-road charging, **b** SOC variation, and **c** accumulated wirelessly delivered energy consumption during one cycle

For stationary and dynamic wireless charging efficiency, 80% and 60% are applied as discussed earlier for stationary and dynamic charging, respectively, with 60 kW of wireless power transfer capacity. In addition, the installed ranges are selected for lower vehicle speed region to maximize the transmitted energy since the dwelling duration is an important parameter for total energy transferred on the same length of installed powered track. Considering the energy balance between expected consumed and required accumulated energy, two locations are selected with 125 m in the length of powered track, of which distance is based upon the typical 6 segmented track length with one power supply inverter.

HHDT drive cycle analysis on OLEV
For HHDT drive cycle analysis, the required power ranges are a lot higher than the NY cycle, and the full range of driving distance has the on-road power supply.

On-campus drive cycle analysis of OLEV
For measured on-campus drive cycle, the stop locations are defined as shown in Fig. 21.7, in addition to considering the measured altitude information of actual driving pattern as in Fig. 21.7c. In addition, the simulations for this case are performed similarly as in NY cycles, and the results are shown in Fig. 21.8, for only stationary charging at five stops.

(1) Only stationary charging at stops
In the simulation of the on-campus drive cycle for OLEV driving case, it is assumed to have five stops while driving about 4.7 km of distance, and the duration of each stop is assumed as 25 s considering typical public transit bus operation. The red circles in Fig. 21.7 indicate the dynamic wireless charging spot at stops during the on-campus drive cycle in the simulation model.
(2) A partial range of wireless power transfer system including stops
For a partial range of dynamic wireless charging, the hilly area of initial 800 m distance (300 s of driving time duration) is selected for powered track installation as shown in Fig. 21.8a, in terms of the required power for driving the on-campus drive cycle and the dynamic wireless delivered power. In Fig. 21.9, the calculated results of SOC variation and the accumulated consumed energy of wirelessly delivered energy are described in (b) and (c), respectively.

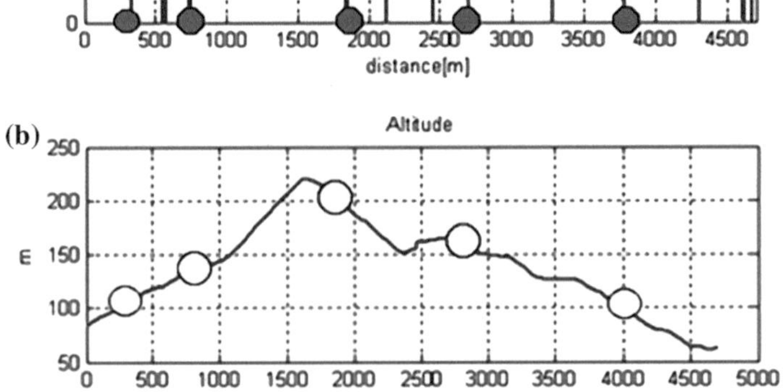

Fig. 21.7 Measured on-campus drive cycle with stop locations. **a** Stationary charging spots over on-campus cycle, **b** stationary charging spots in view of location and altitude

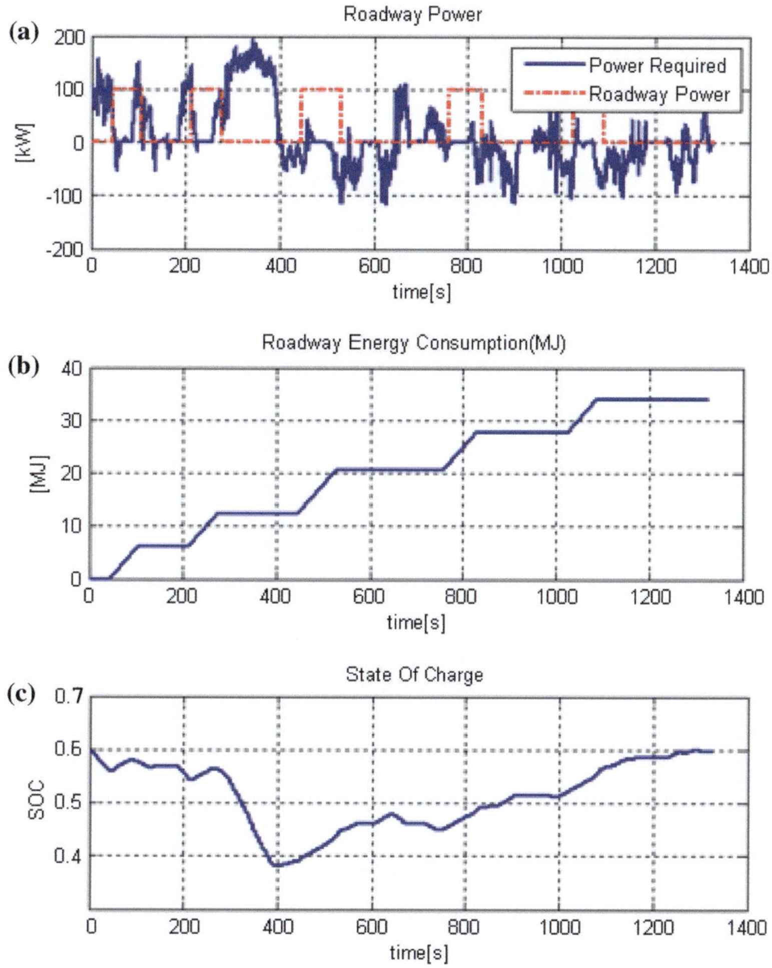

Fig. 21.8 Simulation results of OLEV over on-campus drive cycle with wireless power system at stops only. **a** Required power for driving and wirelessly delivered power from on-road charging, **b** SOC variation, and **c** accumulated wirelessly delivered energy consumption during operation

The energy required for an OLEV operation in this case is supplied in real time from the power supply infrastructure. The OLEV operation efficiency in view of driving distance per energy consumption (km/kWh) is compared under three different configurations:

(1) with the on-road power supply installed only at stops, representing static wireless charging,
(2) installed at stops with additional 'partial' coverage of dynamic wireless power transfer system over the driven route, and

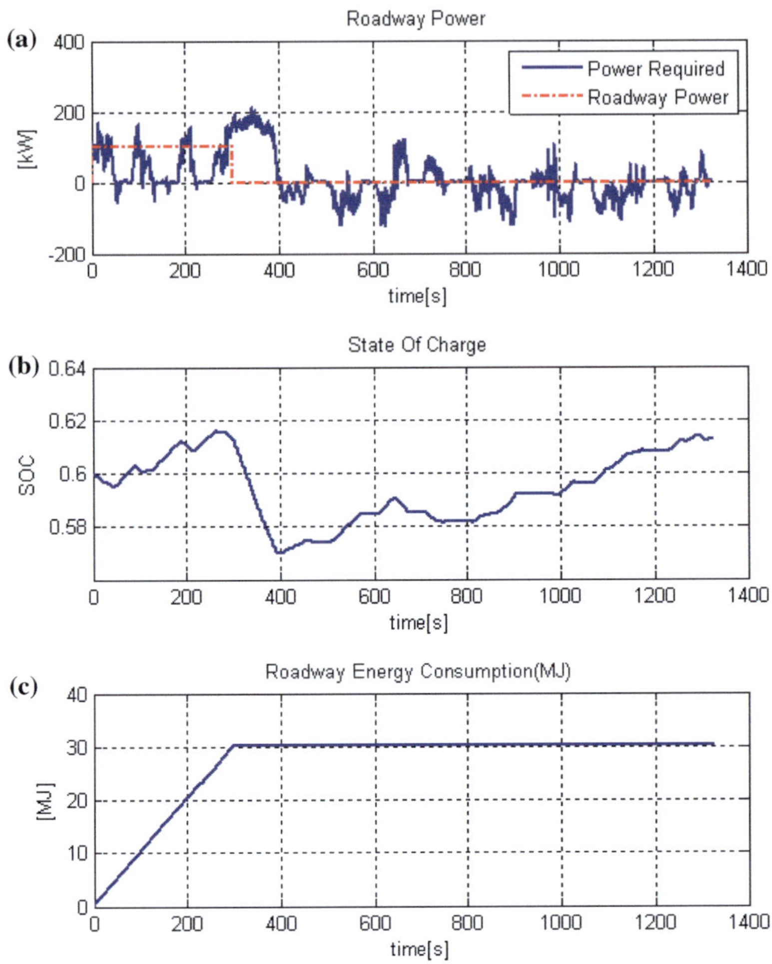

Fig. 21.9 Simulation results of OLEV over on-campus drive cycle with a partial range of wireless power supply. **a** Required power for driving and wirelessly delivered power from on-road charging, **b** SOC variation, and **c** accumulated wirelessly delivered energy consumption during operation

(3) finally, full coverage of the driven route. The efficiency results for three different driving cycles are summarized in Table 21.3 below, as denoted 'At Stops,' 'Partial,' and 'Full Range,' respectively, with different number of onboard battery packs. From the results, it is possible to estimate how much the operational efficiency can be diminished when the dynamic charging distance is increased due to the wireless transmission efficiency loss for all three driving cycles.

Table 21.3 Driving energy economy (km/kWh)—OLEV

Battery units	NY			HHDT		On-campus		
	At stops	Partial	Full range	At stops	Full range	At stops	Partial	Full range
1	0.936	0.785	0.316		0.609	1.114	0.802	0.370
2	0.898	0.768	0.314	0.738	0.607	1.091	1.045	0.369
3	0.862	0.752	0.311	0.777	0.600	1.078	1.024	0.366

21.5 Operational Efficiency Comparison

Overall operational efficiencies for the three types of eco-friendly vehicle systems being compared are summarized in Table 21.4. As can be seen in the table, the OLEV with only stationary charging and dynamic charging shows higher energy economy than a BEV, mainly because of the higher weight of the larger battery packs in the latter. In addition, it should be noted that the energy economy with stationary charging is only superior to that of dynamic charging; however, this requires larger battery sizes and longer wait times at stopping points.

The OLEV and BEV vehicles compared here use the same baseline battery pack, the difference being that the BEV has 20 pack units and the OLEV is only equipped with two packs and partial dynamic charging coverage as well as stationary charging locations. For CNG, a fuel price of \$0.70/L is applied; the electricity price considered as \$0.10/kWh. From Table 21.5, it can be seen that the cost of energy per km is quite different when comparing the OLEV and SHEV systems. The numbers in brackets are the energy cost reduction ratios from the CNG baseline (non-hybrid). The three eco-friendly vehicle systems compared show a 17–70% cost reduction when compared with the baseline CNG engine-driven vehicle. A qualitative analysis of the technologies is presented in Table 21.6, based upon the above observations.

Table 21.4 Operational efficiency comparison

	NY	HHDT	On-campus
SHEV (km/L)	2.41	2.70	2.19
BEV (km/kWh)	0.655	0.759	0.680
OLEV (km/kWh)	0.768	0.666	0.786
OLEV[a] (km/kWh)	0.974	–	0.822

[a]Static wireless charging only

Table 21.5 Operation cost comparison (\$/km)

	NY	HHDT	On-campus
SHEV	0.290 (17%)	0.260 (26%)	0.320 (10%)
BEV	0.153 (56%)	0.132 (62%)	0.147 (60%)
OLEV	0.130 (63%)	0.150 (57%)	0.127 (64%)
OLEV[a]	0.102 (71%)	–	0.121 (65%)

[a]Static wireless charging only

Table 21.6 Overall summary on future vehicle system comparison [8]

	Charging time	Driving efficiency	Driving range	Vehicle cost	Infrastructure cost
SHEV	★★	★	★★★	★★★	★★★
EV	★	★★	★	★	★★
OLEV	★★★	★★★	★★	★★★	★

★★★: favorable ★★: average ★: less favorable

The OLEV with a partial range of dynamic wireless charging installation combined with stationary charging showed better efficiency than a BEV, which requires an impracticably heavy battery to have a driving range equivalent to the baseline CNG bus.

21.6 Conclusions

In this chapter, the operational efficiency of OLEV compared with SHEV and PEV were simulated and discussed. Eco-friendly vehicle technologies for commercial bus applications, such as a plug-in series hybrid electric vehicle (SHEV), a pure battery electric vehicle (PEV), and an electric vehicle (EV) with dynamic wireless charging electric vehicle (OLEV), were analyzed in view of operational efficiency by calculating the optimized energy consumption from a simulation based upon derived specifications of each power drive system.

By developing proper vehicle driving modeling and controllers to achieve the required performance of different types of commercial buses, an energy consumption and efficiency analyses over the NY Manhattan (NY) cycle, the High-Speed Heavy Duty Diesel Truck (HHDT) cycle, and the actual measured 'on-campus' drive cycles were performed for an equivalent and practical comparison. The overall operational efficiency of the PEV and the OLEV, in view of energy cost, was higher than that of the SHEV. Compared with the baseline CNG engine bus, the energy consumption reductions, in terms of energy cost, were found to be 10% for the SHEV, 60% for the PEV, and 65% for the OLEV.

References

1. Suh NP (2011) Design of on-line electric vehicle. In: Proceedings of CIRP design conference. Plenary Speech, Daejeon, Korea 27–29 March 2011
2. Suh IS (2011) Intelligent wireless EV fast charging with SMFIR technology. J Integr Des Process Sci 15(3):3–12
3. Suh IS (2011) On-road electrification for optimized power supply in OLEV application. J Integr Des Process Sci 15(3):13–27

4. Ehsani M, Gao Y, Emadi Ali (2005) Modern electric, hybrid electric, and fuel cell vehicles-fundamentals, theory, and design. CRC Press, Boca Raton
5. Boukehili A et al (2012) Hybrid vehicle power management modeling and refinement. Int J Automot Technol 13(6):987–998
6. Moura SJ et al (2011) A stochastic optimal control approach for power management in plug-in hybrid electric vehicles. IEEE Trans Control Syst Technol 19(3):545–555
7. Kemper P, Suh IS (2014) Implementation cost comparison of different types of EV energy replenishment technologies for public transit bus systems. Transport Res Rec
8. Yi K, Kim J, Hwang K, Lee M, Kim K, Kim H, Suh IS (2015) Operational efficiency comparison on eco-friendly vehicles including dynamic wireless charging. Int. J. Automot Technol 16(6):1017–1030

Chapter 22
The Economics of Wireless Charging on the Road

Jong Han Park and Yong Hoon Jeong

Abstract To accelerate market penetration of EVs, installing a proper charging infrastructure is necessary. The total cost of operating electric vehicles includes vehicle cost, infrastructure cost, and energy consumption cost. We compared the total cost associated with plug-in hybrid (PHEV) and On-Line Electric Vehicles (OLEV), i.e., plug-in versus wireless charging. Because of the high cost of the battery, OLEV is more cost efficient in a mega-city where many vehicles can share the roadways' wireless charging infrastructure. As battery costs fall, EVs with larger batteries will become more cost competitive. Even then, the infrastructure for OLEV can be used as charging infrastructure for PHEVs or PEVs by adding onboard pickup devices to the vehicles.

22.1 Introduction

Increasing concerns about global warming and soaring, energy costs are stimulating technological innovation in the transportation sector to make it greener. Among the eco-friendly vehicles recently developed or under development, the electric vehicle (EV) is the most feasible and economical alternative with potential to displace a significant portion of petroleum consumption by using electricity for all or a portion of given trips [1]. According to [2], hybrid-type EVs will grow rapidly (global CAGR of 16.6% between 2008 and 2015). To accelerate market penetration of EVs, installing proper charging infrastructure is required. Depending on the charging infrastructure type, vehicle cost, infrastructure cost for EVs, and consumers' charging behavior should be determined. The most well-known charging

J.H. Park (✉)
Dongwonolev, 275 Yangae, Seocho, Seoul, South Korea
e-mail: john@dongwon.com

Y.H. Jeong
Nuclear and Quantum Engineering, Korea Advanced Institute of Science and Technology (KAIST), 291 Daehak-ro, Yuseong-gu, Daejeon 34141, South Korea
e-mail: jeongyh@kaist.ac.kr

N.P. Suh and D.H. Cho (eds.), *The On-line Electric Vehicle*,
DOI 10.1007/978-3-319-51183-2_22

infrastructure type in the current market is the plug-in charging system. Under the plug-in-type infrastructure, the EV is only able to recharge at locations with plug-in charging stations and charging takes a longer time. Overnight charging at a home garage or apartment complex is a generally accepted charging scenario for plug-in EVs. To achieve longer trips without recharging, plug-in EVs require larger batteries, which are expensive and heavy.

Another charging infrastructure for EVs is the so called roadway power system [3], OLEV system [4], and dynamic connection [1]. Under these systems, EVs can be connected to an external source of energy along a roadway while moving. While there are many different terminologies for EVs under roadway power systems, currently only the OLEV is commercialized in Korea. Compared to plug-in EVs, the OLEV requires a smaller battery and consumers do not need to allocate time for charging, although it requires infrastructure along a small percentage of heavily travelled roadways.

As mentioned above, charging infrastructure type affects many aspects of EV cost and market penetration. It is therefore necessary to compare the total cost associated with EV market penetration considering certain charging infrastructure type. In this chapter, we consider vehicle cost, infrastructure cost, and energy cost in Seoul, Korea, a case of particular interest because Seoul is a typical mega-city, which is the most appropriate venue for an EV test bed. By evaluating the total cost related to EV infrastructure and utilization, a more cost-efficient system can be determined. The results of the comparison provide useful insight into installing EV infrastructure country wide, a task that requires large investment. Nevertheless, the infrastructure can be used for a long time once installed.

22.2 Comparison OLEV with IC Engine

Before we compare the economic effect of charging infrastructure type of electric vehicle, we would like to mention the economic benefit of OLEV bus compared to CNG bus based on real operation data. From March of 2014, we start operate two OLEV buses as '7' line bus in Gumi city. Each OLEV bus operated about 250 km per day.

The major economic benefit of OLEV comes from energy cost saving. Energy costs differ from different countries. So, this comparison result is based on Korea and the results will be different in other country. Table 22.1 shows the energy cost and efficiency of different vehicle types.

Table 22.1 Energy cost comparison

	Energy efficiency	Energy cost	Cost per month	Cost per year
CNG bus	1.7 km/ℓ	1.0$/ℓ	4411$	52,932$
OLEV bus	0.7 km/kWh	0.11$/kWh	1178&	14,136$

Table 22.2 Life time cost comparison

	Bus cost	Government subsidy	Energy cost (10 years)	Total cost
CNG bus	200,000$	100,000$	529,320$	629,320$
OLEV bus	450,000$	300,000$	141,360$	291,360$

As results, OLEV bus can save 73% of energy cost comparing to CNG bus. The energy cost is calculated with 250 km driving per day. If the driving distance increases, the cost saving effect will increase accordingly.

Based on energy cost comparison results, we can compare two different types of buses as life time cost perspectives. We assume that each bus can be operated for 10 years and basic operating cost and maintenance cost are same. Table 22.2 summarizes the life time cost comparison results.

After ten years operation with government subsidy, OLEV bus can save 337,960$ comparing to CNG buses, 54% of total cost. If we consider CO_2 emission as cost, the saving amount will increase. By using OLEV bus, the city can save 118 ton•CO_2/year compared to CNG bus.

The results shows that OLEV bus is more efficient system compared to CNG without considering government subsidy. Also, OLEV bus cost will be decreased with the increasing the production volume, while CNG bus cost is almost saturated.

22.3 Vehicle Technologies and Cost

The vehicle type considered is a four passenger compact car that is the same size as popular HEVs and PHEVs in the current market. At present, pure EVs have limitations in use because of the limited driving range from the lack of infrastructure and the high battery cost. Hence, we use a hybrid-type EV in the comparison, which uses electricity as the primary power source. Hybrid-type EVs have the same efficiency as HEVs after the charge-depleting range. To compare the total cost under different EV charging systems, we use two types of hybrid-type EVs in the comparison: a PHEV(Plug-in Hybrid Electric Vehicle) and an OLEV (On-Line Electric Vehicle).

22.3.1 PHEV

A PHEV is an HEV with the ability to recharge its energy storage system with electricity from a plug-in charging station. With a fully charged energy storage system, a PHEV will bias toward using electricity over liquid fuels. The secondary energy carrier for the PHEV is a liquid fuel, such as gasoline, stored in the vehicle.

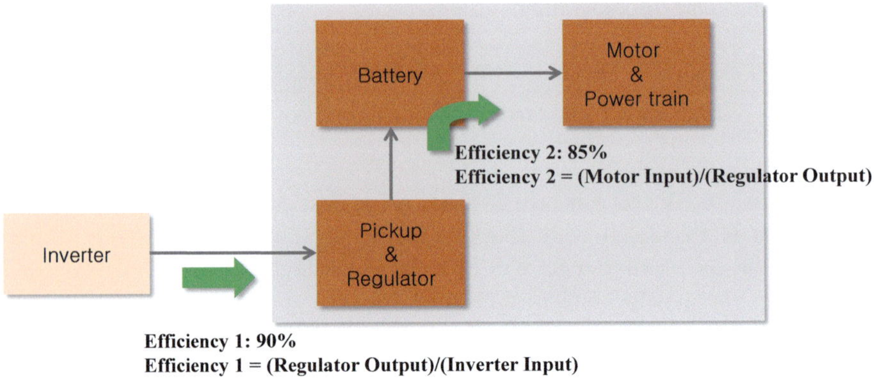

Fig. 22.1 Electricity power transfer efficiency of PHEV

In EV mode, the driving efficiency (motor input energy/distance) of the PHEV is 0.2 kWh/km which is an average of commercialized PHEVs and EVs in the market. In HEV mode, the efficiency is the same as that of a HEV. The efficiency of currently available HEVs in the market, such as the Toyota Prius and GM Volt, is 50 mpg. The PHEV is assumed to have a 16 kWh Lithium-ion battery, as used in the GM Volt.

For the PHEV, the electricity power transfer efficiency from the inverter to the motor is presented in Fig. 22.1.

To calculate overall power transmission efficiency in charging plug-in-type EVs, power transmission efficiency from inverter to pickup (Efficiency 1) and from battery to motor (Efficiency 2) need to be considered. The overall power transfer efficiency is 76%. This means 24% energy loss occurs during the transfer at the current technology state. The total efficiency for the PHEV in EV mode including driving efficiency and power transfer efficiency is 0.26 kWh/km.

22.3.2 OLEV

The On-Line EV (OLEV), a so called Roadway-Powered EV [5–8], uses wireless charging technology based on the concept of shaped magnetic field in resonance (SMFIR) charging. The transmitted electric energy based on magnetic resonance is used for powering motors and recharging batteries on the vehicle. It also has a relatively smaller battery than PHEVs to minimize the installation of underground power supplier lines. In the study, the typical OLEV is assumed to have a 5 kWh lithium-ion battery and the capacity of the onboard pickup device is assumed as 10 kWh/h, which means 10 kWh can be transmitted to the OLEV in one hour, through the magnetic resonance charging technology.

An OLEV is an HEV with the ability to recharge its energy storage system, as is the case of the PHEV. On a power supply road or with a fully charged energy storage system, the OLEV will bias toward using electricity over liquid fuels. In EV mode and HEV mode, the driving efficiency is the same as that of the PHEV. The only difference is the charging mechanism.

For the OLEV, the electricity power transfer efficiency from the roadway inverter to the motor is described in Fig. 22.2. The overall power transfer efficiency is 64%. Because of the loss that occurs in the air gap, the overall efficiency of the OLEV is lower than that of the PHEV. The total efficiency of the OLEV in the EV mode including driving efficiency and power transfer efficiency is 0.31 kWh/km.

22.3.3 Vehicle Cost

The PHEV and the OLEV are based on the HEV paradigm. The main differences are the onboard pickup charging device and the size of the battery due to the distinct charging mechanism. Hence, we estimate the vehicle costs based on a currently available HEV, the Toyota Prius, and add the cost of the onboard charging device and battery. The average Prius manufacturer suggested retail price (MARP) in 2013 was $24,200 including a 1 kWh lithium-ion battery. The PHEV is assumed to have a 16 kWh lithium-ion battery, as used in the GM Volt, which is most dominant PHEV model in the market. For the OLEV, a 5 kWh lithium-ion battery is assumed. To estimate the vehicle cost, we use the estimated battery cost from [9].

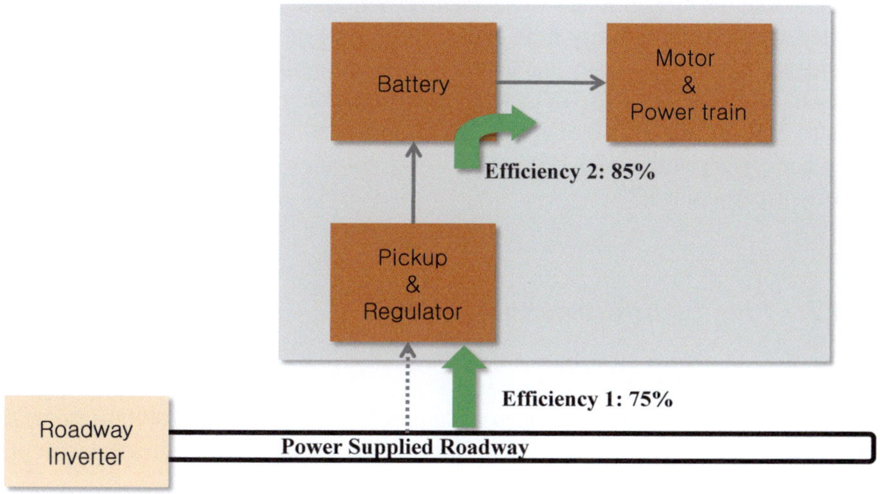

Fig. 22.2 Electricity power transfer efficiency of OLEV

Table 22.3 Estimated vehicle costs in 2013

	HEV base	Onboard Pickup	Battery	Total cost
PHEV	$23,285	$1000	$12,177	$36,462
OLEV	$23,285	$5000	$3805	$32,090

According to a medium cost scenario [9], the battery cost in 2011 is $915/kWh and it will decline at a CAGR (Compound Annual Growth Rate) of 8.8% until 2019 and at 3.5% from 2020 to 2030. The estimated vehicle costs in 2013 are summarized in Table 22.3.

HEV base cost means the cost of the HEV excluding the battery. We assumed the onboard pickup device cost for both PHEV types is $1000 [1]. The onboard pickup for the OLEV is assumed as $5000 according to the OLEV manufacturer with the assumption of mass production. We also estimated the vehicle cost until 2022. The only underlying assumption is that the battery cost will decline according to [9] and other costs such as motor, chassis and body will be same because those are quite matured component and no much change to decline as time pass. The reason for the cost difference between two types of EV stems from the size of the battery and the others are shared. With declining battery cost, the cost difference between the PHEV and the OLEV decreases.

22.4 Charging Infrastructure Types

For EVs, the infrastructure is a necessary condition for their penetration. To install the infrastructure, huge social investment is required. Once the infrastructure is installed, it will undergo very little change. Various types of infrastructure for EVs are under evaluation by many stakeholders to better understand their capability and potential benefits. The cost associated with providing certain type of charging infrastructure for the EV, along with additional costs for onboard electronics and added battery capacities with different EV technologies, will be a key factor in the success of particular EVs.

22.4.1 Plug-in Infrastructure for PHEV

The standalone EV charger for the home is around $5000, and the construction cost is about $2000 in Korea. The Korean government supports up to $7000 per home to install the charger. Therefore, we assumed $7000 for the PHEV infrastructure per vehicle.

22.4.2 On-Line Infrastructure for OLEV

To estimate the On-Line infrastructure cost for the OLEV, we need to determine the length of power supplied roadway and the unit cost. To estimate the length, we consider the Seoul case as a representative mega-city.

(1) **OLEV energy harvesting and consumption pattern**
We assume that the power supplied roadway is installed only in the inner city where most traffic occurs to maximize the cost efficiency of the infrastructure. Outside of the city, the OLEV operates in HEV mode. To decide the length of power supplied roadway installed in the inner city, we test the OLEV energy harvesting from the power supplied roadway and consumption pattern according to the typical driving cycle in the mega-city. In the test, we assume that the power supplied roadway is only installed where vehicles stop and idle on the roadway. Vehicles usually spend much time stopping in front of traffic lights. We use the Japan 10–15 case which shows typical vehicle driving patterns in the mega-city. It involves an average speed of 22.7 km/h and 32% idling time during the total driving time with a total distance of 4.2 km. The case is very similar to Seoul's driving cycle. The average speed of vehicles in Seoul is 23 km/h and the idling time is 27% (during the total driving time). For the OLEV, 20% of the battery capacity(1 kWh) is always restored and not usable and another 1 kWh is restored to be used in HEV mode. Hence, 3 kWh is available to supply power to the motor during driving. To investigate the energy harvesting from the power supplied road and energy consumption, we test three different cases based on the initial battery SOC (State of Charge): 4 kWh (full charge), 2.5, 1 kWh (minimum charge).

- Initial battery SOC: 4 kWh (Full charge)
  Figure 22.3 shows the battery SOC and energy consumption. Energy is harvested when the vehicle stops and the onboard pickup capacity of the OLEV is 10 kWh/h. With a fully charged battery, the OLEV is operated fully in EV mode during the test by using 1 kWh while harvesting 0.45 kWh energy from power supplied roadway. After the test, the battery SOC is 3.45 kWh.
- Initial battery SOC: 2.5 kWh (Full charge)
  Figure 22.4 shows the OLEV is also fully operated in EV mode when the initial battery SOC is 2.5. During the test, the OLEV harvests 0.6 kWh while consuming 1 kWh. After the test, the battery SOC is 2.1 kWh.
- Initial battery SOC: 1 kWh (Minimum charge)
  According to Fig. 22.5, the OLEV uses 0.6 kWh, which is harvested while idling in EV mode. From the battery to the motor, the power transfer efficiency (Efficiency 1) is 85% and the driving efficiency is 0.235 kWh/km. With 0.6 kWh, the OLEV travels 2.52 km of the 4.2 km test bed. Therefore, with the minimum battery SOC, the OLEV travels 60% of the total distance in EV mode.

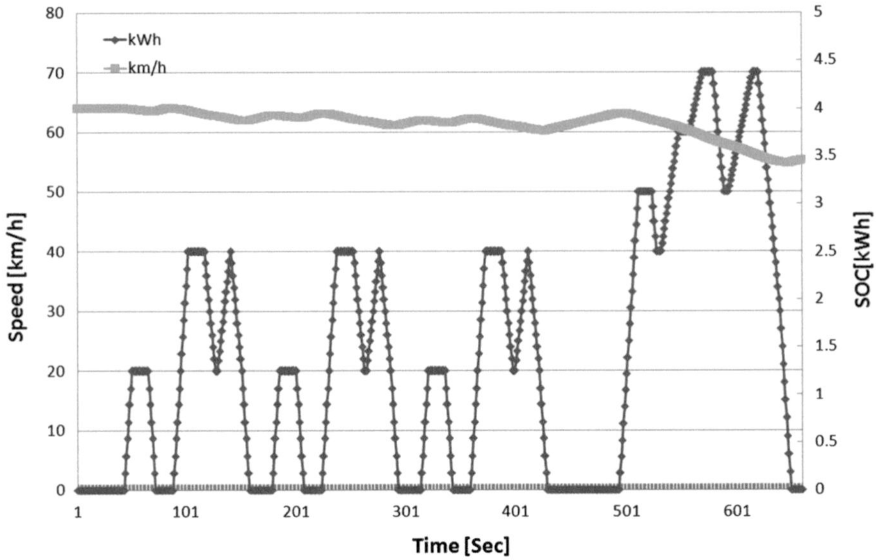

Fig. 22.3 OLEV battery SOC and energy consumption: 4 kWh

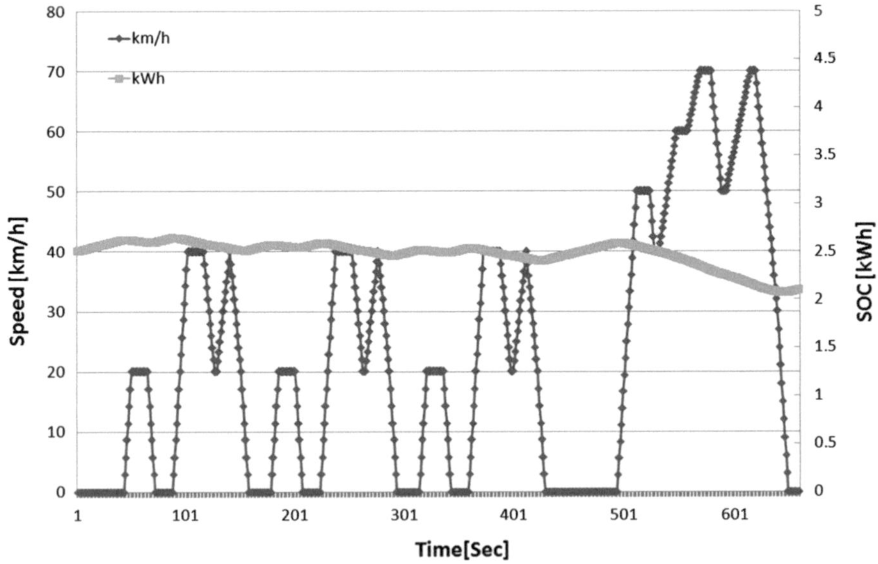

Fig. 22.4 OLEV battery SOC and energy consumption: 2.5 kWh

From the three cases, it is shown that the OLEV travels at least 60% of the total distance in EV mode in the worst case and in other cases it is fully operated in EV mode. In the comparison, we assume that he OLEV travels 80% of the total

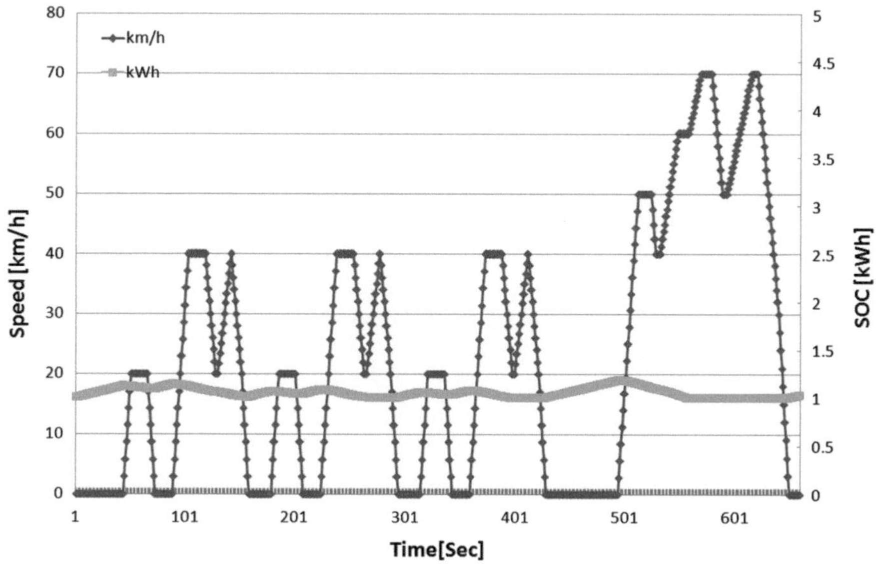

Fig. 22.5 OLEV battery SOC and energy consumption: 1 kWh

Table 22.4 Roadway profile in Seoul and power supplied roadway

Roadway type	Length (km)	Lane-km	No. of target lanes	Power supplied roadway (lane-km)
2-lane	6492	12,984	0	0
4-lane	652	2608	4	391
6-lane	578	3468	6	520
8-lane	300	2400	8	360
10-lane	121	1210	10	182
Total	8143	22,669	–	1453

distance in EV mode in the inner city where the power supplied roadway is installed in case vehicles are idling and the OLEV starts to operate with full charge.

(2) **Length of power supplied roadway**
To estimate the length of power supplied roadway in Seoul, we assume that the distance of idling is 15% of the total distance. Mostly, vehicles idle in front of traffic lights. Table 22.4 shows the roadway profile of Seoul and the length of power supplied roadway that should be installed.
We exclude 2-lane roadways for installation of the power supplied roadway. The length of 2-lane highways is quite substantial but in most cases 2-lane roads are used to travel very short distances, and thus can be disregarded. As a result, a power supplied roadway of 1453 km is constructed in the inner city

and it covers 6% of the total lane-km. By installing 6% of the total lanes, the OLEV can travel 60% of the total distance by using electricity, which is cheap and emits less CO_2 than the use of fossil fuels.

(3) **Unit cost per kilometer for power supplied roadway**
Cost associated with the power supplied roadway includes the roadway inductor, the power conditioners and distribution system, and labor involved with installation as well as engineering. According to the PATH project, the estimated roadway cost under a conservative scenario is $1,950,000 per lane-mile and $1,180,000 per lane-mile under an optimistic scenario. In the comparison, we assume $0.5 m/km for the power supplied roadway cost for an OLEV with mass production.
In installing the power supplied roadway for the OLEV in Seoul, $726 M is needed. The cost is the same as the infrastructure cost for the PHEV when the number of PHEVs in the market reaches 103,772, which is 4% of the total number of vehicles in Seoul. In Seoul, there are 2.5 million vehicles.

22.5 Energy Cost

PHEV and OLEV use gasoline in HEV mode and electricity in EV mode. The different infrastructure systems affect the energy consumption patterns and this results in a difference in energy costs for the two vehicle types.

22.5.1 Key Assumptions

To estimate the energy cost for PHEVs and OLEVs in Seoul, we employ the underlying assumptions presented in Table 22.5.

- Projected gasoline price
  To estimate the energy cost during the vehicle life time, the gasoline price in the next 10 years is required. To project the gasoline price, we use the annual world oil price from 2005 to 2035 by the U.S. Energy Information Administration (EIA) in 2010. We use the reference case among three cases. According to the

Table 22.5 Key assumptions

Vehicle kilometers travelled per day (Year)	41.8 km (15,257 km)	Average from 2000 to 2008 (KTSA)
Vehicle life (years)	10	(Funk and Rabl 1999)
Gasoline price ($/l)	1.65$	Average price in 2013
Electricity prices ($/kW)	0.1$ won	Average electricity price for EV in 2013 (KEPCO)

projected oil price in the reference case, we project the gasoline price by 2035 because gasoline price is mainly influenced by the material price 5.

– Projected electricity price
  To project the electricity price, we also use the reference case of the average annual U.S. retail electricity prices from EIA (2010). Electricity price is influenced by many factors such as fuel, generation method, and distribution. Fuel cost accounts for most of the cost and EIA uses the projected worldwide fuel cost to estimate the electricity price in the future. Similar to the U.S.A., Korea plans to increase renewable and nuclear energy dependence in the future. Hence, we project the electricity price based on electricity prices from EIA (2010) by 2035.

22.5.2 PHEV Energy Cost

To estimate the energy cost for the PHEV, we use the distribution of driving distance per trip based on national statistics (KTSA 2009) rather than a constant distance assumption. The distribution of driving distance has an important impact on fuel economy. Figure 22.6 shows the ratio of driving distance per trip in terms of frequency and distance. Most trips involve short driving distance. While long trips are infrequent, they are important because their length makes them a significant portion of the total distance travelled.

Long trips reduce the PHEV energy efficiency because the PHEV travels in HEV mode after the charge-depleting range. In the study, we assume that the PHEV has a fully charged battery when it starts each trip. This is an optimistic assumption because it assumes higher fuel cost saving by using more electricity rather than gasoline, which is more expensive.

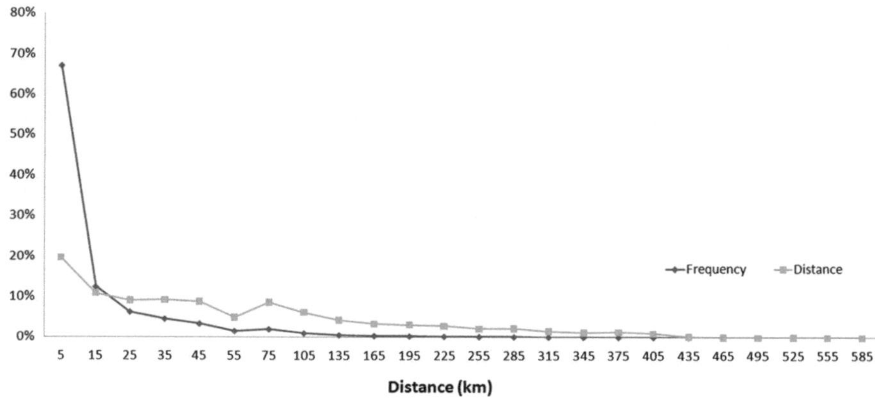

Fig. 22.6 Driving distribution per trip

A PHEV with a 16 kWh battery can use 12 kWh for power. 3 kWh is always restored for the minimum battery SOC and 1 kWh for the HEV mode. With a fully charged battery, the PHEV travels 50 km with 0.23 kWh/km driving efficiency considering 85% of power transmission efficiency from inverter to pickup. The PHEV travels in HEV mode after 50 km. According to Fig. 22.6, the PHEV travels 24% of the total distance in HEV mode.

Estimated energy cost for the PHEV in 2013 under the assumption is $589. Energy cost during the vehicle life (10 years) can be calculated according to the projected gasoline and electricity prices.

22.5.3 OLEV Energy Cost

The OLEV travels in EV mode in the inner city where the power supplied roadway is installed. During the inner city travel, 80% of the total distance is travelled in EV mode. According to a national household travel survey by the Ministry of Land, Transport and Marine affairs (2009), trips to and from work account for 81% of total driving. Therefore, we assume that 81% of total vehicle travel distance occurs in the city. Consequently, the OLEV is assumed to travel 64% of the total distance in EV mode.

Estimated energy cost for the OLEV in 2013 under these assumptions is $803. The OLEV energy cost for 10 years can be calculated according to the projected gasoline and electricity prices.

22.6 Results of Comparison Analysis

To compare the effects of infrastructure type in EV penetration in a mega-city, we compare the total cost between the PHEV and OLEV. Lower total cost will allow faster and more efficient market penetration. The OLEV has a lower vehicle price due to its smaller battery but has higher energy cost. The whole infrastructure for the OLEV must be installed before a single OLEV appears. The PHEV has a higher vehicle cost due to the larger battery size and lower energy cost. Its infrastructure increases as the PHEV increases. These differences are mainly due to the different charging infrastructure. Hence, the economic impact of the different charging infrastructure is analyzed to predict the EV market penetration potential.

22.6.1 Comparison Results

In the comparison, the total cost includes vehicle cost, energy cost for the vehicle life time, and infrastructure cost. Since the PHEV and OLEV are in the same category of an advanced hybrid vehicle, tax, insurance, and repairs have not been considered.

- **Total cost comparison**
  Table 22.6 shows the associated costs in 2013. Energy cost reflects the next projected gasoline and electricity costs for the next 10 years.
  The infrastructure cost for the PHEV is a variable cost whereas the OLEV has a fixed infrastructure cost and the cost is shared by the number of OLEVs distributed. Hence, we need to compare the total cost according to the number of vehicles. In the comparison, we assume that the number of vehicles is assigned to each type of vehicle and compare the total cost along the given number of vehicles. In 2013, if the number of vehicles exceeds 75,573 the total cost of the OLEV becomes less than that of the PHEV. 75,573 vehicles cover only 3% of the total number of vehicles in Seoul. The total number of vehicles in Seoul is 2,500,000. Economic benefit by adopting OLEV compared to PHEV is mainly due to the high vehicle cost of PHEVs compared to OLEVs in 2013. If 10% of the total number of vehicles is changed to hybrid based EVs in 2014, the total cost for the OLEV is $10,947 while the PHEV's total cost is $12,324. By adopting the OLEV, $1377 M can be saved. If the number of vehicles increases, the amount of cost savings will increase.
- **Total cost comparison from 2013 to 2022**
  Each year, total cost changes according to the estimated vehicle cost and project energy cost. We compare the total cost between the PHEV and OLEV from 2013 to 2022, when the hybrid type of EV is expected to be dominant in the market [2]. Figure 22.7 illustrates the breakeven point of the total cost for the OLEV compared to the PHEV regarding the number of vehicles.
  Until 2022, the total cost of the OLEV is always less than that of the PHEV. In the earlier stage, the breakeven point for the OLEV is lower because the vehicle cost of the PHEV is higher due to its expensive battery. The large fixed cost of infrastructure for the OLEV is compensated by the high vehicle cost of the PHEV with a small number of vehicles. As time passes, the breakeven point increases because the vehicle cost decreases according to the cost change of batteries and increased gasoline price. In 2020, the OLEV reaches the breakeven point where 6% of conventional vehicles are converted to hybrid-type EVs.

Table 22.6 Total cost comparison in 2011

Cost factor	Vehicle type	Cost (Thousand $)
Vehicle cost	PHEV	36.4
	OLEV	32.1
Energy cost (10 years)	PHEV	6.8
	OLEV	8.9
Infrastructure cost	PHEV (per vehicle)	7.0
	OLEV	726,401.1

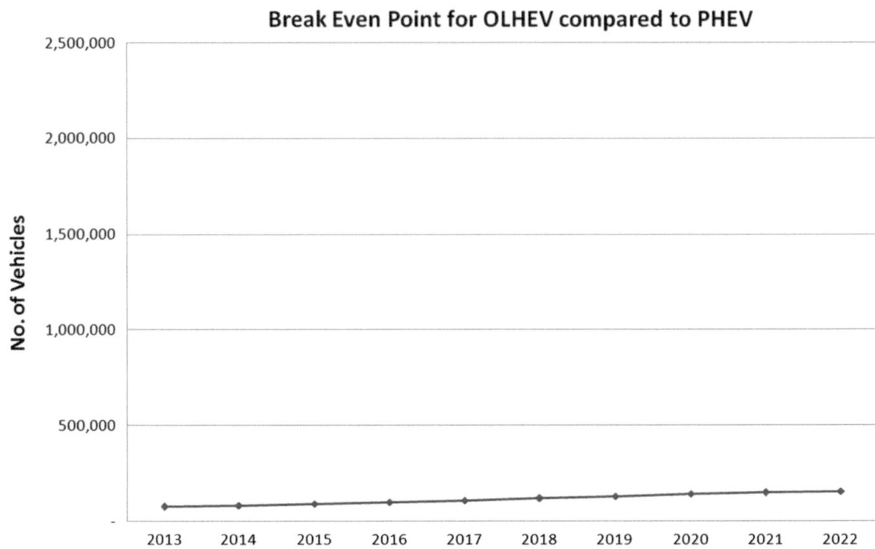

Fig. 22.7 Breakeven point for OLEV compared to PHEV

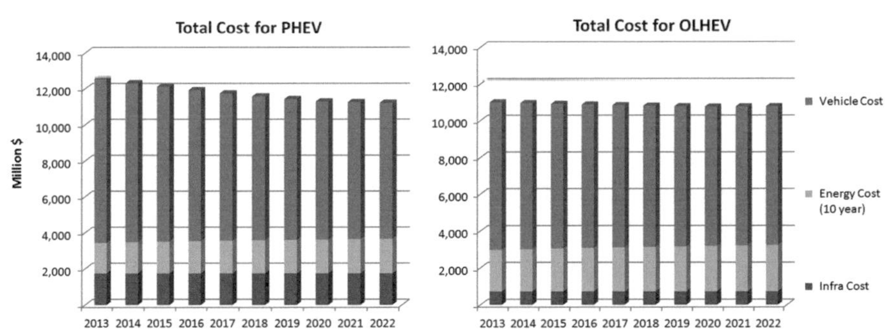

Fig. 22.8 Comparison of cost factors

– **Comparison of cost factor**

To analyze each cost factor's impact on the total cost along the time period, we fixed the number of converted vehicles each year at 250,000 (10% of total number of vehicles). Figure 22.8 shows the comparison results.

In both vehicle types, the vehicle cost occupies most of the total cost. The vehicle cost for the PHEV is relatively higher and its energy cost is lower than that of the OLEV. The total cost of the PHEV decreases due to the diminishing vehicle cost even though the energy cost increases slightly. The OHEV's energy cost increases at a relatively steeper rate because it depends on gasoline more than the PHEV. The total cost for the PHEV is higher all the time because the vehicle price itself is higher and the infrastructure cost increases as the number of vehicles increases. Even though the OLEV requires a huge investment for initially installing the infrastructure, if the number of vehicles converted to HEV based EVs is over 103,772 (4% of total passenger cars in Seoul), the total infrastructure cost for the OLEV falls.

- **Oil price change effect to total cost**
  According to our assumptions, OLEV is driving in EV mode for 64% of its total mileage, while PHEV for 76%. Hence, OLEV's energy cost is more sensitive to the oil price. In our comparison, we assumed the reference oil price is 1.65$/l in Fig. 22.5. If the oil price increase 30%, and 10% of the total number of vehicles is changed to hybrid based EVs in 2014, the total cost for the OLEV is $11,398 while the PHEV's total cost is $12,609. By adopting the OLEV, $1211 M can be saved and the saving is decreased by $166 M, 12% of total saving comparing to the original Oil price. Even though, OLEV depends more on Oil resource, the sensitivity is not crucial to affect the total cost comparison because the oil resource cost is very small portion in total cost.

22.6.2 *Discussion*

According to the results, the OLEV is more cost efficient when the battery cost is high. To advance EVs to the market to ensure a secure energy source and address environmental issues, the charging infrastructure decision should be made to adopt the On-Line system for OLEVs in the future. When the battery cost becomes sufficiently low, the PHEV or a pure EV also can be an alternative. Even at that time, the infrastructure for the OLEV can be used by adding an onboard pickup device to the PHEV or a pure EV. Hence, the investment to the power supplied roadway will not be wasted if any type of EV is ultimately used, as they always require charging.

In addition to advancing the OLEV technology, there are strong implications regarding the commercial value of this technology. Researchers from the National Renewable Energy Laboratory in the U.S.A. assessed the economics of the various types of electric vehicles based on U.S. market data-based assumptions for the battery, motor, and engine [1]. Under the U.S. environment, they compared the economics of conventional vehicles, hybrid electric vehicles (HEVs), plug-in hybrid electric vehicles (PHEVs) with various all-electric driving ranges, pure electric vehicles (EVs) with an 80 mile range, and electrified hybrid electric vehicles (EHEVs). The EHEV can connect to an external source of energy along some roadways while moving. The OLEV technology is a candidate for this kind of dynamic plug-in. According to their results, using current battery cost and life, PHEVs and EVs were not cost effective for many different configurations. The EHEV is the only alternative to provide a cost-effective pathway to vehicle electrification. Therefore, the outcome of [1] strongly supports our results from a different angle.

22.7 Conclusions

To accelerate market penetration of EVs, installing a proper charging infrastructure is necessary. Depending on the charging infrastructure system, vehicle cost, infrastructure cost, and energy consumption patterns are determined. In the study, we compared the total cost associated with PHEVs and OLEVs regarding the infrastructure system. Due to the expensive battery, the OLEV is more cost efficient in a mega-city where lots of vehicles share the roadways in the city. As battery costs fall, EVs with larger batteries will be more efficient. Even at that time, the infrastructure for OLEVs can be used as charging infrastructure for PHEVs or PEVs by adding onboard pickup devices to the vehicles.

The current analysis results are based on various assumptions regarding the technology of vehicles and battery cost trend and projected energy cost. These assumptions accordingly impact the results. Another limitation is that the study considers only the total cost minimization for EV penetration. The perspective of consumers who make a decision of purchasing vehicles is not included. Another limitation is that the comparison analysis is done based on a mega-city, Seoul, Korea. Smaller cities or different countries will have different driving patterns and energy costs. Hence, to generalize the results, it is necessary to expand our study to different city sizes and different countries.

In further comparison, the optimal system for PHEV and OLEV should be identified. By adapting the battery size, onboard pickup capacity, and the ratio of power supplied roadway, optimal PHEV and OLEV systems that minimize the total cost can be established. Based on optimal systems, the comparison results and related conclusions can be different. Also, the optimal PHEV and OLEV should be compared for other green car options such as fuel cell vehicles and pure electric vehicles.

References

1. Brooker A, Thornton M, Rugh J (2010) Technology improvement pathways to cost-effective vehicle electrification. NREL/CP-540-47454
2. Pike Research (2010) Hybrid electric vehicles for fleet markets. Available via http://www.pikeresearch.com/research/hybrid-electric-vehicles-for-fleet-markets. Accessed 08 Dec 2010
3. California PATH Research Report (1996) Roadway powered electric vehicle project, parametric studies: phase 3D final report. UCB-ITS-PRR-96-28
4. Lee SL, Park CB, Cho GH, Huh J, Choi NS, Rim CT (2010) On-line electric vehicle using inductive power transfer system. Energy Convers Congr Exposition (ECCE): 1598–1602
5. Bolger JG, Kirsten FA, Ng LS (1978) Inductive power coupling for an electric highway system. IEEE Veh Technol Conf 28:137–144
6. Zell CE, Bolger JG (1982) Development of an engineering prototype of a roadway powered electric transit vehicle system: A public/private Sector Program. IEEE Veh Technol Conf 32:435–438
7. Shladover SE (1987) Roadway-powered electric transit vehicle—progress and prospects. Transp Res Rec 1155:28–36

8. Covic GA, Boys JT, Kissin MLG (2007) A three-phase inductive power transfer system for roadway-powered vehicles. IEEE Trans Industr Electron 54(6):3370–3378
9. Hensley R, Knupfer S, Pinner D (2009) Electrifying cars: how three industries will evolve. McKinsey quarterly 3:87–96

Chapter 23
Regulatory and Safety Issues

Dong Ho Cho

Abstract This chapter deals with safety issues related to the high electric power, electromagnetic interference (EMI), and electromagnetic force (EMF) of power cable module and pickup module. We propose a safety-test method for road-embedded power cable and pickup module with 20 kHz high current. Based on the test of voltage and current induced by 20 kHz high current, regulator safety verification method is proposed based on conventional electric safety certification procedure for electric bus. The test environment, test metric, and the procedure for EMI generated by 20 kHz magnetic resonated power cable module pickup module, and inverter system are proposed to prevent the malfunction of electronic device or equipment in close proximity. We also propose specifications for certification tests for EMF generated by combination of power cable module and pickup module.

23.1 Introduction

The regulatory issues related to a high-power electric bus and electromagnetic compatibility are introduced. For safety issues related to a high electric power supply system, safety tests for electronic shock, insulation distance, temperature rise, insulation resistance, leakage current, dielectric withstanding voltage, ground continuity, IP, appearance, and power cable are explained in detail. Also, functional tests of light load and function, ripple voltage and current measurement, sudden change of system voltage, sudden phase change of system voltage, over current capacity, load, efficiency, power factor measurement, output side short circuit, line side harmonics, and load of a power supplied road are described. With regard to safety issues for an electric bus with wireless charging, voltage ratings of electric circuits, basic protective measures, an insulation resistance test, a withstanding voltage test, a connecting parts connectivity test, and safety standards of high-power electric devices of

D.H. Cho (✉)
School of Electrical Engineering KAIST, Korea Advanced Institute of Science and Technology (KAIST), 291 Daehak-ro, Yuseong-gu, Daejeon 34141, South Korea
e-mail: dhcho@kaista.ac.kr

N.P. Suh and D.H. Cho (eds.), *The On-line Electric Vehicle*,
DOI 10.1007/978-3-319-51183-2_23

electric vehicles are introduced. In addition, in view of electromagnetic compatibility safety issues for the power supply system, the requirements and the measurement method of electromagnetic emission are explained. Finally, regarding electromagnetic compatibility safety issues for the power collector system, the requirements and the test method of electromagnetic immunity are explained.

23.2 Safety Issues for High Power Supply System

23.2.1 Safety Test

(1) Electric Shock Protection Test

In order to test protection against electric shock caused by contact with a live electric part, tests with a test finger and a test pin defined in IEC 61032 shall be conducted to declare a pass/fail decision. The test with a test finger shall be conducted at a force of 50 N [1].

[Pass/Fail Criterion]

- In the tests with a test finger and a test pin, the finger or the pin should not contact a live part of 25 Vac or 60 Vdc or higher.
- The live electric part should have an external box or at least a protective wall conforming to the requirements of IP2X (Protective Rating against Infiltration of Solid) of IEC 60529 [2]. The surface of an easily accessible external box or the protective wall should be IP20 or higher for the indoor type and IP44 or higher in the case of an outdoor type.

(2) Insulation Distance Test

The insulation distance test is divided into the spatial distance measurement test and creepage distance measurement test.

(a) The results of the spatial distance measurement test should be bigger than the spatial distances listed in Tables 23.1 or Table 23.2 depending on the pollution level standard of IEC 60664-1, as shown below, and there should be no dielectric breakdown during an impulse voltage test [3].

  - Pollution Level 1: Major environmental condition—a dry place free of pollution and accumulated pollution
  - Pollution Level 2: Major environmental condition—a normal place in which pollution may temporarily accumulate (conductive)
  - Pollution Level 3: Major environmental condition—a place with accumulated pollution and humidity
  - Pollution Level 4: Major environmental condition—a place where pollution accumulates as the place is exposed to dust, rain, snow, etc.

Table 23.1 Spatial distance depending on the pollution level and test impulse voltage (between the main circuit and the external box)

Rated insulation voltage Vac standard (Vdc = Vac/$\sqrt{2}$)	Pollution level				Impulse voltage 1.2/50 μs (kV)
	1 (mm)	2 (mm)	3 (mm)	4 (mm)	
$50 < x \leq 100$	0.5	0.5	0.8	1.6	1.5
$100 < x \leq 150$	1.5	1.5	1.5	1.6	2.5
$150 < x \leq 300$	3.0	3.0	3.0	3.0	4.0
$300 < x \leq 600$	5.5	5.5	5.5	5.5	6.0
$600 < x \leq 1000$	8.0	8.0	8.0	8.0	8.0

Table 23.2 Spatial distance depending on the pollution level and test impulse voltage (within the main circuit)

Rated insulation voltage Vac standard (Vdc = Vac/$\sqrt{2}$)	Pollution level				Impulse voltage 1.2/50 μs (kV)
	1 (mm)	2 (mm)	3 (mm)	4 (mm)	
$50 < x \leq 100$	0.1	0.2	0.8	1.6	0.8
$100 < x \leq 150$	0.5	0.5	0.8	1.6	1.5
$150 < x \leq 300$	1.5	1.5	1.5	1.6	2.5
$300 < x \leq 600$	3.0	3.0	3.0	3.0	4.0
$600 < x \leq 1000$	5.5	5.5	5.5	5.5	6.0

(b) In the creepage distance measurement test, the creepage distance is measured in accordance with the following CTI (Comparative Tracking Index) classification criteria of IEC 60664-1:

- Insulation Material Group I: $600 \leq$ CTI;
- Insulation Material Group II: $400 \leq$ CTI < 600;
- Insulation Material Group IIIa: $175 \leq$ CTI < 400;
- Insulation Material Group IIIb: $100 \leq$ CTI < 175;

[Pass/Fail Criteria]

- The spatial distances should be larger than the values in Table 23.1 or 23.2, and the power supply device should have resistance to the impulse voltage.
- The creepage distances should be larger than the values listed in Table 23.3.

(3) Temperature Rise Test

The power shall be operated at the rated voltage and rated frequency. The power supply device shall be set to have the rated output. The test temperature shall be 30 ± 5 °C when the test is conducted indoors and the temperature correction shall be done to 30 °C. The temperature shall be corrected based on 40 °C in an outdoor test.

Table 23.3 Creepage distance depending on the insulation material

Rated insulation voltage (V_{rms})	Pollution level								
	1	2				3			
		Insulation material				Insulation material			
	(mm)	I (mm)	II (mm)	IIIa (mm)	IIIb (mm)	I (mm)	II (mm)	IIIa (mm)	IIIb (mm)
50 < x ≤ 80	0.22	0.67	0.95	1.3	1.3	1.7	1.9	2.1	2.1
80 < x ≤ 125	0.28	0.75	1.05	1.5	1.5	1.9	2.1	2.4	2.4
125 x ≤ 250	0.56	1.25	1.8	2.5	2.5	3.2	3.6	4.0	4.0
250 < x ≤ 500	1.30	2.5	3.6	5.0	5.0	6.3	5.1	8.0	8.0
500 < x ≤ 1000	3.2	5.0	5.1	10.0	10.0	12.5	14.0	16.0	16.0

Table 23.4 Pass/fail criteria

Location	Metal surface °C	Non-metal surface °C
A handle or knob held to move	50	60
A handle or knob not held to move	60	85
Surface normally touched	70	95

The temperature at each part of the element should be within the specified temperature presented by the manufacturer

(a) Operation shall proceed until the temperature rise in each part reaches the saturation state as shown in Table 23.4. The points at which the temperatures are measured shall be the upper part of the external box, transformer, reactors, the DC link capacitor, and the switching element heat sink.
(b) The temperature of each part shall be measured when the temperature rise reaches the saturation state.

(4) Insulation Resistance Test

The insulation resistance shall be measured at the test points in Fig. 23.1 applying 500 V DC. The insulation resistance shall be measured immediately after completing the dielectric withstanding voltage test and thermostatic test.

[Pass/Fail Criteria]

- The insulation resistance of equipment type 1 shall be 1 MΩ or higher.
- The insulation resistance of equipment type 2 shall be 7 MΩ or higher.

(5) Leakage Current Test

The AC power shall be operated at the rated voltage and rated frequency. The power supply device shall be set to have the rated output. U_2 shall be measured by building a circuit in the power supply/power collector, as shown in Fig. 23.2. The frequency applied is 15 Hz–1 MHz.

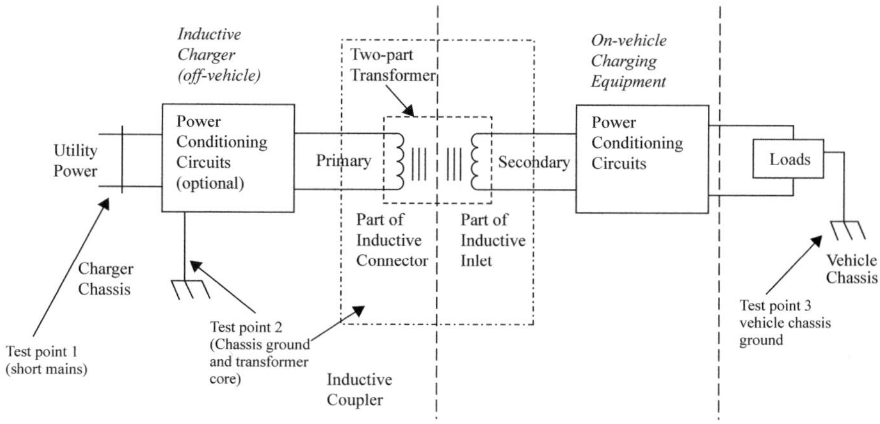

Fig. 23.1 Insulation resistance test

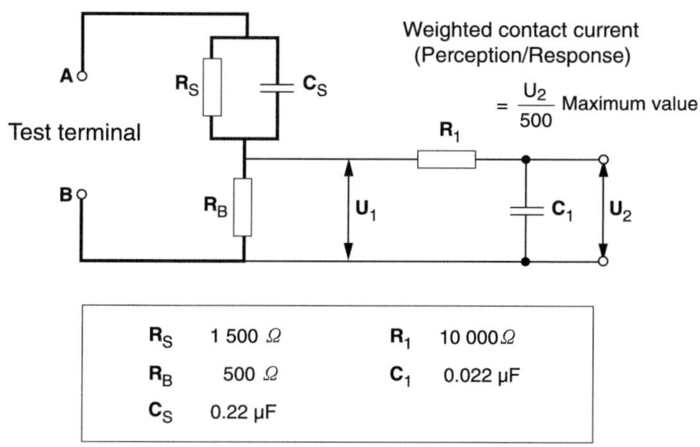

Fig. 23.2 Leakage current test

[Pass/Fail Criteria]

- Equipment type 1: 3.5 mA or lower
- Equipment type 2: 0.7 mA or lower

(6) Dielectric Withstanding Voltage Test

The dielectric withstanding voltage test shall be conducted applying the following voltages for 1 min at the rated frequency. The test shall first be conducted between A and B, and then between B and C, as shown in Fig. 23.3.

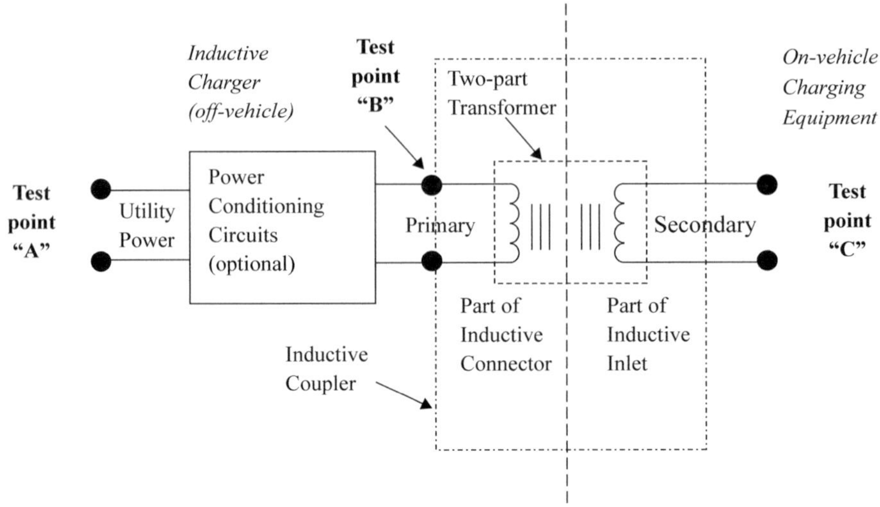

Fig. 23.3 Dielectric withstanding voltage test

(a) Equipment type 1: 2000 V rms shall be applied to the common mode and the differential mode.
(b) Equipment type 2: 4000 V rms shall be applied to the common mode, and 2000 V rms shall be applied to the differential mode.

[Pass/Fail Criterion]

- The device should run smoothly in the operation after the test.

(7) Ground Continuity Test

A DC current of 16 A shall be applied between the ground terminal or the ground electrode and a metal part accessible to the human body using AC or DC power with a no-load voltage of 12 V or lower.

[Pass/Fail Criterion]

- The resistance between the metal part and the ground terminal or the ground electrode shall be 0.1 Ω or lower.

(8) IP Test

The cover, cap, and the protective grill should be firmly secured. In addition, the outer coverings that can be opened such as a door or a drawer must be separated from the key, device, or the live parts.

[Pass/Fail Criterion]

- The power supply system should be of IP54 or higher.

(9) Appearance Test

The connectors and other transfer parts should have no sharp edges when installed.

(10) Power Cable

The specifications of the cable for a wirelessly recharged electric vehicle and the reason for its use, waterproofing of the Mechanical Contact (MC) box switching segment inside the underground box, the conductor area, the conductor resistance, the hot set, and the test results of allowable current and flame retardancy should be provided to enable the special cable to be safely installed in the real road operation environment.

(a) **Conductor Area and Conductor Resistance**

0.204 Ω/km or lower (in the case of 90 mm^2 cable).

[Pass/Fail Criteria]

As shown in Table 23.5, the following conductor resistance per conductor area should be satisfied when measured using the Double Bridge measurement technique:

(b) **Hot set**

- With load: The elongation shall be within 175%.
- After cooling: The elongation shall be within 15%.

[Pass/Fail Criteria]

- As shown in Table 23.6, the elongation is measured after applying the given temperature and the load in a thermo-hydrostat and after cooling, respectively.
  - Insulator: Elongation within 175% when load is applied at 200 °C and within 15% after cooling
  - Sheath: Elongation within 175% when load is applied at 140 °C and within 15% after cooling

Table 23.5 Nominal cross-sectional area and maximum conductor resistance table

Nominal cross-sectional area (mm^2)	Conductor resistance (Ohm/km)
70	0.268
90	0.204
95	0.193

※ *Double Bridge measurement technique* A total of 5 strands forming a 1-m-long cable shall be bound together to make a terminal and the resistance shall be measured after terminal connection

Table 23.6 Hot set experiment of the insulator

Insulator	Room temperature	Tensile strength	125 N/mm^2 or higher	N/mm^2	12.51
		Elongation	200% or higher	%	425
	Heated	Tensile strength	Within ± 25% of the value before heating	%	6.0
		Elongation	Within ± 25% of the value before heating	%	−6.0
	Hot set		Elongation of 175% or lower when load is applied	%	90
			Elongation of 15% or lower after cooling	%	2.5
	Shrinkage test		4% or lower	%	1.0

(c) Allowable Current of Cable [Japanese Automotive Standard Organization (JASO) D609]

- It is calculated by entering the maximum allowable temperature for the cable sheath and the general air temperature.
- It is calculated by entering the allowable temperature for the insulator specified by National Electrical Manufactures Association (NEMA) [example: 75 °C in the case of Thermoplastic Elastomer (TPE)].

[Pass/Fail Criteria]

- The temperatures of the sheath and the conductor shall be measured at an interval of 2 min or 10 min until the temperature saturation phenomenon occurs, applying the maximum allowable current (300A) at the rated working frequency (20 kHz) as the current source, the result of which shall be used as the pass/fail criterion.

The equation specified by JASO (Japanese Automotive Standard Organization) for the electric vehicle cable is as follows:

$$I = K\sqrt{\frac{T_C - T_A}{Y^{R_{th}}}},$$

$$Y = Y_0\{1 + \alpha(T_C - 20)\},$$

$$R_{th} = R_1 + R_2,$$

$$R_1 = \frac{P_1}{2\pi} log \frac{d_2}{d_1},$$

$$R_2 = \frac{10P_1}{\pi d_2}$$

I: Allowable current (A),
D_1: Outer diameter of conductor (mm),
Y: Effective conductor resistance of cable at $T_C(\Omega/cm)$,
D_2: Insulation outer diameter (mm),
T_C: Maximum allowable temperature of cable (°C)
P_1: Specific thermal resistance of insulation sheath (°Ccm/W),
T_A: Ambient temperature (°C),
R_{th}: Thermal coefficient of resistance (°Ccm/W),
α: Ambient temperature (°C),
K: Allowable current reduction factor in the Case of multi-core cable

(d) Flame Retardancy of Cable

- It is calculated by entering the maximum allowable temperature for the cable sheath and the general air temperature.
- It is calculated by entering the allowable temperature for the insulator specified by NEMA (example: 75 °C in the case of TPE).

18.14 Flame Retardancy Test
18.14.1 Combustion Test of a Single Cable
This test shall be applied only to the cables sheathed with ST_1, ST_2, or SE_1, and shall be conducted only when specially required.
18.14.2 Combustion Test of Bundle Cable
This test shall be applied to the low toxic cables sheathed with ST_8. As to the test method and requirements, the provisions of KS C IEC 60332-3-24 shall be followed.

[Pass/Fail Criteria]

- The cable shall be left in room temperature for 16 h or longer before burning it and measuring in the mm length from the bottom of the upper support to the point where carbonization has started. The cable should not burn beyond 540 mm away from the lower part of the clamp.

(e) Waterproofing of MC(Magnetic Contactor) Box (IEC 60529:2002) [2]

- The water resistance shall be checked in the IPX7 test conditions.
- A specimen of size 640 mm * 410 mm * 232 mm shall be tested in the conditions of 23.5°, 63% RH, and 98 kPa.

[Pass/Fail Criteria]

- The temperature difference between the water and the specimen must be within 5 K
- The water resistance test shall be conducted for 30 min, placing the MC box in 1-m-deep water.

(f) Input/Output Differential Current

In relation to the current transmission route, if the line connected to the secondary side of the transformer is buried underground over a long distance, the current of the output (+) side and that of the return (−) side may differ due to the self-capacitance existing as a uniformly distributed capacitance. As this is different from the leakage current resulting from a ground fault or insulation defect, it is called the 'Input/Output Differential Current'. Based on the IEC 60479 standard, which specifies that the maximum safe amperage increases by 100 mA when a maximum safe amperage of 5 mA, which has no harmful electro-physiological effect at 60 Hz and a frequency factor of 10 kHz or higher (the ratio of the limiting current related to physiological effect at frequency f to the limiting current at 60 Hz, 20 times), is applied, if the input/output differential current per 1 m does not exceed 100 mA at 20 kHz AC, which is the normal standard for step voltage, the specimen passes the test.

[Pass/Fail Criteria]

- Occurrence of Input/Output Differential Current of 100 mA per 1 m step voltage or less
- 7 A or less per segment including the common electricity supply line {Segment power supply cable length (70 m) × Threshold of differential current (100 mA)}

(g) Induced Voltage

Although the current resulting from the induced voltage generated during operation of a wirelessly recharged electric vehicle should be 5 mA or lower, which is the lower limit of the electric shock current standard that is predicted not to cause any damage to human organs in general, if the actual detection threshold depending on frequency is applied to the case of 20 kHz by referring to 60 Hz, the current threshold, which is about 20 times larger compared to 60 Hz, should be the electric shock current level that is predicted not to cause any damage to human organs at 20 kHz. Accordingly, considering the safety of general people, the standard is set to 100 mA in the case of using 20 kHz at an energizing cycle of 10 s, which is enough time to perceive and respond to currents, and the standard of the maximum induced voltage is set to 57.5 V applying the total human body impedance, which is the minimum asymptotic value of the 575 Ω.

[Pass/Fail Criteria]

- The voltage between the ground and the metal facility shall be 57.5 V or lower, and the current converted based on the total human body impedance minimum asymptotic value of 575 Ω shall be 100 mA or lower.

(h) Marking

The specimen should have the following markings: manufacturer, device reference, manufacture date, serial number, input voltage, input frequency, and input current.

23.2.2 *Functional Test*

(1) Light Load and Functional Test

The power supply system shall be tested being connected to the rated input voltage, and in the maximum and minimum input voltage conditions in accordance with the provisions of IEC 60146-1-1 [4]. The functional test and the load should be selected so as to satisfy the required performance. During this test, it should be checked whether the control equipment, auxiliary devices, protective equipment, and the main circuit are properly operating. This test may be conducted by different methods depending on the shape of the equipment.

[Pass/Fail Criteria]

- All parts of the electric circuit and the cooling system of the power supply system should properly operate together with the main circuit.
- The control equipment, auxiliary devices, protective equipment, and the main circuit should operate properly.

(2) Ripple Voltage and Ripple Current Measurement Test

In accordance with the test specified in IEC 60146-1-1, measurements of the superimposed AC voltage, superimposed AC current, and noise voltage or noise current at the DC side should be separately specified [4].

[Pass/Fail Criterion]

- The ripple voltage and the ripple current shall be within the specified values presented by the manufacturer.

(3) System Voltage Sudden Change Test

The AC power shall be operated at the rated voltage and the rated frequency. The output of the power supply system shall be set to the rated output.

(a) The power supply system shall be operated at the rated output.
(b) The system voltage shall be rapidly changed to the value of the rated voltage +15% in the form of a step function (1 cycle or less rise time), which shall be maintained for 10 s before being returned to the rated voltage.

(c) The system shall be operated at the rated voltage.
(d) The system voltage shall be rapidly changed to the value of the rated voltage −15% in the form of a step function (1 cycle or less fall time), which shall be maintained for 10 s before being returned to the rated voltage.
(e) The voltage waveform and the current waveform of the input and the output shall be recorded.

[Pass/Fail Criteria]

- The power supply system should stably operate following the sudden change in the system voltage.
- The change in the AC output current should be 150% of the rated rms current or less and 0.5 s or shorter.

(4) Sudden Change Test of System Voltage Phase

The AC power shall be operated at the rated voltage and the rated frequency. The output of the power supply system shall be set to the rated output.

(a) The output voltage phase of the power supply system in normal operation conditions shall be set as 0°.
(b) The system voltage phase shall be changed from 0° to +10° in the form of a step function, which shall be maintained for 10 s before being returned to 0° again in the form of a step function.
(c) The system voltage phase shall be changed from 0° to −10° in the form of a step function, which shall be maintained for 10 s before being returned to 0° again in the form of a step function.
(d) The output voltage waveform and the output current waveform shall be recorded.
(e) The test in (b) shall be repeated after changing the phase change value of +10° to +120°. The output voltage and current waveforms shall be recorded.

[Pass/Fail Criteria]

- During the sudden change of phase within ±10°, the power supply system should stably operate following the rapidly changing phase of the system voltage.
- During the sudden change of phase of the system voltage up to +120°, the power supply system should stably operate following the rapidly changing phase of the system voltage, or stably stop without damage to any parts, in which case the system should be automatically started.

(5) Overcurrent Capacity Test

In the overcurrent capacity test, short-duration overcurrent of the value specified for the load test or continuous input of an actual load should be applied at the specified interval in accordance with the provisions of IEC 60146-1-1 [4].

[Pass/Fail Criterion]

- The power supply system should normally operate in the allowable overcurrent range presented by the manufacturer.

(6) Load Test

The following shall be measured at 25, 50, 75, and 100% of the nominal speed and in the relevant maximum load condition of each speed:

- Voltage Uo, current Io, and power Po at the input side of the power supply system
- Voltage U_1, current I_1, and power P_1 at the input side of the motor.

[Pass/Fail criteria]

- As soon as load is disconnected, switch shall be opened.
- In case that the load of one phase is adjusted to zero in 3 phase inverter, normal stable operation time must be above 30 min.

(7) Efficiency Test

The efficiency shall be measured applying a load to the output side of the power supply system to enable it to have the rated output.

[Pass/Fail Criteria]

- The inverter efficiency shall be 95% or higher.

(8) Power Factor Measurement Test

The measurements shall be made in accordance with the provisions of IEC 61800-2 under the rated operation conditions of the power supply system [5].

[Pass/Fail Criterion]

- The inverter power factor shall be 0.9 or higher.

(9) Output Side Short-circuit Test

The test shall be conducted by/ short-circuiting the output side in accordance with the provisions of IEC 61558-1. A dielectric strength test shall be conducted when the specimen is cooled to room temperature after the short-circuit test [6].

[Pass/Fail Criteria]

- No flame, metal solution, or toxic or ignitable gas shall be generated during the test.
- The temperature shall not exceed the value listed in Table 3 of IEC 61558-1 [6].
- The specimen shall withstand the dielectric strength test.

(10) Line Side Harmonic Test

The amount of harmonics at the input side of the inverter shall be measured under the rated operation conditions. IEC 61000-4-7 shall be followed [7].

[Pass/Fail Criterion]

- The amount of input harmonics shall be within the specified value presented by the manufacturer.

(11) Load Test of Power Supplied Road

As there is a regulation on overloaded vehicles in Paragraph 1, Article 54 of the current Road Act, and traffic of vehicles of which the axial load exceeds 10 tons or the total weight exceeds 40 tons is restricted in accordance with Paragraph 3, Article 28 of its enforcement decree, a dynamic load of 17 tons designed for compressive strength is applied by estimating the design axial load of 10 tons and applying the dynamic load coefficient of 1.7–10 tons, which is the axial load used for restriction of vehicle traffic.

(a) Compressive Strength Test [KS F 2403/KS F 2405 (revised from ISO 1920:1976)] [8]

[Pass/Fail Criteria]

- The Standard Compressive Strength shall be 30 MPa or higher.
  - The compressive strength test shall be conducted as shown in Fig. 23.4 using a UTM tester that can apply a load up to 50 tons.
  - The size of the specimen shall be 800 * 500 * 300 mm, and the size of the UTM press plate shall be 200 * 500 mm.
  - The UTM equipment shall be capable of applying a load of up to 50 tons, but a load of up to 45 tons shall be used considering the reliability of the equipment.

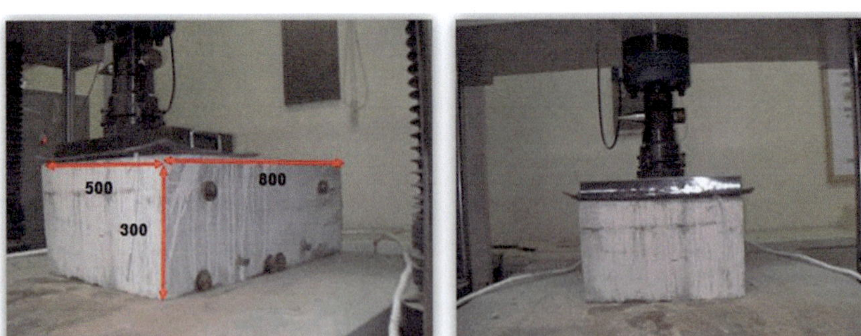

Fig. 23.4 Compressive strength Load Test

Plain concrete | Concrete containing a power supply cable line

Fig. 23.5 Trisection point Load Test of specimen

- A first grade or higher compressive tester specified in KS B 5533 shall be used.
- The thickness of the upper and lower press plates shall be 25 mm, the size of the press plates shall be 200 * 500 mm, and the compression surface shall be finished by grinding.
- The concrete specimen shall be tested immediately after predetermined curing.

(b) Flexural Strength Test (KS F 2408 (ISO 4013:1978), Trisection Point Load Method) [9]

As shown in Fig. 23.5, the test shall be conducted for the concrete specimen containing the power supply cable line structure using the fracture strength test method (ISO 4013) of the hardened concrete specimen in accordance with the trisection point load test.

[Pass/Fail Criteria]

- 1/5–1/8 of the compressive strength
  - As shown in Fig. 23.6, the tester shall be used within the range from the point where the maximum load is 1/5 of the full capacity to the capacity, and if the capacity of the tester is variable, each capacity shall be considered as a separate capacity.
  - The span of the supporting roller shall be 3 times the height of the specimen.
  - The specimen containing a power supply cable line shall be placed at the center of the bearing width and the upper load device shall be positioned in contact with the trisection points of the span. (There should be no gap between the contacting surface of the load device and the specimen surface).

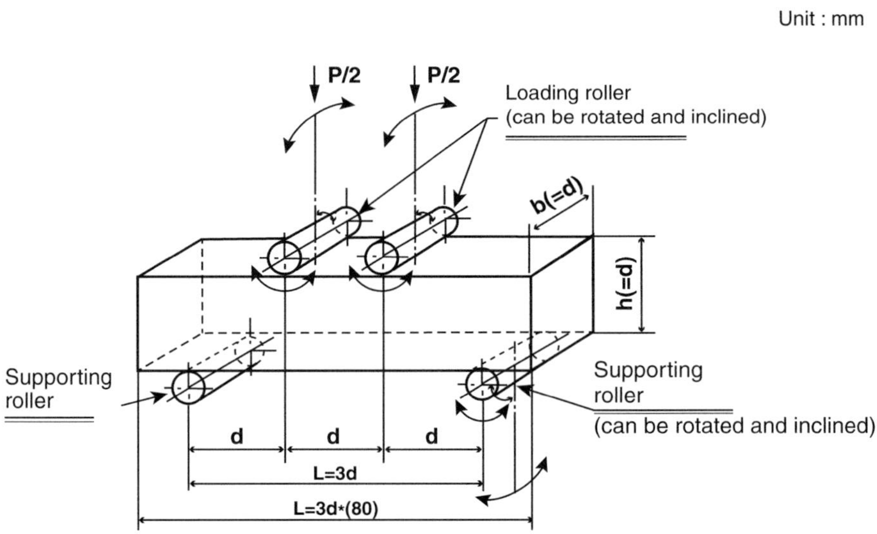

Fig. 23.6 Example of trisection point load device

Table 23.7 Voltage Ratings of Electric Circuits

Voltage rating	Maximum operating voltage	
	DC (V)	AC [V (rms)]
A	$0 < U \leq 60$	$0 < U \leq 30$
B	$60 < U \leq 1500$	$30 \leq U \leq 1000$

Remark DC 60 V and AC 30 V are the values selected considering wet weather condition

- The load shall be applied at a constant speed lest the impact should be applied to the specimen containing the power supply cable line, and the speed shall be adjusted so as to let the increase rate of the edge stress intensity be 0.06 ± 0.04 MPa (=N/mm^2) per second, which shall be maintained until the maximum load is reached.
- The load shall be applied until the specimen containing the power supply cable line breaks, and the value of the maximum load appearing on the tester shall be read by significant three digits.

23.3 Safety Issues for Electric Bus with Wireless Charging

23.3.1 Voltage Ratings of Electric Circuits

Each electric circuit falls under one of the voltage ratings specified in Table 23.7 in accordance with its maximum operating voltage U.

Fig. 23.7 Marking of parts of which the voltage rating is B. marking symbol (Background color: *Yellow*/ Symbol: *Black*)

In the vicinity of the Rechargeable Energy Storage System (RESS) and the fuel cell stack with voltage rating B, the symbol in Fig. 23.7 should be marked conspicuously.

When a live part of a circuit with a voltage rating B is exposed, the symbol should be conspicuously marked on the partition and the external box. If the partition and the external box are accessible or removable, the symbol should be shown without fail.

The symbol shall be drawn in a black on yellow background in accordance with ISO 3864-1 [10].

This warning mark should be located at a position on the energy storage system where it can be easily seen.

23.3.2 Basic Protective Measures

People should be protected from electric hazards caused by direct contact with a live part of an arbitrary electric circuit of which the voltage rating is B. The protective measures against direct contact should include one or more of the following:

- Basic insulation of the live part
- Prevention of access to the partition/external box and the live part.

23.3.3 Insulation Resistance Test

As shown in Table 23.8, before measurement, the equipment to be measured shall be left at a temperature of 5 ±2 °C, the preliminary condition range, at least for 8 h, after which it shall be left for 8 h at a temperature of 23 ± 5 °C, humidity of 90 ± 5%, and atmospheric pressure of 86–106 kPa, which are the condition ranges for arrival at the dew point.

The preliminary conditions and the main conditions can be optionally changed. After start of the main condition range, there is a short transition period where dew

Table 23.8 Pass/Fail criteria

Insulation type	Minimum transient insulation resistance (kΩ/V)	Minimum final insulation resistance (kΩ/V)
Equipment Type 1	0.1	1
Equipment Type 2	0.5	5

The resistance shall be measured based on the rated voltage of the relevant device

Table 23.9 Pass/Fail criteria

Test voltage		
Equipment type 1	Equipment type 2	
Basic insulation AC. V	Additional insulation AC. V	Double or reinforced insulation ACa.c. V
2U + 1000, Min. 1500 V	2U + 2250, Min. 2750 V	2U + 3250, min. 3750 V

U is the maximum operating voltage of the equipment

drops are generated. The insulation resistance shall be measured in this range. The insulation resistance shall be measured between the conductive part and the car chassis and between the conductive part and the added electric circuit.

23.3.4 Withstanding Voltage Test

As shown in Table 23.9, the withstanding voltage test shall be conducted after separating the traction battery and connecting a different electric circuit electrically to the chassis. AC voltage shall be applied between 50 Hz and 60 Hz for 1 min.

No dielectric breakdown or flashover should occur during or after the test.

23.3.5 Connecting Parts Connectivity Test (Equipotential Connection)

The current from the power source of which the no-load voltage does not exceed DC 60 V (greater current between 1.5 times the maximum current of the relevant power circuit or 25 A) shall be applied between two exposed conductive parts selected at random for 5 s at minimum. The voltage drop between the two exposed conductive parts selected as the objects of the test shall be measured. The resistance calculated from the relevant current and the voltage drop should not exceed 0.1 Ω.

[Pass/Fail Criterion]

- The resistance should not exceed 0.1 Ω.

23.3.6 Safety Standard for High-Power Electric Device of Motor Vehicle

The following should be fulfilled in accordance with Article 18-2 (High-Power Electric Device) of the Rules for Motor Vehicle Safety Standard.

※ Objects of application: All motor vehicles directly driven using one of the following power sources:

(1) Direct Current (DC): 60–1500 V
(2) Alternating Current (AC): 30–1000 V

- In order to ensure general structural safety of the wirelessly recharged electric vehicle power collector, the cabin should be isolated from the power collector and the system should have an excessive voltage/excessive current cutoff function.
- The insulation resistance standard of the wirelessly recharged electric vehicle power collector, a minimum of 500 Ω/V, shall be satisfied.

[Pass/Fail Criteria]

- Whether the cabin is isolated from the power collector and whether the system has excessive voltage/excessive cutoff function
- The insulation resistance standard of the power collector, a minimum of 500 Ω/V, shall be met.

23.4 Electromagnetic Compatibility Safety Issues for Power Supply System

23.4.1 Requirements for Electromagnetic Emission

(1) General Matters

The EMI in a frequency range between 0 Hz and 1 GHz is covered.

The requirements for electromagnetic emission are to ensure that the EMI generated by normal operation of the power supply system of the on-line electric vehicle does not reach a level where the intended operation of other equipment is interfered with.

Remark 1: The thresholds in this standard cannot provide perfect protective measures against radio disturbance caused by a power supply system of an on-line electric vehicle occurring when the receiving antenna of a radio set or TV set is used within a 10 m radius.
Remark 2: In a special case, for example, when equipment with high susceptibility is used at a close distance, additional reinforcement actions should be taken to reduce the electromagnetic emission below the specified level.

(2) Conducted Emission Threshold

As to the requirements for conducted emission of the on-line electric vehicle power supply system, IEC 61000-6-4 shall be complied with as follows [11]:

(a) Threshold of Interference Voltage at AC Power Input Port

The threshold of the power supply device of an on-line electric vehicle should not exceed the values in Tables 23.10 and 23.11. The power supply system of an on-line electric vehicle should satisfy both the mean threshold and the quasi-peak threshold when measured using an average detection receiver and a quasi-peak

Table 23.10 Interference voltage threshold of AC power input port at frequency of 0.15–30 MHz

Output current (A)	Frequency range (MHz)	Threshold [dB(μV)]	
		Quasi-peak	Average
>16–100	0.15 0.5[b]	100	90
	0.50–5.0[b]	86	76
	5.0–30.0	90–70[a]	80–60[a]
>100	0.15–0.5[b]	130	120
	0.50–5.0[b]	125	115
	5.0–30.0	115	105

[a]The threshold decreases linearly in accordance with an algebraic increase in the frequency
[b]A low threshold is applied to a band-edge frequency

Table 23.11 Interference voltage threshold of communication port at frequency between 0.15 and 30 MHz

Frequency range (MHz)	Quasi-peak	Average
0.15–0.50[a, b, c]	97 dB (μV)–87 dB (μV) 53 dB (μA)–43 dB (μA)	84 dB (μV)–74 dB (μV) 40 dB (μA)–30 dB (μA)
0.50–30[a, c]	87 dB (μV) 43 dB (μA)	74 dB (μV) 30 dB (μA)

[a]A lower threshold is applied to a band-edge (changing stage) frequency
[b]The threshold linearly decreases between 0.15 and 0.5 MHz in accordance with the algebraic change in the frequency
[c]The threshold for voltage and current interference shall be measured using an ISN made of 150 Ω common mode impedance (asymmetric mode) at the tested communication terminal (conversion factor is $20\ log_{10}150/I = 44dB$)

Table 23.12 Threshold of radiated emission at frequency between 30 and 1000 MHz

Frequency range (MHz)	Quasi-peak threshold, @10 m [dB(μV/m)]
30–230	40
230–1000	47

detection receiver, respectively. If the mean threshold is satisfied when measured using a quasi-peak detection receiver, the equipment under testing shall be regarded as having satisfied both thresholds. Measurement is not required using an average detection receiver. If the readings of the detection receiver fluctuate near the threshold, the record should be observed for at least 15 s at each measurement frequency. The highest values excluding fluctuating values that are far from the threshold and thus should be disregarded should be recorded.

(b) Threshold of Interference Voltage at AC Output

Unless the AC output terminal is connected to other equipment in a conductive manner, the threshold is not applied.

(c) Threshold of Communication Port

(3) Radiated Emission Threshold

As to the requirements for radiated emission of an on-line electric vehicle power supply system, IEC 61000-6-4 shall be complied with as follows [11]:

The threshold of an on-line electric vehicle power supply device should satisfy the value in Table 23.12. If the readings of the detection receiver fluctuate near the threshold, the record should be observed for at least 15 s at each measurement frequency. The highest values excluding fluctuating values that are far from the threshold and thus should be disregarded should be recorded.

The threshold shall not be applied to the radiated emission in the frequency band below 30 MHz.

23.4.2 Requirements for Electromagnetic Immunity

Regarding the requirements for electromagnetic immunity of an on-line electric vehicle power supply system, IEC 61000-6-2 shall be complied with as follows [12]:

(1) General Matters

The requirements for electromagnetic immunity at a frequency range of 0 Hz–1 GHz are covered.

These test requirements represent the basic EMC requirements for electromagnetic immunity. The test requirements are specified for each port that is considered.

The test should be conducted in a well defined and reproducible manner.

The tests should be conducted as individual and consecutive tests. The sequence of the tests can be selected.

A detailed explanation of the tests, proper testers to be used, and the test setup are given in the basic standard stated in the following tables.

The content of this basic standard shall not be repeated here. However, the corrections to be applied to the actual tests and additional information are provided in this standard.

The levels given in this section cannot satisfy the extreme cases of very low probability that can occur anywhere. In such cases, higher levels may be required.

(a) Pass/Fail Criteria for Performance

If the test results defined in this standard render the equipment dangerous or unsafe, the equipment shall be regarded as having failed the test.

The definition and functional explanation of the pass/fail criteria for performance to be used during an EMC test or for evaluation of the test results should be prepared by the manufacturer, and a test report should be prepared based on one of the following pass/fail criteria for each test, as specified in Tables 23.13, 23.14 and 23.15.

Table 23.13 Electromagnetic immunity requirements for external box port [14, 15, 18]

	Environmental phenomenon	Test condition	Unit	Basic standard	Remark	Pass/Fail criterion
1.1	Power frequency Electromagnetic field	50, 60 30	Hz A/m	IEC61000-4-8	The test should be conducted at the frequency of the electric power supplied[a]	A
1.2	Radio frequency Electromagnetic field Amplitude modulation	80–1 000 10 80	MHz V/m % AM (1 kHz)	IEC61000-4-3	The specified test level is the r.m.s. value of the unmodulated carrier wave[b]	A
1.3	Radio frequency Electromagnetic field Amplitude modulation	1.4–2.0 3 80	GHz V/m % AM (1 kHz)	IEC61000-4-3	The specified test level is the r.m.s. value of the unmodulated carrier wave	A
1.4	Radio frequency Electromagnetic field Amplitude modulation	2.0–2.7 1 80	GHz V/m % AM (1 kHz)	IEC61000-4-3	The specified test level is the r.m.s. value of the unmodulated carrier wave	A

(continued)

Table 23.13 (continued)

	Environmental phenomenon		Test condition	Unit	Basic standard	Remark	Pass/Fail criterion
1.5	Electrostatic discharge	Contact-discharge	±4 (Charging voltage)	kV	IEC61000-4-2	The basic standard shall be referred to when making the decision on test methods, such as contact discharge or aerial discharge	B
		Aerial discharge	±8 (Charging voltage)	kV			B

[a]This applies only to equipment containing a device susceptible to a magnetic field
[b]The level is 3 V/m excluding the ITU broadcasting frequency bands 87–108, 174–230, and 470–790 MHz

Table 23.14 Electromagnetic immunity requirements for AC power input port [16, 17, 21]

	Environmental phenomenon	Test condition	Unit	Basic standard	Remark	Pass/Fail criterion
2.1	Radio frequency common mode	0.15–80 10 80	MHz V % AM (1 kHz)	IEC61000-4-6	The specified test level is the r.m.s. value of the unmodulated carrier wave[a, b]	A
2.2	Surge	1.2/50 (8/20)	Tr/Th us	IEC61000-4-5		B
	Line and ground	±2	kV (Open circuit test voltage)			
	Between lines	±1	kV (Open circuit test voltage)			
2.3	Fast transient	± 2	kV (Open circuit test voltage)	IEC61000-4-4		B
	Phenomenon	5/50	Tr/Th ns			
		5	Repetition frequency kHz			

[a]The test level can be defined as the equivalent current for 150 Ω load
[b]The level is 3 V excluding the ITU broadcasting frequency band 47–68 MHz

① Performance Pass/Fail Criterion A

The equipment should continuously operate according to the intended function during and after the test. When the equipment is used normally, deterioration of performance or loss of function below the performance level specified by the manufacturer is not allowed. The performance level may be replaced by the

Table 23.15 Electromagnetic immunity requirements for signal line/control line ports [16, 17, 21]

	Environmental phenomenon	Test condition	Unit	Basic standard	Remark	Pass/fail criterion
3.1	Radio frequency common mode	0.15–80 3 80	MHz V % AM (1 kHz)	IEC61000-4-6	The specified test level is the r.m.s. value of the unmodulated carrier wave[a, b, c]	A
3.2	Fast transient	±0.5	kV (Open circuit test voltage)	IEC61000-4-4	A capacitive clamp is used[c]	B
	Phenomenon	5/50	Tr/Th ns			
		5	Repetition frequency kHz			
3.3	Surge	1,2/50 (8/20)	Tr/Th us	IEC61000-4-5	d, e	B
	Line and ground	±1	kV(Open circuit test voltage)			

[a]The test level can be defined as the equivalent current for the 150 Ω load
[b]The level is 3 V excluding the ITU broadcasting frequency band 47–68 MHz
[c]It can be applied only to the terminals connected to a cable of which the total length is 3 meters or longer in accordance with the functional conditions of the manufacturer
[d]It can be applied only to the terminals connected to a cable of which the total length is 30 meters or longer in accordance with the functional conditions of the manufacturer
[e]If a normal function of the equipment under testing does not operate due to the effect of coupler/decoupler network(CDN), this test is unnecessary

allowable performance loss. If the minimum performance level or the allowable performance deterioration is not designated by the manufacturer, the information can be obtained from the product specifications and manual; otherwise, if the equipment is used normally, it can be reasonably predicted by the user.

② Performance Pass/Fail Criterion B

The equipment should continuously operate normally after the test. When the equipment is used normally, deterioration of performance or loss of function below the performance level specified by the manufacturer is not allowed. The performance level may be replaced by the allowable performance loss. However, performance deterioration is allowed during the test. Change in the actual operation state or the stored data is not allowed. If the minimum performance level or the allowable performance deterioration is not designated by the manufacturer, the information can be obtained from the product specifications and manual; otherwise, if the equipment is used normally, it can be reasonably predicted by the user.

③ Performance Pass/Fail Criterion C

If the functions can be recovered through self-recovery ability or by a control operation, temporary loss of function is allowed.

(b) Operation Conditions during Test

The equipment under testing should be tested in the most susceptible operation mode expected by conducting limited advance tests and other factors. This operation mode should have consistency through normal application. The setup of the equipment under testing should be changed so that the equipment will be in the most susceptible conditions through practical application and the installation state.

If the specifications of the manufacturer require an external protective device or measurement clearly stated in the user manual, the test required in this standard should be conducted in the relevant place or by applying an external protective device.

The test setup and the operation mode during the test should be accurately described in the test report. It is not always possible to test all the functions of the equipment. In such a case, the most decisive operation mode should be selected.

If the equipment has terminals used for similar connections or many similar terminals, a sufficient number of terminals should be selected simulating the actual operation conditions, and it should be verified that terminations of all different forms have been achieved.

Unless otherwise instructed in this standard, the test should be conducted at the specified temperature, humidity, pressure, and operation range, and at the rated voltage for the product.

(2) Basic Requirements for Electromagnetic Immunity

The application of tests for evaluation of immunity depends on the particular equipment, its configuration, its ports, its technology and its operating conditions. Tests shall be applied to the relevant ports of the equipment. Tests shall only be carried out where the relevant ports exist.

23.4.3 Electromagnetic Emission Measurement Method

(1) General Matters

Electromagnetic emission shall be measured in accordance with Comité International Spécial des Perturbations Radioélectriques (CISPR 11). The following exceptions shall be preferentially applied [13].

(2) Conducted Emission Measurement Method

Measurement of the conducted emission shall be done in an electromagnetically shielded room in principle. If site measurement is required, the measurement should be made considering the measurement conditions of CISPR 11 [13].

(3) Radiated Emission Measurement Method

Measurement of radiated emission shall be done in an OATS (Open Area Test Site) or a SAC (Semi Anechoic Chamber) in principle. If site measurement is required, the measurement should be made considering the measurement conditions of CISPR 11 [13].

23.4.4 Electromagnetic Immunity Test Method

(1) General Matters

An electromagnetic immunity test shall be conducted basically in accordance with the basic standard specified in IEC 61000-6-4 [11].

(2) Electromagnetic Immunity Test Method

(a) Test Method for Immunity against Electrostatic Discharge

The test shall be conducted in accordance with IEC 61000-4-2 [14].

(b) Test Method for Immunity against Radiated Emission

The test shall be conducted in accordance with IEC 61000-4-3 [15].

(c) Test Method for Immunity against EFT/Burst

The test shall be conducted in accordance with IEC 61000-4-4 [16].

(d) Test Method for Immunity against Surge

The test shall be conducted in accordance with IEC 61000-4-5 [17].

(e) Test Method for Immunity against Power Frequency Magnetic Field

The test shall be conducted in accordance with IEC 61000-4-8 [18].

23.5 EMC Safety Issues for Power Pickup System

23.5.1 Requirements for Electromagnetic Emission

(1) General Matters

The EMI in a frequency range of 0 Hz–1 GHz is covered.

The requirements for electromagnetic emission are to ensure that the EMI generated by normal operation of the power supply system of an on-line electric vehicle does not reach a level where the intended operation of other equipment is interfered with.

(2) Radiated Emission Threshold

(a) Electromagnetic Wave Emission Standard

Table 23.16 Peak and quasi-peak values for radiated disturbance—method of using an electromagnetically shielded room equipped with an electromagnetic wave absorber

Service/band	Frequency (MHz)	Level [Unit: Db(μV/m)]									
		Rating 1		Rating 2		Rating 3		Rating 4		Rating 5	
		Peak value	Quasi-peak value	Peak value	Quasi-peak value	Peak value	Quasi-peak value	Peak value	Quasi-peak value	Peak value	Quasi-peak value
Broadcasting											
Long wave	0.15–0.30	86	73	86	63	66	53	56	43	46	33
Medium wave	0.53–1.8	72	59	64	51	56	43	48	35	40	27
Short wave	5.9–6.2	64	51	58	45	52	39	46	33	40	27
FM	76–108	62	49	56	43	50	37	44	31	38	25
TV band I	41–88	52	–	46	–	40	–	34	–	28	–
TV band III	174–230	56	–	50	–	44	–	38	–	32	–
Digital audio broadcastingIII	171–245	50	–	44	–	38	–	32	–	26	–
TV band IV	468–944	65	–	59	–	53	–	47	–	41	–
Digital terrestrial TV broadcasting	470–770	69	–	63	–	57	–	51	–	45	–
Digital audio broadcasting Lband	1447–1494	52	–	46	–	40	–	34	–	28	–
Digital satellite radio	2320–2345	58	–	52	–	46	–	40	–	34	–
Mobile communication service											
Civil band	26–28	64	51	58	45	52	39	46	33	40	27
VHF	30–54	64	51	58	45	52	39	46	33	40	27
VHF	68–87	59	46	53	40	47	34	41	28	35	22
VHF	142–175	59	46	53	40	47	34	41	28	35	22
Analog UHF	380–512	62	49	56	43	50	37	44	31	38	25

(continued)

Table 23.16 (continued)

Service/band	Frequency (MHz)	Level [Unit: Db(μV/m)]									
		Rating 1		Rating 2		Rating 3		Rating 4		Rating 5	
		Peak value	Quasi-peak value	Peak value	Quasi-peak value	Peak value	Quasi-peak value	Peak value	Quasi-peak value	Peak value	Quasi-peak value
RKE wireless key	300–330	56	–	50	–	44	–	38	–	32	–
RKE wireless key	420–450	56	–	50	–	44	–	38	–	32	31
Analog UHF	820–960	68	55	62	49	56	43	50	37	44	–
GSM800	860–895	68	–	62	–	56	–	50	–	44	–
EGSM/GSM900	925–960	68	–	62	–	56	–	50	–	44	–
Civil GPS L1	1567–1583	–	–		–		–	–	–		–
GSM1800 (PCN Personal communication network)	1803–1882	68	–	62	–	56	–	50	–	44	–
GSM1900	1850–1990	68	–	62	–	56	–	50	–	44	–
3G/IMT 2000	1900–1992	68	–	62	–	56	–	50	–	44	–
3G/IMT 2000	2010–2025	68	–	62	–	56	–	50	–	44	–
3G/IMT 2000	2108–2172	68	–	62	–	56	–	50	–	44	–
Bluetooth/802.11	2400–2500	68	–	62	–	56	–	50	–	44	–

Table 23.17 Mean threshold for radiated disturbance—method of using an electromagnetically shielded room equipped with an electromagnetic wave absorber

Service/band	Frequency (MHz)	Level [Unit: Db(μ*V*/m)]				
		Rating 1	Rating 2	Rating 3	Rating 4	Rating 5
		Average	Average	Average	Average	Average
Broadcasting						
Long wave	0.15–0.30	66	56	46	36	26
Medium wave	0.53–1.8	52	44	36	28	20
Short wave	5.9–6.2	44	38	32	26	20
FM	76–108	42	36	30	24	18
TV band I	41–88	42	36	30	24	18
TV band III	174–230	46	40	34	28	22
Digital audio broadcasting III	171–245	40	34	28	22	16
TV band IV	468–944	55	49	43	37	31
Digital terrestrial TV broadcasting	470–770	59	53	47	41	35
Digital audio broadcasting Lband	1447–1494	42	36	30	24	18
Digital satellite radio	2320–2345	48	42	36	30	24
Mobile communication service						
Civil band	26–28	44	38	32	26	20
VHF	30–54	44	38	32	26	20
VHF	68–87	39	33	27	21	15
VHF	142–175	39	33	27	21	15
Analog UHF	380–512	42	36	30	24	18
RKE wireless key	300–330	42	36	30	24	18
RKE wireless key	420–450	42	36	30	24	18
Analog UHF	820–960	48	42	36	30	24
GSM800	860–895	48	42	36	30	24
EGSM/GSM900	925–960	48	42	36	30	24
Civil GPS L1	1567–1583	34	28	22	16	10
GSM1800 (PCN Personal communication network)	1803–1882	48	42	36	30	24
GSM1900	1850–1990	48	42	36	30	24
3G/IMT 2000	1900–1992	48	42	36	30	24
3G/IMT 2000	2010–2025	48	42	36	30	24
3G/IMT 2000	2108–2172	48	42	36	30	24
Bluetooth/802.11	2400–2500	48	42	36	30	24

The standard of CISPR 25 shall be followed, and the peak value, quasi-peak value, and mean value measured in an electromagnetic shield room attached with an electromagnetic wave absorber should meet the requirements in Tables 23.16 and 23.17 [19].

23.5.2 Requirements for Electromagnetic Immunity

(1) General Matters

The requirements for electromagnetic immunity at a frequency range of 0 Hz–1 GHz are covered.

These test requirements represent the basic EMC requirements for electromagnetic immunity. The test requirements are specified for each port considered.

The test should be conducted in a well defined and reproducible manner.

The tests should be conducted as individual and consecutive tests. The sequence of the tests can be selected.

A detailed explanation of the tests, proper testers to be used, and the test setup are given in the basic standard stated in the following tables.

The content of this basic standard shall not be repeated here. However, corrections to be applied to the actual tests and additional information are provided in this standard.

The levels given in this section cannot satisfy the extreme cases of very low probability that can occur anywhere. In such cases, higher levels may be required.

(a) Pass/Fail Criteria for Performance

Various types and diversity of the equipment within the application scope of this standard make it difficult to define accurate pass/fail criteria for evaluation of the immunity test results.

If the test results defined in this standard make the equipment dangerous or unsafe, the equipment shall be regarded as having failed the test.

The definition and functional explanation of the pass/fail criteria for the performance to be used during an EMC test or for evaluation of the test results should be prepared by the manufacturer, and a test report should be prepared based on one of the following pass/fail criteria for each test as specified in the table.

① Performance Pass/Fail Criterion A

The equipment should continuously operate normally during and after the test. When the equipment is used normally, deterioration of performance or loss of function below the performance level specified by the manufacturer is not allowed. The performance level may be replaced by the allowable performance loss. If the minimum performance level or the allowable performance deterioration is not designated by the manufacturer, the information can be obtained from the product specifications and manual; otherwise, if the equipment is used normally, it can be reasonably predicted by the user.

② Performance Pass/Fail Criterion B

The equipment should continuously operate normally after the test. When the equipment is used normally, deterioration of performance or loss of function below the performance level specified by the manufacturer is not allowed. The performance level may be replaced by the allowable performance loss. However, performance deterioration is allowed during the test. Change in the actual operation state or the stored data is not allowed. If the minimum performance level or the allowable performance deterioration is not designated by the manufacturer, the information can be obtained from the product specifications and manual; otherwise, if the equipment is used normally, it can be reasonably predicted by the user.

③ Performance Pass/Fail Criterion C

If the functions can be recovered through self-recovery or by a control operation, temporary loss of function is allowed.

(b) Operation Condition during Test

The equipment under testing should be tested in the most susceptible operation mode expected by conducting limited advance tests and other factors. This operation mode should have consistency through normal application. The setup of the equipment under testing should be changed so that the equipment can be operated in the most susceptible conditions through practical application and the installation state.

If the specifications of the manufacturer require an external protective device or measurement clearly stated in the user manual, the test required in this standard should be conducted in the relevant place or by applying an external protective device.

The test setup and the operation mode during the test should be accurately described in the test report. It is not always possible to test all functions of the equipment. In this case, the most decisive operation mode should be selected.

Unless otherwise instructed in this standard, the test should be conducted at the specified temperature, humidity, pressure, and operation range, and at the rated voltage for the product.

(2) Basic Requirements for Immunity

The basic immunity requirements for external box are shown in Table 23.18

Table 23.18 Immunity requirements for external box port [14, 20]

Test item	Allowable level	Criterion
Immunity against electromagnetic radiation (ISO 11452-2)	80–1000 MHz, 24 V/m	A
Immunity against electrostatic discharge (IEC 61000-4-2)	Contact discharge (± 4 kV), Aerial discharge (±8 kV)	B

23.5.3 Electromagnetic Emission Measurement Method

(1) General Matters

Electromagnetic emission shall be measured in accordance with CISPR 25 [19]. The following exceptions shall be preferentially applied.

(2) Radiated Emission Measurement Method

Measurement of radiated emission shall be done in an OATS (Open Area Test Site) or a SAC (Semi Anechoic Chamber) in principle. If site measurement is required, the measurement should be made considering the measurement conditions of CISPR 11 [13].

23.5.4 Electromagnetic Immunity Test Method

(1) Test Method for Immunity against Electrostatic Discharge

The test shall be conducted in accordance with IEC 61000-4-2 [14].

(2) Test Method for Immunity against Radiated Emission

The test shall be conducted in accordance with ISO 11452-2 and IEC 61000-4-2 [14, 20].

23.6 Conclusions

In OLEV project, there have been many important regulatory safety issues in the area of electric high power, electric bus with pickup module, and electromagnetic compatibility.

To verify high power safety of road-embedded power cable module in case of using 20 kHz frequency, we suggested test environment, test metric, and procedure because there is no IEC safety standard. And road-embedded power cable module was certified to be safe. In addition, load test of road-embedded power cable module embedded in the road was made in view of compressive strength test and flexural strength test, and our road-embedded power cable module was proven to be safe.

Also, to support the safety of electric bus with pickup module, we proposed safety certification standard. And, our OLEV bus system was certified to be safe by certification aurhority.

Moreover, to prove that there are no EMF, EMI, and EMC problems, electric bus with wireless charging and road-embedded power cable module were tested based on international regulatory standard in view of safety and were certified to be safe. Especially, to acquire c certification, we used optimum structure and power variables for road-embedded power cable module and pickup module.

References

1. IEC 61032 (2005) Protection of persons and equipment by enclosures—probes for verification, 9 Dec 2005
2. IEC 60529 (2002) Degrees of protection provided by enclosures (IP Code), 30 Sep 2002
3. IEC 60664-1 (2002) Insulation coordination for equipment within low-voltage systems—part 1: principles, requirements and tests, 30 Nov 2002
4. IEC 60146-1-1 (2002) Semiconductor convertors General requirements and line commutated convertors part 1-1: specifications of basic requirements, 18 Jun 2002
5. IEC 61800-2 (2002) Adjustable speed electrical power drive systems—part 2: general requirements—Rating specifications for low voltage adjustable frequency A.C. power drive systems, 29 Oct 2002
6. IEC 61558-1 (2002) Safety of power transformers, power supplies, reactors and similar products—part 1: general requirements and tests, 31 Jul 2002
7. IEC 61000-4-7 (2004) Part 4: testing and measurement techniques—section 7: general guide on harmonics and interharmonics measurements and instrumentation, for power supply systems and equipment connected thereto, 11 Aug 2004
8. KS F 2403/ KS F 2405 (revised from ISO 1920:1976) (1965) Standard test method for making and curing concrete specimens, 10 Apr. 1965
9. KS F 2408(ISO 4013:1978) (1968) Standard test method for flexural strength of concrete, 10 Jul 1968
10. ISO 3864-1 (2002) Graphical symbols—safety colours and safety symbols—part 1: design principles for safety symbols and safety markings, 31 Dec 2002
11. IEC 61000-6-4 (2002) Electromagnetic compatibility (EMC)—part 6-4: generic standards—emission standard for industrial environments, 26 Oct 2002
12. IEC 61000-6-2 (2002) Electromagnetic compatibility (EMC)—part 6-2: generic standards: immunity standard for industrial environments, 26 Oct 2002
13. CISPR 11 (2002) Electromagnetic compatibility (EMC)—industrial, scientific and medical equipment—radio-frequency disturbance characteristics—limits and methods of measurement, 26 Oct 2002
14. IEC 61000-4-2 (1995) Electromagnetic compatibility (EMC)—part 4: testing and measure-ment techniques—section 2: electrostatic discharge immunity test, 29 Dec 1995
15. IEC 61000-4-3 (2003) Electromagnetic compatibility (EMC)—part 4: testing and measure ment techniques—section 3: radiated radio-frequency, electromagnetic field immunity test, 30 Aug 2003
16. IEC 61000-4-4 (2003) Electromagnetic compatibility (EMC)—part 4: testing and measurement techniques—section 4: electrical fast transient/burst immunity test, 30 Aug 2003
17. IEC 61000-4-5 (2003) Electromagnetic compatibility (EMC)—part 4: testing and measure ment techniques—section 5: surge immunity test, 30 Aug 2003
18. IEC 61000-4-8 (2003) Electromagnetic compatibility (EMC)—part 4: test and measurement method—section 8: power frequency magnetic field immunity test, 30 Aug 2003
19. CISPR 25 (2002) Limits and methods of measurement of radio disturbance characteristics for the protection of receivers used on board vehicles, 27 Jul 2002
20. ISO 11452-2 (2002) Road vehicles—electrical disturbances from narrowband radiated electromagnetic energy—component test methods—part 2: absorber-lined shielded enclosure, 22 May 2002
21. IEC 61000-4-6 (2003) Electromagnetic compatibility (EMC)—part 4: testing and measure-ment techniques—section 6: immunity to conducted disturbances, induced by radio-frequency fields, 30 Aug 2003

Chapter 24
Energy Revolution: Journey towards a Greener Planet

Kon Fah Loh

Abstract Alarming rise in the level of anthropogenic CO_2 and increase in temperature of Earth are the two most symptomatic evidence of climate change. The approach to combat this impending catastrophe should be multi-pronged, requiring technological, social, economic and even political forces to work in harmony to protect our planet. This chapter focuses on the technology tools that can potentially replace the traditional energy sources and addresses the root case of harmful GHG emissions that created global warming in the first place. It also quantifies the potential benefits of the new approach and charts out a clear path towards creating a greener planet.

24.1 Global Energy Consumption

In Chap. 1, the overall energy consumption of the world and the source of green house gases (GHG) were presented. The use of petroleum to power internal combustion (IC) engines and the use of coal for electricity generation were two primary anthropogenic emission of CO_2. The use of renewable energy, solar and wind in particular, to generate electricity to power electric automobiles and replace coal-burning electric power plants was emphasized. Many other chapters of this book were devoted on the On-Line Electric Vehicle (OLEV). This chapter discusses specific means of replacing the traditional energy sources in the twenty-first century.

Global energy consumption and economic growth go hand in hand. Advanced industrialized countries use more energy per unit of economic output and per capita than poorer countries. Our world's energy demand will continue to grow in highly populated countries such as China and India (Fig. 24.1).

K.F. Loh (✉)
Independent Environmental Evangelist, 10, Jalan Tun Dr. Ismail,
30350 Ipoh, Perak, Malaysia
e-mail: konfahloh@yahoo.com

N.P. Suh and D.H. Cho (eds.), *The On-line Electric Vehicle*,
DOI 10.1007/978-3-319-51183-2_24

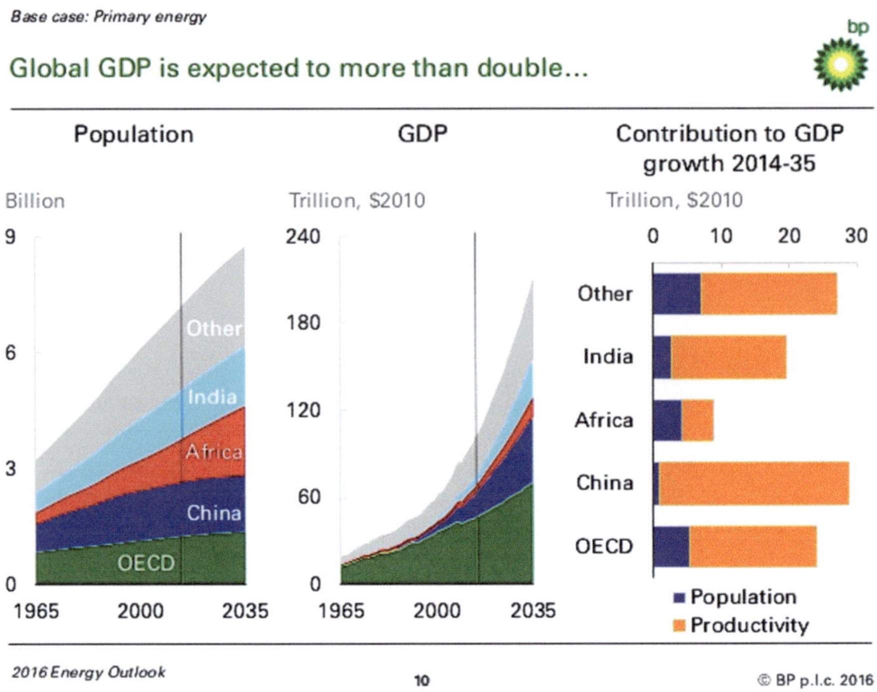

Fig. 24.1 World primary energy driven by strong growth in emerging Asia (BP)

According to International Energy Outlook 2016 (IE02016) study, by the US Energy Information Administration (EIA), the total global energy consumption is set to rise 48% from 549 quadrillion British Thermal Units (BTU) in 2012 to 815 quadrillion BTU in 2040. Fossil fuels will still account for 78% of world energy use in 2040 [1] India) will account for more than half of the increase. China will be central to the demand growth due to the underlying rise of oil demand and its build-up of strategic petroleum reserves, which will reach at least 500 million barrels by 2020 (Fig. 24.2).

As stated in Chap. 1, climate scientists have observed that carbon dioxide (CO_2) concentrations in the atmosphere have been rising significantly over the past century, compared to the pre-industrial era of about 280 parts per million (ppm) by volume. In 2014, the concentration was 397 ppm, an increase of about 40% higher than in mid-1800s [2]. The big percentage increase in CO_2 is largely responsible for global warming. Among the many human activities that produce greenhouse gases transportation, production of electricity and energy for cement and steel production contribute to maximum CO_2 emission (Fig. 24.3).

The International Energy Agency (IEA) preliminary data suggest that electricity generated by renewables plays a critical role in stemming the rise in CO_2 emissions, having accounted for about 90% of new electricity generation in 2015; wind farm alone produced more than half of new electricity generation. Global power

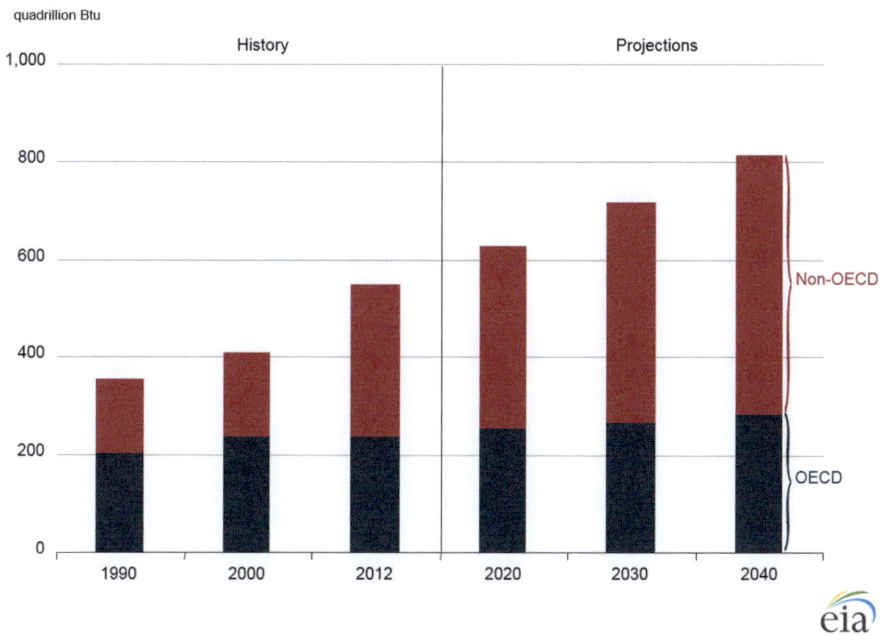

Fig. 24.2 Shows world energy consumption and outlook, 1990–2040 (*Source* EIA)

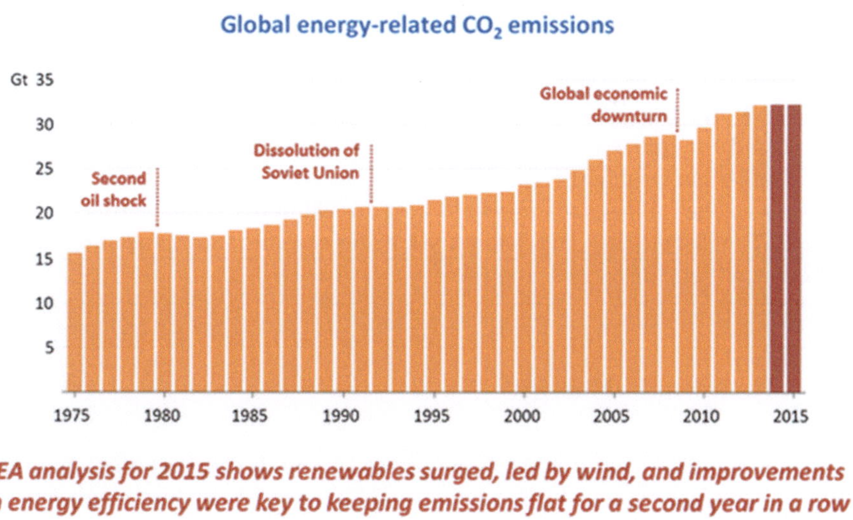

Fig. 24.3 IEA analysis for 2015 shows renewables surged, led by wind and improvement in energy efficiency were key to keeping emissions flat for a second year in a row

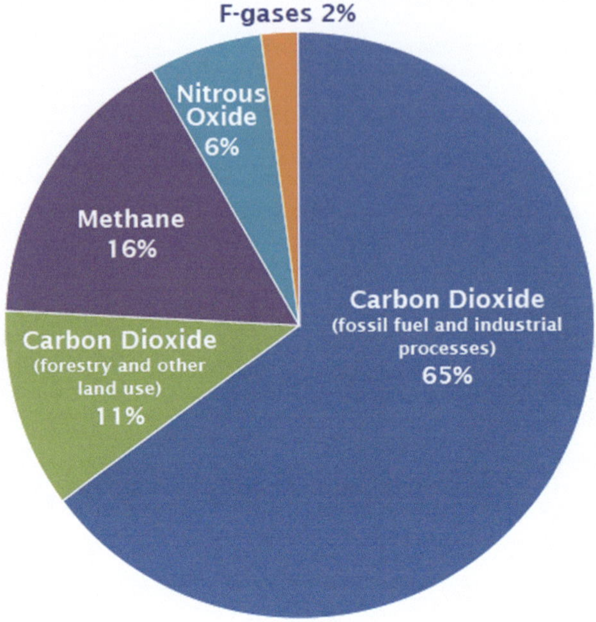

Fig. 24.4 Shows GHG emissions, and majority is made up of CO_2

generation is set to shift away from coal, the share of which will fall from 40% at the present to 30% in 2040.

IEA's World Energy Outlook 2011 report estimated the potential for Carbon Capture and Storage (CCS) to contribute 22% of global CO_2 mitigation through 2035. Further analysis by IEA in the Energy Technology Perspectives 2010 report also showed that climate action would cost an additional US$4.7 trillion without CCS. CCS will be costly and has not yet being commercially proven on an integrated basis (Figs. 24.4, 24.5 and 24.6).

24.2 Renewable Energy

Phasing out nuclear and fossil fuels entirely is still a revolutionary idea, and many energy experts are sceptical. However, more and more scientists, engineers and activists actively promote a 100% renewable energy vision. According to the IPCC's latest assessment report, we have already used up almost 2/3 of our carbon budget, if we are to have a reasonable chance of limiting global mean temperature rise to no more than 2 °C. At the current and projected rate of consumption,

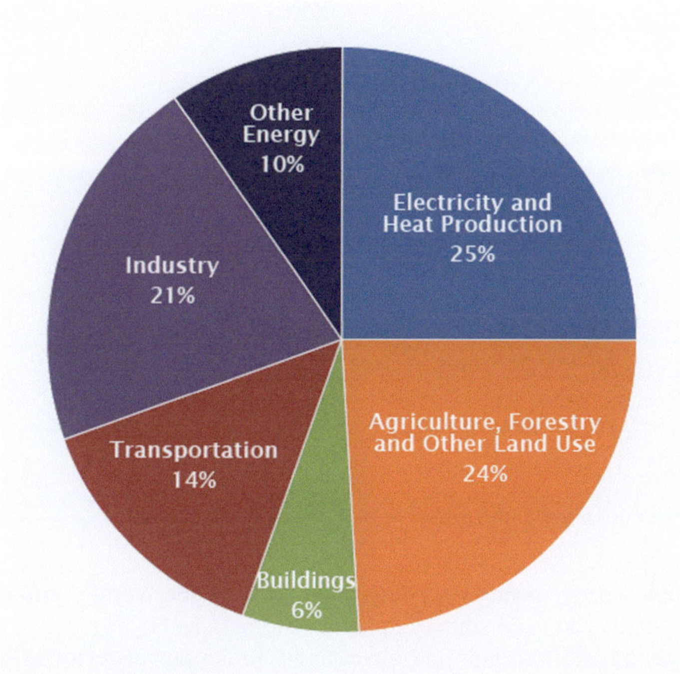

Fig. 24.5 Shows world's CO_2 emissions by sector in 2013

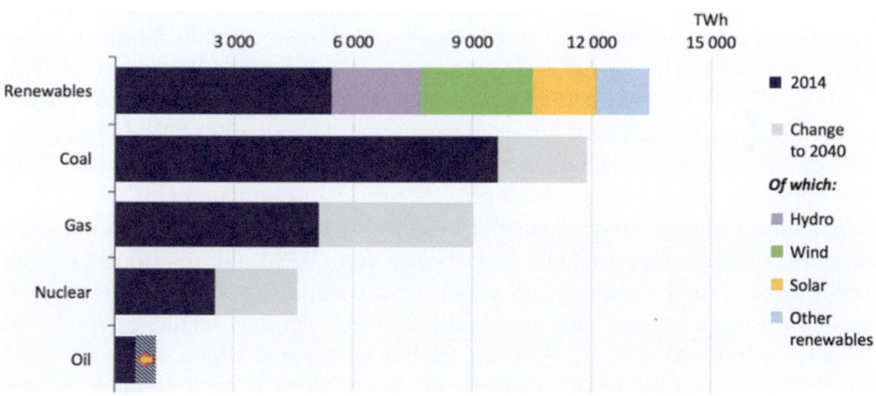

Fig. 24.6 Global electricity generation by source in 2014 (terawatt hours, dark blue bars) and change to 2040 (*grey*, *lilac*, *green*, *yellow* and *light blue bars*). *Source* IEA World Energy Outlook

this entire budget will be used up by 2040. So it is essential that we move rapidly towards a cleaner form of energy supply—one that delivers 100% renewable energy by 2050.

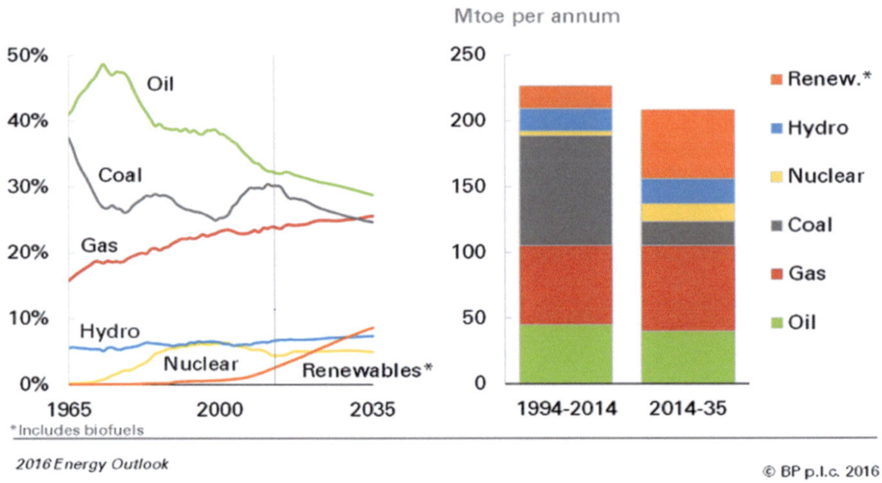

Fig. 24.7 Shares of primary energy, fossil fuels (%) used (BP Outlook)

Renewable energy sources accounted for 12% of the world's primary energy demand in 2012. The main sources today are biomass and hydro, which are mostly used for heating and transport and increasingly in the power sector as well. More than one billion people or 15% of the global population still lack access to a power grid, and more than 2.6 billion people rely on traditional biomass for cooking and heating, with the related negative health impacts.

Solar energy is the most abundant permanent energy resource on earth, and it is available for the use in its direct (solar radiation) and indirect (wind, biomass, hydro, ocean, etc.) forms. The sun emits energy at a rate of 3.8×10^{23} kW per second. Even if only 0.1% of this energy could be converted at an efficiency of only 10%, it would be four times the world's total generating capacity of about 3000 GW. Looking at it another way, the total annual solar radiation falling on the earth is more than 7500 times the world's total annual primary energy consumption [3].

Wind is a complex energy source and estimating the global wind resource size is extremely difficult. One estimate (Cole 1992) from 1992, put total onshore wind energy to be around a million GW of potential generating capacity: if only 1% of the area was utilized, and allowance made for the lower load factors of wind plant (15–40%, compared with 75–85% for thermal plant) that would still correspond, roughly to the total worldwide capacity of all electricity-generating plant in operation. The offshore wind resource is also vast, with European resources, for example, capable of supplying all the European Union's electricity needs, without going further than 30 km offshore (Fig. 24.7).

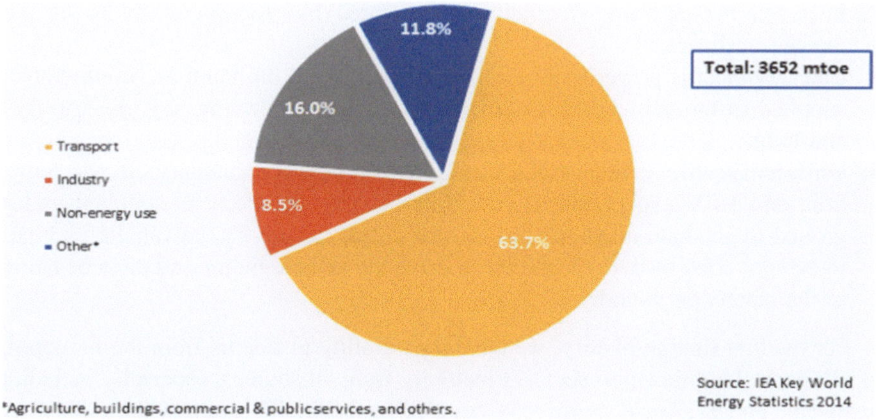

Fig. 24.8 Global crude oil consumption in 2012 by sector (*Source* IEA 2014)

24.3 Fossil Fuels

In 2015, the world consumed about 94 million barrels of crude oil per day (b/d). About 63.7% of that (see chart 4), or 60 million b/d, was used for transportation (Fig. 24.8).

24.4 Energy Efficiency

Energy efficiency is important as it helps to decrease the use of primary energy resources and achieve considerable savings. The implementation of OLEV technology will drastically reduce the use of fossil fuels and CO_2 emissions in the land transport sector. It can completely replace it with renewable energy, and the implication is huge on the environment and climate warming.

Oil has the monopoly in the land transport sector and commanded a premium over other fossil fuels. Oil has the greatest energy density, that is, oil gives us the greatest amount of energy per unit volume or per unit weight.

Almost all land transport is based on internal combustion engine (ICE) except for tram and train, which are connected to supply of electricity as power train. ICE vehicles are very inefficient. The efficiency from well to wheel is between 20%, whereas if the oil is converted to electricity using the combined cycle system, the efficiency is 60% plus some 'waste heat' which can be used for heating.

Linking electric cars to a renewable energy grid is the best possible option to reduce emissions from the transport sector. The electricity may be fed into the electric motor via several means both on or off road.

1. Battery electric (BEV) power train: electricity is generated offload and is stored on board.

2. Fuel cell power train: electricity is generated by reaction of hydrogen and oxygen on board.
3. Catenary electric power train: electricity is generated off board and continuously supplied to the vehicle via external catenary during driving, e.g. electric train and tram.
4. On-Line Electric Vehicle (OLEV) is a hybrid between catenary electric power train and BEV except that it is 'cordless'. The electricity is supplied underground to generate magnetic flux, which is 'picked up' by the vehicle on board to be converted back to electricity to drive the vehicle motor and or to be stored in the battery on board.

For the first time in history, we have a possibility to free us from the monopoly of oil for the land transport sector. Electricity from all sources, especially including renewable energy, could be directly coupled to the OLEV through the grid.

24.5 Reduction in Energy and CO_2 Due to Efficiency

The Kyoto Protocol calls for a short-term action and set out a clear long-term pathway. The long-term goal is to phase out of fossil fuels and nuclear power by 2050 through a just transition towards 100% renewable energy, thereby protecting and restoring forests (Table 24.1).

Table 24.1 Development of efficiency and investment costs for selected new power plant technologies exemplary data for OECD Europe

	2010	2020	2030	2040	2050
Coal fired condensing power plant					
Efficiency (%)	43	44	45	47	49
Investment costs ($2012/кW)	2000	2000	2000	2000	2000
Lignite fired condensing power plant					
Efficiency (%)	41	43	44	45	45
Investment costs ($2012/kW)	2200	2200	2200	2200	2200
Gas turbine					
Efficiency (%)	39	40	40	41	42
Investment costs ($201 2/KW)	500	500	500	500	500
Gas fired combined cycle power plant					
Efficiency (%)	60	61	62	63	64
Investment costs ($2012/lM)	1000	1000	1000	1000	1000
Gas fired combined cycle CHP plant					
Efficiency (%)	83	84	85	86	88
Investment cocts ($2012 kW)	1300	1300	1300	1300	1300

Source Based on WEO 2014 and own assumptions note CO_2 emissions refer to power station outputs only; life-cycle emissions are not considered

As stated above, the efficiency has improved tremendously from 20% to 60%. Thus, we need to consume half the amount of oil if we are to use oil to generate electricity. The reduction is about 30 million barrels per day or 10,950 million barrels per year. This works out to be about US$ 493 billion savings per year (based on US$45/barrel) on a worldwide basis.

Fossil fuels are sold on different units such as barrel, tons or volume. The price comparison based on the common denominator of British Thermal Units (BTU) or per Giga Joule (GJ) is shown below.

Energy prices per million BTU [4]

- Coal—Powder River Basin—$0.56
- Coal—Northern Appalachia—$2.08
- Natural gas—$5.69
- Ethanol tax credit—$5.92
- Propane—$13.28
- Petroleum—$13.43
- #2 Heating oil—$14.74
- Jet fuel—$15.48
- Diesel—$15.59
- Gasoline—$17.81
- Wood pellets—$18.57
- Corn ethanol—$23.46
- Electricity—$26.31
- Cellulosic ethanol from corn cobs—$30.92

Furthermore, if instead of using oil, we replace it with gas to generate electricity, there are additional savings as gas price is about 2–4 times cheaper than oil. In 2012, the price of crude oil in North America was almost ten times more than the natural gas in energy equivalent terms. Therefore, there are further savings of at least 60%. This is translated to be about $300 billion per year. The total savings are about $800 billion per year!

(Note: the price of oil and gas depends on supply and demand like other commodities. It is greatly influenced by weather and geopolitical situation in the oil producing countries. Since there is a cartel (OPEC), the price remains high over the years until in recent year where the gap between supply and demand is getting greater. With the introduction of OLEV, the 'black gold' will lose its shine. Over time, the price will ease and only the cheapest cost will have a chance to survive.)

The emission of carbon dioxide would be reduced by about 70% by switching from ICE to OLEV using gas for power generation to electricity.

24.6 Insurance for Future Generation from Saving Energy

With such huge amount of potential savings, we have a realistic opportunity to solve some of the ills in the world, such as climate warming, carbon sequestration to remove some of the CO_2 in the air, increase renewable energy gradually but surely so that we will be fully dependable on renewable energy for the future. The net positive impact will be increase in water supply ensuring food security for the world's growing population and providing jobs and alleviating a billion of people out of extreme poverty. With this goal, we propose the following:

Assuming worldwide the savings is $800 billion per year, we should invest the saved fund as follows:

1. 25% to offset with carbon sequestration in reforestation and afforestation allocated to the third world, especially those within the tropics and semi-tropics where carbon sequestration per given area is high. Instead of contributing to carbon dioxide emission, it could reverse the effect by carbon sequestration. The net result is more carbon sequestrated (fixed) than carbon emitted.
2. 25% to subsidize the installation of renewable energy such as but not limited to wind and solar. This is to slow down and eventually stop carbon dioxide emission by replacement with renewable source.
3. 40% to divert water from plentiful to places of scarcity within the country.
4. 10% to assist poor countries to divert water for food and other vegetation.

The only drawback of OLEV is the initial infrastructure cost. This could be financed in many ways, and we suggest a carbon tax for all ICE vehicles. It should be based on energy consumed. The carbon tax is to be increased gradually so that people have the incentives to change over to electric vehicles (EVs). Assuming on the average, we get $20 per month. The annual tax will be $240, and assuming there are more than 200 million cars in a country like in the USA. The annual fund available for the infrastructure is $50 billion. Over a period of 10–15 years, the whole country is electrified (except some pockets in the country side).

This programme will have to start in cities and later to be connected to inter cities. Concurrently, the introduction of OLEV into all the cities will generate employment. Electric vehicles using OLEV carry a smaller battery (1/3–1/4 compared to BEV vehicles). It is lighter and uses less energy and material to build in comparison with BEV vehicles. BEV cars can be modified for OLEV use. The manufacturing and maintenance cost of OLEV cars is reasonably low. Future policy may limit ICE vehicles into the city making the status car 'incapacitated'.

In the past agriculture, deforestation and other land use (AFOLU) had contributed to carbon emission. Millions of hectares of forests were logged, cleared and burnt like those in Malaysia, Indonesia and Brazil where large areas were cleared annually to grow oil palm, soya and other crops. Many countries in the humid tropic also practiced 'shifting cultivation', where a forested area was cleared and burnt to provide nutrient to plant a crop for the subsistent farmers. It caused soil erosion and when it soon ran out of fertility, it would then abandon the area and started clearing another patch to repeat the process all over again. Temperate and tropical forested areas were cleared for agriculture and other land use for the growing population. It is high time to conserve whatever remaining forests we have and also engage in reforestation, afforestation and enrichment.

To decrease the CO_2 in the air, it is proposed to capture the CO_2 (CCS) and store it in deep well. This methodology has not been proven yet. However, this process is expensive and can be dangerous with the sudden release or leak from the well. A total of 1400 people were killed due to the sudden escape of CO_2 from Lake Nyos in 1986 in Cameroon, Africa. Instead, CO_2 could be released slowly on a calm and sunny day in a dense forest to enhance carbon sequestration. It is a known fact that when CO_2 concentration in an agriculture greenhouse is increased to about 900–1000 parts per million, the yield of tomato increases by about 30% and likewise the growth of the forest can be increased in a similar way. More research has done to confirm financial viability. This method is relatively cheaper and easier to implement.

In the humid tropic alone, about 300 million to 1 billion hectares of land are available for reforestation and afforestation. The amount of carbon dioxide sequestered over 100 years could range from 60 to 140 billion tons. Protecting 130 million hectares of land from deforestation in the Amazon alone has the potential to 'lock' in 62Gt of CO_2.

Reforestation and afforestation in developing countries provide employment to the poor and alleviate them from extreme poverty. Using the 'taungya' system, it will provide them with a source of food. It is a marriage between shifting cultivation and the establishment of the forest plantation. Once the canopy is closing, they move to another patch of newly planted trees to grow their crops. Crops are planted between the rows of trees and are weeded. We could also grow agroforest trees such as: Galip nut (Canarium idicum), Pili nut (Canarium ovatum), Okari nut (Terminalia kaernbachii), Sea almond (Terminalia catappa), Pacific litchi (Pometia pinnata). Another way is to grow forest trees to be intercropped with coffee, cocoa, peach palm (Bactris gasipaes for food), rattan, vanilla and sunn hemp (Crotalaria juncea) as 'green manure' cover crop.

24.7 Carbon Sink in the Soil

Forest clearing and previous agricultural practices have played a major role in reduction in the suitability of soils for agricultural and forest production in large parts of the world. The results of this change include drying up of soil and loss of soil surface due to wind and erosion by rain, but more importantly reduction in nutrients in the soil and a severe reduction in the soil micro-organisms and soil carbon.

The massive use of chemicals to control weeds, pests and plant disease is an increasingly major reason for soil deterioration as many of the chemicals have adverse effects on the essential micro-organisms.

The provision of biological 'waste' material, particularly plants, is very important for soil improvement. Carbon, not as ash, but as black carbon, is also critical for the role micro-organisms play. These biological items are essential for retention of moisture in the upper soil layer and for the preservation of nutrients in a form that can be absorbed by the plant roots.

In severely depleted soils, it would be necessary to reintroduce the micro-organisms by taking soil from high-quality soil areas to the depleted soils. The soil used in nurseries to produce the plants to be established in the field would be an important source of the micro-organisms.

Where degraded areas have been created by removal of mangroves it would be beneficial to replace them. Mangroves are known to be the most efficient carbon absorbing plants. They are also a critical site for reproduction of marine species that are a significant source of food species for marine life. They constitute an important course of income for coastal people who rely on fishing for livelihood.

There is a growing concern regarding the present energy systems because of inherent risks connected with security of supply and potential international conflicts and on account of the potential damage they can do to the natural environment. Public opinion worldwide is increasingly aware of these risks, and they are pointing to an urgent need to fundamentally transform present energy systems into a more sustainable basis. A major contribution to this transformation can be expected to come from solar radiation, the prime energy source. This is to protect the natural support systems on which humanity depends and to eradicate energy poverty in developing countries.

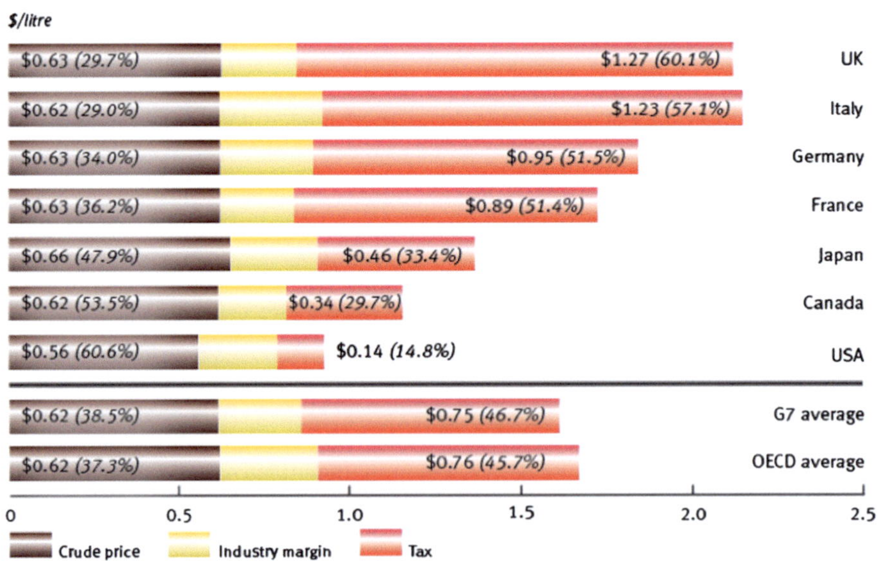

Fig. 24.9 Who gets what from a litre of oil in 2014

24.8 Barrier and Determination

We have the 'tools' and the funding from savings to restructure our world's economy and lower the income disparity among the world population. Many countries are addicted to oil taxes. The USA, Japan, UK and many of the OECD governments earn taxes from fuel consumers as shown in Fig. 24.9.

According to Ernst and Young's report on oil and gas taxes, the EU collects €250 billion in taxes from gasoline and diesel. Taxes on gasoline range as much as between 40 and 65%. The original intention of the taxes was to discourage its uses through higher prices. But over time the consumers have become 'immune' to higher prices. In fact, higher taxes made the consumers less sensitive to changes in oil price.

What would happen if natural gas or electric vehicles start to take away the market share from oil products? Surely the governments will look for ways to replace those oil taxes. This conflict of interest explains why many OECD governments have cynically been promoting alternative fuels 'on the surface' but not in practice.

The above proposal provides plenty of jobs, food security, prevent disastrous climate warming, providing clean water for all, alleviating poverty and protecting our biodiversity, with clean sustainable renewable energy and restructure our world economy for the benefit of mankind and the environment. The trillion dollar question is: do we have the political will to do the right thing for the future?

In September 2000, at the United Nations, the world leaders adopted the Millennium Declaration that expressed the goals of the world on the eve of the new millennium. These commitments included eight Millennium Development Goals (MDGs), adopted as specific, time-bound objectives to improve the conditions of the poorest of the poor by the year 2015 in the areas of income, hunger, diseases control, education and environmental sustainability. The MDGs were subsequently given financial impetus in the Monterrey Consensus of 2002 and several summits of the so-called G8, the eight richest large economies. Taken together, the Rio treaties, the Plan of Action on Population and Development, and the Millennium Development Goals can be called our Millennium Promises for sustainable development.

What have we achieved so far? Not much we know of. It is time we give ourselves a second chance and ensure a better world that future generations deserve.

References

1. US Energy Information Administration, International Energy Outlook (2016) http://www.eia.gov/pressroom/presentations/sieminski_05112016.pdf
2. International Energy Agency, CO_2 Emissions from Fuel Combustion - Highlights (2015) https://www.iea.org/publications/freepublications/publication/CO2EmissionsFromFuelCombustionHighlights2015.pdf

3. World Energy Council (2013) World Energy Resources https://www.worldenergy.org/wpcontent/uploads/2013/09/Complete_WER_2013_Survey.pdf
4. Rapier, R. (2010) The price of energy. http://www.forbes.com/sites/energysource/2010/01/26/the-price-of-energy/#366e089e7780

Epilogue

Electrification of ground transportation systems (EGSTs) is one of the most promising solutions for alleviating the problems associated with global warming. As the first step in realizing EGTS, we created on-line electric vehicle (OLEV). It uses SMFIR (shaped magnetic field in resonance) technology, which transmits hundreds of kW of electric power wirelessly over a significant distance to moving vehicles. This OLEV/SMFIR solution eliminates the need to carry a large number of batteries on board of electric vehicles (EVs) and ameliorates the concerns related to manual charging, long charging times, and range anxiety of EVs. This technology in combination with electric power generated from renewable sources, i.e., solar and wind energy, should reduce about 2/3 of anthropogenic greenhouse gas (GHG) emission worldwide, which is close to the goal established by COP21. To completely replace fossil fuel burning power plants with electricity generated from renewable sources, we will need a cost-effective technology that stores excess electric energy.

Large-scale implementation of EGTS is inevitable for the following six drivers for change: (1) the need to reduce CO_2 emission, (2) continuing advances in electricity generation using renewables, (3) technological and business innovations that will create new industries, (4) complete automation of transportation systems using information and communications technology (ICT), (5) elimination of human intervention to refuel or recharge, and (6) availability of cost-effective electric power storage technologies. Just as the breakup of old AT&T in the USA spurred new businesses and technological innovations in ICT (information and communications technology) worldwide, EGTS is most likely to herald a new era of innovation in transportation and energy-related industries.

OLEV buses were installed in cities in Korea. This inaugural installation in urban areas was well chosen since these areas have more environmental and transportation problems than rural areas. However, in the future, the technology should be extended to all regions and to all forms of EGTS, including passenger cars, forklifts, trains, trucks, and coastal ships. These additional applications should reduce the overall cost of all vehicles, including infrastructure cost, with the increased production volume for hardware. What is presented in this book on OLEV and SMFIR should provide the requisite fundamental knowledge. The ideas

N.P. Suh and D.H. Cho (eds.), *The On-line Electric Vehicle*,
DOI 10.1007/978-3-319-51183-2

imbedded in SMFIR and OLEV can be expanded and applied to a variety of new applications that involve wireless electric power transfer.

OLEV and SMFIR were implemented in two years from the initiation of the project. This is shorter than the duration of typical system development projects. The basic secret behind such rapid development and deployment is good design, i.e., a design of the system that is *uncoupled*, as discussed in the first few chapters of this book. Many examples of successful system development are given in Chap. 3, which were also completed in short times. In the case of OLEV, there are three additional factors that made this rapid progress possible: (1) outstanding designers, researchers, engineers, and managers at KAIST, who dedicated themselves to make this project successful, (2) systematic monitoring of the design and development processes by system architects to prevent coupling of functional requirements (i.e., coupled designs), and (3) support of collaborating industrial firms and the Korean government. To advance the field of system design and development, universities should strengthen their curriculum to generate more engineers and managers who know how to design large systems based on the fundamental principles of design.

During the next few decades, new energy and transportation industries will emerge in order to satisfy the need of ever-increasing world's population without harming the environment. The ideas presented in this book should help those who wish to change the transportation and energy industries. To realize the full potential of an enlightened energy–environment–transportation nexus, we will need a renaissance in the triad of *public policy–industry–universities* so as to reduce the GHG emission before the dire predictions made by IPCC (U.N. Intergovernmental Panel on Climate Change) overwhelm human ability to cope with the crisis. The problem will get compounded if we miss the finite window for taking actions. For technologists, next few decades will be both challenging and intellectually exciting and rewarding.

Policy makers should facilitate a smooth transition from the fossil-fuel-based economy to a renewable-energy-based economy. The industrial firms that are manufacturing automobiles with IC engine should be given incentives for timely transition into businesses related to EGTS. Oil companies should become energy companies that deal with renewable energies. Some of the oil-rich nations in the Middle East might become "solar energy exporters" by harnessing their abundant solar energy in the form of electricity or exporters of materials converting the oil into new high-value-added materials. They may export electricity to Europe, Asia, and Africa. Ultimately, through innovation of new technologies, all nations can strive to improve their economy. This transition into EGTS should create new opportunities that improve the quality of life and enhance the security of our habitat.

Index

N.P. Suh and D.H. Cho (eds.), *The On-line Electric Vehicle*,
DOI 10.1007/978-3-319-51183-2

MIX
Papier aus verantwortungsvollen Quellen
Paper from responsible sources
FSC® C105338

If you have any concerns about our products,
you can contact us on
ProductSafety@springernature.com

In case Publisher is established outside the EU,
the EU authorized representative is:
Springer Nature Customer Service Center GmbH
Europaplatz 3, 69115 Heidelberg, Germany

Printed by Libri Plureos GmbH
in Hamburg, Germany